WEALTH, ENERGY, AND HUMAN VALUES

The Dynamics of Decaying Civilizations from Ancient Greece to America

Thomas P. Wallace, Ph.D.

authorHOUSE®

AuthorHouse™
1663 Liberty Drive
Bloomington, IN 47403
www.authorhouse.com
Phone: 1-800-839-8640

© 2009 Thomas P. Wallace, Ph.D.. All rights reserved.

No part of this book may be reproduced, stored in a retrieval system, or transmitted by any means without the written permission of the author.

First published by AuthorHouse   5/20/2009

ISBN: 978-1-4389-7629-7 (e)
ISBN: 978-1-4389-7627-3 (sc)
ISBN: 978-1-4389-7628-0 (hc)

Library of Congress Control Number: 2009903688

Printed in the United States of America
Bloomington, Indiana

This book is printed on acid-free paper.

This volume is dedicated to my wonderful and successful children
Andy, Mark, Mike, and Sarah,
in recognition and appreciation for the joy and contentment they have brought to their parents.

"You may give them your love but not your thoughts,
For they have their own thoughts.
You may house their bodies but not their souls,
For their souls dwell in the house of tomorrow, which you cannot visit, not even in your dreams.
You may strive to be like them, but seek not to make them like you,
For life goes not backward nor tarries with yesterday.
You are the bows from which your children as living arrows are sent forth."

Kahlil Gibran (1883-1931)

Table of Contents

Foreword:	The Inevitable Exhaustion of Civilizations	ix
Preface:	The Mysteries of Decaying Civilizations: A Scientific Perspective	xv
Chapter 1	The Role of Energy in Society's Organizational and Functional Systems	1
Chapter 2	Wealth, Energy, and Science-based Economics	23
Chapter 3	The Driving Force of Cultural Complexity and Disorder	50
Chapter 4	Socio-economics: A Wealth-Energy Based Perspective	74
Chapter 5	Socio-economic Transformations: Labor Power and Social Energy	98
Chapter 6	The Mechanisms and Energetics of the Human Social Order	129
Chapter 7	The Characteristics and Dynamics of Sociocultural Transitions	163
Chapter 8	Periodicity of Human Advancement: Ancient Greece to the Twenty-First Century	191
Chapter 9	Properties and Characteristics of Sociocultural Stagnation	229
Chapter 10	Ecological Ramifications of Modern Socioeconomic Progress: Chesapeake Bay and the World's Oceans	270
Chapter 11	Sociopolitical Evolution of Economic Power: The Stagnation of Western Civilization	290
Chapter 12	Degradation of the American Culture: Abuse of Founding Principles	319

Chapter 13	Shifting Twenty-First-Century Civilization Patterns: The Asian and Islamic Resurgence	404
Chapter 14	The Mechanistic-Thermodynamic Paradigm: A Unifying Perspective of Civilization Prosperity and Failure	440
Appendix A	The Fundamentals of Thermodynamics Applied to Socioeconomics	469

Foreword:
The Inevitable Exhaustion of Civilizations

> "The two greatest problems in history are how to account for the rise of Rome, and how to account for her fall."
>
> J. S. Reid[1]

> "We may come nearer to understanding them if we remember that the fall of Rome, like her rise, had not one cause but many, and was not an event but a process spread over 300 years."
>
> Will Durant[2]

The degradation of the modern American culture, including its 2008 financial and economic crisis, and the modern rejuvenation of Asian cultures are best understood within the context of 4,000 years of human history and the lessons of failed civilizations. Such are the consequences of the dynamics of societal growth, stagnation, and deterioration, responding to the cultural variables of wealth, energy, and human values and behavior.

The concept that all civilizations inherently and inevitably face deterioration and collapse is an unsettling, complex notion for most people to seriously contemplate. Nevertheless, prominent historians and social thinkers of the twentieth century acknowledge the reality of this unexplained human experience that has occurred many times over the last 4,000 years. The Greek poets Homer and Hesiod viewed history as a regressive process from the Golden Age through the Bronze and Iron Ages. But perhaps the most impressive and recognized analyses are the extensive works of Oswald Spengler, Arnold Toynbee, A. L. Kroeber, and Pitirim Sorokin, who trace the sociocultural development of primitive societies through their mature civilization stage leading to cultural deterioration and stagnation.

Models, concepts, and theories have been proposed relative to the historic cultural degradation of past great civilizations. Many factors and variables have been identified and analyzed that influence social, economic, and political systems, contributing to the human existence within the framework of a social order. However, a complete, unified theory that satisfactorily accounts for the accepted historical signature pattern or periodic fluctuations of cultural growth, decay, and stagnation has remained elusive.

This elusiveness of a unified concept of human societal development has been due to the absence of a science-based, multidisciplinary perspective that incorporates the disciplines of history, economics, and sociology. The civilization studies literature does not appropriately incorporate scientific definitions and principles pertaining to the consumption of societal wealth and available energy resources. Specifically, the science of energy consumption or transfer (i.e., thermodynamics) regulates society's acquisition, refinement, and utilization of Mother Nature's resources to provide the food and fuels necessary for the vital processes of human existence.

Thus, the much-respected sociological and historical analyses of human behavior and events associated with cultural development have neglected the primary role of wealth-energy resources and the fundamental influences of basic science, while often utilizing scientific methodologies. Historically, this perspective may be traced to the 19th century, when a dualism of traditional intellectual thought prevailed whereby knowledge was categorized as either appropriate to the natural sciences or to cultural and social behavior. Consequently, the natural sciences were viewed as restricted to the closed system of the world of nature, which relied, at the time, on classical mechanics. The natural sciences were not viewed as conceptually applicable to sociocultural studies. Max Weber's adopted methodology for his social science investigations eroded this artificial historic barrier, thereby promoting functional linkages among the natural sciences, human behavior, economics, and cultural development.[3] It is now recognized that the long sought unified theory of cultural growth and decay must strictly and rigorously integrate scientifically defined concepts and parameters of wealth and energy within fundamental socioeconomic processes.

The ultimate success of a civilization is influenced by human values, priorities, and behavior and by the randomness and probabilities associated with the activities of nature and society. Consequently, the dynamics of cultural growth and decay are primarily affected by wealth, energy, and human values.

It will be demonstrated that human socioeconomic aspirations and accomplishments, while fundamentally influenced by values and behavior, will be restricted and controlled by the availability and effective utility of wealth and energy resources. Typically, inadequate resources have inspired and precipitated negative human behavior of both the rulers and the ruled, particularly as mature civilizations evolve into the phases of cultural deterioration and stagnation. While adequate wealth-energy resources may appear to be an obvious prerequisite for cultural advancement, thermodynamic principles and applications have been excluded from the most respected analyses of sociocultural development and transitions. Clearly, the science governing energy consumption associated with the nourishment of people and the fueling of machines is also applicable to the vital processes of a social order's existence. This science is the key to comprehending the inevitable exhaustion of civilizations as well as to potential socioeconomic rejuvenation.

The scientific principles governing and restricting all processes of nature and society that contribute to the inevitable decline of civilizations may be illustrated by an analogy with the life span of a car battery. The purchaser of an automobile battery recognizes that the useful life of the battery is limited and approximately five years, depending on its use and abuse. The question is, what happens operationally within a car battery: Why does it have a limited life span (i.e., what causes the *inevitable decline and collapse* of a car battery)? The answer is that Mother Nature's laws of science control and restrict the battery's chemical processes, which produce electricity. The car battery contains chemicals that provide, on command, the electrical energy necessary to start the car's engine. This is referred to as an electrochemical process whereby an electrical current is produced that powers a starter motor.

Once the engine is running, the engine's rotating alternator electromechanically generates an electrical current that reverses the chemical reactions within the battery and recharges it. An ongoing internal view of the battery would reveal that during the many charging-discharging cycles, physical and chemical changes occur that are referred to as the *aging of battery cells*. While the battery is performing useful work by starting the car, the original lead plates gradually deteriorate, producing *undesirable contaminants* and water, which dilutes the sulfuric acid electrolyte. Irreversible material and energy degradation inherently occurs as useful work is being performed. This effect of the deterioration of the lead plates and the weakening electrolyte solution increases the *internal resistance* of the battery, which, over time, reduces the battery's capacity to produce an electrical current. Energy is consumed (i.e., wasted), in order to overcome the increasing *internal resistance* of the battery. All batteries, whether rechargeable or not, will inevitably reach the point where the internal resistance of the cells is sufficiently large and consumes sufficient energy that inadequate current remains to perform useful work. This gradual net degradation of the battery's chemical system over time is an irreversible and spontaneous transition, an inherent and inevitable mandate of chemical thermodynamics.

Likewise, the normal operational mechanisms and energetics of cultural existence and advancement (e.g., economic systems) generate *undesirable contaminants* or societal consequences that constitute an accumulating *internal resistance* toward human progress that gradually reduces the rate of cultural growth. Such undesirable cultural by-products of socioeconomic advancement include escalating societal complexities, disorder, pollution, and social conflict as well as bureaucratic ineffectiveness and expense. These dysfunctional cultural ramifications are a predictable consequence of the second law of thermodynamics, a *societal entropy concept*, which mandates the spontaneous direction of all processes of nature and society and the generation of greater cultural complexity and disorder. Human actions, whether controlling the fate of an automobile battery or pursuing cultural advancement, will consume wealth-energy resources and employ human values, but Mother Nature's scientific principles will limit and direct outcomes and produce consequences.

The *mechanistic-thermodynamic paradigm* is a controlling principle for thermodynamic systems, including the organization and function of a social order as well as the operation of an automobile battery. This concept will be developed in the following chapters, providing a unifying perspective for the twentieth-century sociological, historical, and economic literature of civilization prosperity and failure from ancient Greece to twenty-first-century America.

The "inevitable exhaustion of civilizations" has become quite visible in the first decade of the twenty-first century as the world witnesses perhaps the most revealing and significant evidence of American society's self-destructive journey of cultural degradation. The massive failure of the American financial and economic systems in the latter part of 2008, the nation's most serious financial crisis since the 1930s, shook the world's financial markets and will continue to have massive and unprecedented long-term worldwide ramifications. The nature, scope, and global consequences of these events, reflective of America's mounting cultural degradation, illustrate the self-destructive nature of mature civilizations and more specifically the abuse of wealth and energy resources and the lack of respect for, and responsibility to, one's social order by unethical, greedy, and ruthless human behavior. Arnold Toynbee's comment regarding *cultural self-destruction* is appropriately noted: "A society does not ever die 'from natural causes,' but always does from suicide or murder—and nearly always from the former."[4]

Specific chapters are devoted to the stagnation of Western civilization, the late twentieth century Asian and Islamic resurgence, the deterioration of the American culture, and the ecological destruction of the Chesapeake Bay and the world's oceans; collateral damage or "externality" of socio-economic profitability...an environmental entropy effect.

References and Notes
1. Durant, Will. *The Story of Civilization: Part III: Caesar and Christ*. New York: Simon and Schuster. 1944. 665.
2. Ibid., 665.
3. Weber, Max. *The Theory of Social and Economic Organization*. New York. The Free Press. 1947. 8-29.
4. Toynbee, Arnold J. *A Study of History*, ab. Sommervell, D. C. New York: Oxford University. 1957. I-VI: 273.

Preface
The Mysteries of Decaying Civilizations: A Scientific Perspective

"History ends at least once and occasionally more often in the history of every civilization."   Samuel P. Huntington[1]

The twentieth-century civilization studies literature reveals a number of perplexing themes that have been the subject of continuing interest and debate among authors engaged in comparative historical and behavioral analyses of past civilizations. These topics include (a) the elusive underlying principles explaining the eventual disintegration of all known civilizations; (b) the basis for recurring sequential periodicity of civilization development; (c) the roles and significance of warfare and religion in civilization growth, decay, and rebirth; (d) the identification of inclusive, major variables influencing societal transitions; and (e) the significance of human behavior and cultural traits relative to the effective societal investment of national wealth in the successful development of socioeconomic systems.

Arnold Toynbee points out that each mature civilization enjoys its era of socioeconomic prosperity but ultimately succumbs to "the mirage of immortality," assuming that its good fortunes will continue unabated forever.[2] However, history has never known a civilization that has avoided inevitable cultural breakdown, stagnation, and deterioration leading to ossification or collapse. Cultural degradation may lead to swift extinction or to a rejuvenated status, even if only to be followed by collapse at a later time. The rise and fall of civilizations is a familiar topic in the literature, but the inevitability of cultural disintegration appears to be more of a curiosity than a recognized and significant, unexplained universal phenomenon. Interestingly, civilization studies authors, usually employing narrow perspectives within the social sciences, have been unable to fully explain this historical reality.

The Mysteries of Decaying Civilizations: A Scientific Perspective

Occasionally, if only superficially, a number of authors have noted the importance of energy in maintaining a viable social order. However, energy-based economic models have not historically been appropriately applied in the analyses of deteriorating major civilizations. This is a result of the historic inability or unwillingness of economists to incorporate rigorous, scientifically based models of national wealth in economic, analysis, planning, and policies.

> **The economist studies mental states rather through their manifestations than in themselves; and if he finds they afford evenly balanced incentives to action, he treats them prima facie as for his purposes equal.** [3]

It will be demonstrated that the *mysteries* surrounding the inevitable decline of civilizations become less mysterious if energy concepts, that is, thermodynamic principles, are appropriately applied to social, economic, and political processes. Specifically, the centrality of the intimate relationship of energy to a scientifically defined concept of a national economy is an imperative for the proper analysis of socioeconomic transitions. The need for gasoline to fuel a car and for food to nourish living creatures is universally recognized and accepted. However, the universality of society's dependence on adequate, continuously available wealth-energy resources as the fundamental limiting factor for effective social, economic, and political processes of an organized society is seldom formally and accurately acknowledged. The following questions highlight the breadth and significance of a society's dependence on accessible energy:

- What processes supporting the existence of human life and the lives of other living creatures *do not* require an energy source?
- Which machines and devices *do not* require an energy source?
- Which of society's physical, economic, social, cultural, and political processes and systems *do not* require energy sources for human and machine consumption?

The answer to all three questions is *none*. Now consider the following:

- Can society and its machinery and devices or Mother Nature's processes accomplish any useful work, as in physical,

chemical, and biological activity, without consuming energy?
- Is it possible to consume energy without degrading some of the initial energy content into waste energy and material?

The answer to both questions is *no*.

All processes of *living and doing* conducted by nature and the animal kingdom consist of mechanistic steps and sequences, each having energy requirements and consequences. These physical, chemical, and biological processes of a society's cultural existence involve functional mechanisms that are controlled by thermodynamic principles. This perspective will be fully developed in the following chapters and is referred to as the *mechanistic-thermodynamic paradigm*. Each mechanistic step has energy considerations that are subject to the laws of thermodynamics. While not as generally understood and appreciated as the law of gravity, the laws of thermodynamics are scientifically as valid and as vital to the universe as the principle of gravity.

The nature and ramifications of the mechanistic-thermodynamic paradigm have inherently contributed to a common set of characteristics associated with the cultural degradation of the historic civilizations. Greek, Roman, Spanish, British, and American societies have exhibited such characteristics as they evolved from their more primitive form to more mature civilizations. First and foremost, the primary motivation of a society is economic success, whether for a primitive society struggling for basic survival or for mature civilizations seeking readily available luxury goods and services. Sociopolitical institutions and functions are created by a society in response to adopted economic priorities, goals, and strategies, intended to accomplish the objectives of wealth, power, and status. Materialist goals and rewards are, and have been, people's overriding motivator.

Second, as a civilization matures and becomes more affluent, cultural mentalities migrate from primitive communally oriented altruistic values toward more self-serving, materialistic values. The societal equilibrium of adopted cultural mentalities shifts as a society matures from a *primitive fear for survival* toward *materialistic self-interest and greed*. Such shifting values diminish the sense of community, cooperation, and mutual assistance and, instead, enhance self-oriented priorities and outcomes.

Third, an assumed superiority of race, intellect, social class, moral values, and political and religious ideologies are commonly found within an aspiring or reigning civilization as it maneuvers toward greater dominance and wealth. Such pretentious, self-anointed supremacy is often extended to the role of divine providence and the mission of creating a superior world, which becomes license to inflict inhumane suffering on the conquered in the name of moral superiority and spiritual salvation. Rulers, religious and business leaders, and the elite class have historically found common purpose in their associations to create self-serving, socioeconomic hierarchies within populations. Such co-conspirators have successfully manipulated society's institutions in order to share wealth, power, and glory, usually at the expense and suffering of the masses.

Fourth, successful periods of rapid socioeconomic progress are characterized by (a) times of technological inventiveness; (b) external threats and competitiveness; (c) economic creativity toward new products and markets; (d) innovative tactics employed to confiscate wealth and territory; (e) the acquisition of natural resources and cheap or slave labor; and (f) the creation of a superior military force.

Innovative financial strategies to support territorial expansion and warfare designed to secure new wealth-energy resources have been a key historical factor in the success of past civilizations. In the 1690s, British financial creativity and inventiveness were credited as being a major factor to achieving global military and financial supremacy, while at the same time avoiding the traditional unpopular tactic of raising taxes for imperialist adventures. The unique concept of national debt was introduced in order to raise large sums of money quickly to support overseas expansion by selling shares of stock that would yield an annual dividend. This creative financial strategy has been credited with Britain's overseas economic success by virtue of being able to finance a global military power during the seventeenth century.

Fifth, a civilization's military force often becomes an economic tool for the acquisition of additional wealth, by creating new commercial markets, material resources, and cheap labor. Warfare is a familiar competitive tactic for a young civilization attempting to conquer new territories in order to achieve greater power and to secure the additional wealth required to fuel its expanding economic aspirations and opportunities. Warfare is also employed as a desperate tactic by mature civilizations facing socioeconomic breakdown and financial insolvency resulting from their overinvestment in cultural complexity, expansionism, and the extravagancies of a materialistic culture. Conversely, warfare is occasionally and honorably employed as a necessity for self-defense and altruistic societal outcomes that benefit humankind. America's entry and participation in World War II is an example.

Sixth, as a civilization ages, its social, economic, and political systems intrinsically become more complex and integrated in order to achieve the culture's primary and escalating aspirations for greater socioeconomic prosperity. This objective requires a continual increase in a society's functional and organizational complexity illustrated by the proliferation of more costly and time-consuming bureaucracies, policies, laws, regulations, codes, daily schedules, and family and societal expectations. Such so-called *progressive rudiments* of a successful society are intended to achieve more effective, efficient, productive, and equitable aspects for a society's existence, but at great expense. In addition, the history of maturing civilizations demonstrates that socioeconomic objectives, attuned to accumulating wealth and self-satisfaction, become more highly focused on the profitability of producing luxury goods and services at the expense of more disciplined long-term investments that will generally benefit important long-term economic fundamentals. Thus, national wealth consumption increasingly becomes overly committed to production categories incapable of spawning new forms of wealth for reinvestment in further profitability. Meanwhile, a mature civilization's escalating financial commitments, which are necessary to maintain the momentum of creeping progress, increase exponentially and devour an increasingly larger percentage of available wealth. Over time, the financial requirements to support increasing cultural complexity and misplaced societal priorities for nonessential production ultimately create an unsustainable national budgetary burden, leading to massive debt and financial insolvency. This accurately portrays America in 2009.

Essentially, insufficient resources are invested in the types of production capable of generating the new capital necessary to support human aspirations for continuous socioeconomic growth and advancement. In a thermodynamic analogy, the society's *economic engine* reaches such a low level of effectiveness that it is unable to accomplish sufficient *useful work* (generate new investment capital) to meet the requirements of the *job* (progress and financial solvency). Society's financial burden or load becomes too heavy for the economic engine, and economic productivity becomes insufficient to achieve unrealistic human expectations. The economist would refer to this condition as a *declining marginal productivity*: too little capital being invested in growth potential capable of generating additional wealth, required to maintain a continuous positive rate of cultural growth.

Joseph Tainter points out that "complex societies with large, well-developed economies have historically been able to sustain only rather inferior rates of economic growth."[4] He also notes: "There is in complex societies a recurrent and seemingly inexorable trend toward declining marginal productivity in hierarchical specialization."[5] This socioeconomic phenomenon is analogous to a gasoline-powered vehicle that is increasingly required to take on heavier cargos at various stops during the course of its journey. The vehicle increasingly experiences declining gas mileage (miles per gallon of fuel), lower maximum speeds, and increasing difficulty in climbing mountain roads and eventually becoming immobile. As the load periodically increases during the journey, the portion of the fuel's energy content that accomplishes the desired useful work (i.e., locomotion) declines. Ultimately, the point of immobility is reached, where no useful work is being produced. During this sequence, as the vehicle assumes heavier and heavier cargos, a larger and larger percentage of fuel is consumed to overcome the accumulating load or resistance to further progress in completing the intended journey. These changing conditions during the process increasingly result in the conversion of a given quantity of energy (fuel) into higher percentages of wasted energy (heat). As a smaller portion of the consumed fuel is effectively utilized to accomplish the useful purpose of vehicle motion, the engine's efficiency declines. At some point, motion ceases and the journey ends.

Consider the analogy of this hypothetical vehicle being subjected to an increasing burden that threatens the completion of its journey to a mature society pursuing *its journey* toward socioeconomic affluence. A civilization's economy increasingly achieves a lower efficiency (i.e., a higher ratio of wasted to wisely and effectively utilized wealth-energy resources) as its financial burden escalates due to increasing cultural complexity and to a high priority being placed on the production of nonessential goods and services. The outcome represents a declining marginal return on investment, an increasingly wasteful consumption of resources (albeit perhaps pleasurably), and a declining economic effectiveness of national wealth and energy consumption in achieving societal well-being. Accordingly, increasing cultural complexity, which seldom contributes to economic productivity, diverts resources from useful processes capable of enhancing a society's advancement and longevity. As a result, a civilization possessing a declining marginal return on its investments of national wealth finds itself on a slippery slope toward socioeconomic collapse. As stated by Tainter: "The marginal return on investment in complexity is at present the best explanation of collapse."[6]

Exploring the etiology of cultural deterioration has been an important element in the comparative analyses of past civilizations and constitutes one phase of the recurrent cultural periodicity of cultural growth and decay. B. G. Brander, in *Staring into Chaos: Explorations in the Decline of Western Civilization*, asks the question, "Can we really find laws, patterns, cycles, or consistencies in history?"[7] Oswald Spengler, Arnold Toynbee, and Pitirim Sorokin answer affirmatively and address the periodicity of societal development, whereas others disagree with their concepts and debate the question. Some authors approach cultural evolution and civilization development as simply a chronicle of random historic circumstances and events that humankind has endured, while others rely primarily on innate behavioral traits. Unfortunately, the importance of the interrelatedness of a broad range of factors that influence civilization growth and decline is marginalized, if not totally ignored. However, the prosperity of a given civilization, identified and quantified by its rate of socioeconomic progress, is affected by many interrelated variables including human nature and behavior, the availability of material resources, mathematical probability, and a globally competitive environment.

Overwhelmingly, authors of the civilization studies literature create and label patterns of growth and decay based on assumed limited sociological and psychological capabilities of people. However, regardless of a society's moral, ethical, cultural, emotional, and intellectual capabilities and traditions, energy and other material resources are a universal requirement for the functioning and prosperity of socioeconomic systems. Additionally, human events are influenced by mathematical probability that may randomly affect specific circumstances and outcomes despite the wisest and most noble human intentions, plans, and efforts. Authors addressing an individual's freedom of choice and the randomness of human thought processes have infrequently and only partially acknowledged probability. This has been the basis for reconciling outcomes of closely contested significant historical events as being determined by chance and unpredictable occurrences.

Jacob Burckhardt[8] predicts a low probability that eventually a few spiritually motivated people would emerge and provide the necessary moral, ethical, and creative elements to counteract forces of societal disintegration. On the other hand, Oswald Spengler regards cultures as possessing morphological or organism characteristics with fixed predictable life cycles: "The great Cultures accomplish their majestic wave-cycles. They appear suddenly, swell in splendid lines, flatten again and vanish, and the face of the waters is once more a sleeping waste."[9]

Arnold Toynbee places greater emphasis on choice, free will, and probability: "I do think that the scientific apparatus can be applied fruitfully to human affairs to some extent, e.g., when they are considerably affected by the physical environment and when it is the subconscious part of the psyche more than the will and the intellect that is in command."[10] However, Toynbee's primary view of societal deterioration is that people fail to make wise judgments and not on a predetermined "organism" cycle or pathway. Toynbee and Sorokin appear to agree that poor choices result from the clash of opposing mentalities of materialism versus spiritualism with the former, motivated by individual self-interest, ultimately winning the conflict among free choices.

Toynbee identifies consistent and specific historical patterns of cultural disintegration while professing that societal growth results from responses to challenges with creative solutions.[11] Sorokin's sociological approach is based on the premise that a society's cultural arts reflect a sequence of transitions between spiritualistic and materialistic phases. These are due to shifting values, behavioral characteristics, and differing motivations of a population during major cultural transformations.[12] His methodology produces data that trace, like a barometer, societal change but have proven incapable of revealing specific *dynamic forces* leading to cultural transformations and deterioration.

Subsequent chapters will explore these *dynamic forces* that, since the time of ancient Greece, have driven societal development and are responsible for fluctuating periods of cultural growth and decay, as well as for the ultimate deterioration of civilizations. The productive approach to unraveling the *mysteries of disintegrating civilizations* is to examine, from a science-based perspective, the interrelationships among energy, wealth, and human values and behavior and to assess the resulting dynamic forces of change. This necessitates the integration of the academic disciplines of history, sociology, economics, and science, which will be utilized in three major approaches to unlocking the mystery.

First, an energy-based economic model utilizing scientifically defined concepts of *wealth* and *energy*, in contrast to the traditionally ill-defined *currency-based wealth*, is the appropriate, precise foundation for socioeconomic systems. Wealth-energy resources will be more fully and scientifically defined in a later chapter in terms of usable heat content and/or economic value. In order to maximize accuracy, precision, and usefulness, any analysis of the periodicity of cultural growth and decay of civilizations must be assessed utilizing a wealth-energy based economic perspective. The engine of any economy is fundamentally fueled by a combination of materials possessing economic value and energy (i.e., wealth-energy capital). Thus, any attempted analysis of a society's socioeconomic system that evades this standard is flawed and highly misleading, and ignores major long-term economic pitfalls. The circumstances surrounding the financial mischief of the 2008 collapse of major Wall Street financial institutions provides a colossal, unfortunate illustration of such flawed economics and failed human values.

Second, the wealth-energy, thermodynamic approach of representing the functional operations of a society's socioeconomic system is analogous to the scientific concepts and principles of molecular reactivity. That is, the mechanistic-thermodynamic paradigm represents the functional steps of societal processes and the related energy considerations that comprise economic systems in the same manner as found in chemical, biological, and physical processes. An overall, or a net chemical, reaction, such as the process of photosynthesis, is the net result of many complex mechanistic steps, each of which is associated with an energy transfer. Likewise, the physical, chemical, and biological processes of a social order and nature consist of specific mechanisms, which are enabled, directed, and controlled by energy transfer, i.e., thermodynamics. The same principles that apply to such molecular events are also applicable to the social, economic, and political processes that comprise the organizational and functional systems of developing civilizations.

Perhaps the most remarkable outcome of this science-based approach to cultural development is that it provides a nonempirical foundation for unifying the empirically-based sociological, economic, and historical concepts of civilization development contributed by such eminent pioneers as Pitirim Sorokin, Arnold Toynbee, Oswald Spengler, and A. L. Kroeber. These authors base their work on the analysis of empirical data extracted from sociological and cultural aspects of historical events without benefit of the science of energetics which, it turns out, provides the unimpeachable endorsement of Mother Nature's laws of science to many of their major concepts of civilization growth and decay.

Third, the application of the *mechanistic-thermodynamic paradigm* to national economics reveals that a civilization that is able to maintain a continuous, high degree of economic effectiveness in its conversion of national wealth-energy resources into new forms of wealth-energy capital will increase its probability of prosperity and long-term survival. However, this desirable outcome also requires that society minimizes consumption of wealth-energy resources for nonessential goods and services, avoids excessive debt, and maximizes societal investments that provide for the general well-being of society. Such a principle will not enable a society to freely utilize national wealth to finance every interesting or alluring individual or societal opportunity or desire. Thus far in the history of the world, no civilization has been able to accomplish this objective and avoid the inevitable consequences of cultural degradation.

In summary, the mysteries of the inevitable disintegration of civilizations may be demystified by the application of the mechanistic-thermodynamic paradigm to the socioeconomic processes of cultural growth and decay. This is the theme of this work.

It becomes the primary function of culture, therefore, to harness and control energy so that it may be put to work in man's service. Culture thus confronts us as an elaborate thermodynamic, mechanical system. By means of technological instruments energy is harnessed and put to work. Social and philosophic systems are both adjuncts and expressions of this technologic process. The functioning of culture as a whole therefore rests upon and is determined by the amount of energy harnessed and by the way in which it is put to work.[13]

References and Notes

1. Huntington, Samuel P. *The Clash of Civilizations and the Remaking of World Order.* New York: Simon and Schuster. 1996. 301.
2. Toynbee, Arnold J. *A Study of History,* ab. Sommervell, D. C. New York: Oxford University Press. 1957. XXIV: 4-10.
3. Marshall, Alfred. *Principles of Economics.* 8th ed. London: Macmillan. 1927. 16.
4. Tainter, Joseph A. *The Collapse of Complex Societies.* New York: Cambridge University Press. 2004. 108.
5. Ibid., 106.
6. Ibid., 203.
7. Brander, B. G. *Staring into Chaos: Explorations in the Decline of Western Civilization.* Dallas: Spence. 1998. 367. Also, excellent coverage of the views of Oswald Spengler, Arnold Toynbee, and Pitirim Sorokin.
8. Ibid., 31.
9. Spengler, Oswald. *The Decline of the West,* trans. with notes by Charles Francis Atkinson. New York: Oxford University Press. 1991. 73.
10. Letter of 1949 to Kenneth W. Thompson of the Rockefeller Foundation. *Montagu, Toynbee and History.* 202-203.
11. Brander, B. G. *Staring into Chaos.* 157-228.
12. Ibid., 235-365.
13. White, Leslie A. *The Science of Culture.* Toronto: Doubleday Canada. 1969. 367-368.

CHAPTER 1
The Role of Energy in Society's Organizational and Functional Systems

"The Civilization is the inevitable destiny of the Culture, and in this principle we obtain the viewpoint from which the deepest and gravest problems of historical morphology become capable of solution. Civilizations are the most external and artificial states of which a species of developed humanity is capable. They are a conclusion, the thing-become succeeding the thing-becoming, death following life, rigidity following expansion, intellectual age and the stone-built, petrifying world-city following mother-earth and the spiritual childhood of Doric and Gothic. They are an end, irrevocable, yet by inward necessity reached again and again."   Oswald Spengler[1]

A multidisciplinary review of the literature regarding the evolution of societies, cultures, civilizations, and empires reveals common patterns of formation, growth, maturation, and eventual decline. Historians, political economists, sociologists, cultural anthropologists, and psychologists offer various interpretations and rationales for historic events and cultural attitudes, behaviors, and transformations. However, there is universal agreement that something appears to occur within mature civilizations that ultimately and significantly reduces their power, wealth, and global stature. The literature is abundant with theories, models, and rationalizations regarding the fates of past civilizations including debates as to if, and when, their destiny may appropriately be described as disintegration, stagnation, collapse, or death. Generally, authors attribute the historically observed sequence from a society's genesis to growth followed by decline as a consequence of people's inherent human instincts and cultural influences in an evolutionary process referred to as a sociocultural transition. However, some of the same authors admit that a complete and satisfying understanding of why successful civilizations ultimately decline is lacking and has not been revealed from social science approaches.

Additionally, authors tracing the evolution of cultural transitions employ the descriptive phrase *cultural dynamics,* but with rare exception, such literature is devoid of appropriate models that mention a driving force or an energy source necessary to energize such dynamics. *The Encarta Dictionary*[2] defines "dynamics" as "the forces that tend to produce activity and change in any situation or sphere of existence." Basic physics, as well as intuition, tells us that a force requires an energy source. One wonders what such authors view as the energy source of the forces they refer to as responsible for the dynamic functions of a culture. Perhaps, it is simply a restrictive view of the energy humans absorb by consuming food and the dynamics involved in their daily activities implementing impulses of human nature.

As will be demonstrated, a thermodynamic-based economic model identifies the *dynamics* that drive all human existence including the economic, social, and political activities of a society. Such a model must employ strict science-based definitions of national wealth, wealth-energy resources, and related financial and economic parameters and must not violate the laws of thermodynamics. The *activity and change* resulting from these societal dynamics are the consequence of an almost-infinite number of specific mechanisms and conditions representing society's operational functions. It is instructive to consider a small-scale analogy of a chemical reaction to societal functional processes of existence (e.g., the complex biochemical process of photosynthesis). Such chemical processes consist of many sequential interrelated reaction mechanisms, occurring over time, under various potential sets of conditions. Depending on the nature of the specific reactants and conditions, desirable and undesirable products may be realized, with each potentially producing additional multiple products, thereby propagating a continuous multi-step process. Such mechanisms are fueled by some form of internal or external energy and conform to the laws of thermodynamics. This analogy will be more fully developed in a later chapter in conjunction with a model for civilization development.

Joseph Tainter, in *The Collapse of Complex Societies*, notes the mystery that surrounds the historic decline of all known civilizations:

"Collapse is a recurrent feature of human societies, and indeed it is this fact that makes it worthwhile to explore a general explanation."[3] Further, he states: "Explanations of collapse have tended to be ad hoc, pertaining only to one or a few societies, so that a general understanding remains elusive."[4] The "general explanation" and "understanding" that "remain elusive" are due to a lack of recognition of the role wealth-energy resources play in providing the dynamics that drive the day-to-day functions of a society. More specifically, the second law of thermodynamics, the entropy concept, is the basis for why "collapse is a recurrent feature of human society," and this perspective will be fully explored in later chapters.

Leslie White emphasizes the necessity of available energy to drive cultural systems and, in general, recognizes the fundamental role of thermodynamics in the processes of a society:

Cultural systems, like all material systems, are thermodynamic systems. Their existence and operation require energy. Every cultural event ... involves the expenditure of energy.[5]

It is noteworthy that only when the required amount of wealth-energy resources are available to provide the required dynamics for a given societal function can human judgments, behavior, and capabilities have a role in initiating a process mechanism and proceeding to achieve specific outcomes. Furthermore, the laws of thermodynamics are rules that control and restrict society's attempts to consume Mother Nature's resources in the pursuit of socioeconomic objectives.

While it is obviously useful to have historians and cultural anthropologists report observed social patterns and historical events for understanding, analyses, and for future projections, such literature does not provide a fundamental explanation for the inevitable decline of past civilizations. However, a thermodynamic-based economic model incorporating the mechanistic-thermodynamic paradigm properly represents a social order as it consumes wealth-energy resources. This constitutes the basis for the inevitable decline in the rate of socioeconomic progress, an inevitable thermodynamic consequence of the processes of cultural existence that generates by-products of costly and disruptive societal complexity and disorder. This will be defined as the *societal entropy production effect*, which will be developed further in a later chapter. Interestingly, this concept is conceptually consistent with, and supportive of, the hypothesis of economist Joseph Schumpeter that mature economies are destroyed by their own success.

> **The capitalist process not only destroys its own institutional framework but it also creates the conditions for another. Destruction may not be the right word after all. Perhaps I should have spoken of transformation.... In the end there is not so much difference as one might think between saying that the decay of capitalism is due to its success and saying that it is due to its failure.[6]**

> **(Note: This quote may bring to mind the analogy offered in the Foreword of a car battery's internal chemical system deteriorating as it accomplishes useful work of starting the car engine. The battery's ultimate failure is also due to its success!)**

Schumpeter, in effect, describes an inherent property of a declining socioeconomic system as characteristic of its maturity. Other authors attribute cultural disintegration and collapse to the theory of *faulty behavior and failure to adapt* to increasingly unstable social and economic conditions (i.e., a culture of endlessly expanding complexity and bureaucracies). A more precise description, consistent with Schumpeter's view that "the decay of capitalism is due to its success," is a human *inability to contend* with the inherent accumulation of negative elements of out-of-control, irresolvable social, economic, and political by-products of a successful civilization. In actuality, Schumpeter is reacting to the accumulated *societal entropy production* reaching an intolerable level, precipitating a significant reduction in socioeconomic progress and resulting in what Toynbee refers to as cultural "breakdown" leading to "stagnation." In order to fully develop the important linkages among the works of Schumpeter, Toynbee, Sorokin, and others, the historical development of a society's creation of organizational and functional systems will be reviewed and related to the mechanistic-thermodynamic paradigm.

Cultural development, based on the creation of organizational and functional systems, is often characterized by a variety of conflicting terminology and definitions. Some authors of the civilization studies literature devote many printed pages to painstakingly providing such definitions with supporting rationale. Unfortunately for the reader, such definitions often lack the necessary scientific basis and the precision of mathematics, thus providing inadequate conceptual design for the nature and complexity of the particular matter being discussed. Precise, science-based definitions will be emphasized as concepts and models are developed in this work. This point is illustrated by the definitions and discussions of the concepts of society, culture, empire, civilization, and, most significantly, the laws of nature that apply to the processes of energy consumption.

The generic term *society* will be applied to a group of people who voluntarily organize and function to achieve common purposes of survival, security, and future socioeconomic success. The term implies the existence of some form of social order, morality, common purpose, self-determination, cooperation, and self-regulation, whether describing a primitive or more advanced and mature society. Members of a society possess some degree of centrality of mission and common agenda and thus consider a societal relationship as their most promising alternative for achieving their future objectives and aspirations.

Individual decision-making related to joining and enthusiastically participating in a society is based on the attractiveness of finding unity in the anticipated relationships with others. These judgments are influenced by ideologies, issues, and objectives pertaining to religion, governance, and politics; values; socioeconomics; and the nature and quality of leadership. Additionally, individuals may define themselves and create a given social order based upon such hereditary and cultural characteristics as ethnicity, race, tribe, language, organized religion, social practices, and some form of political ideology and nationalism. The same cultural elements are also the basis for people defining who and what they oppose, which is often founded on myth, ideology, and prejudice rather than on fact, reason, fairness, and the Golden Rule.

A composite of such characteristics begins to define the cultural traits of a given society. Leslie White notes that: "Culture, the cultural process, is an interactive process; it is composed of cultural traits that interact with one another, forming new permutations, combinations, and syntheses.... Cultural systems must be explained in terms of themselves."[7] In addition, Durkheim states that cultural traits "attract each other, repel each other, unite, divide themselves, and multiply."[8] Thus, the inherent distinctiveness, properties, and interactions of cultural traits within a society are responsible for observed cause-and-effect phenomena whose outcomes and consequences are dependent on human decision-making. Also, ethnicity denotes a unique group of people who are perceived to possess a cultural distinctiveness, expressed by their language, art, values, literature, religion, rituals, food, etc.

Depending on circumstances, cultural traits are capable of being divisive as well as unifying. Huntington points out that people separated by ideology but united by culture may eventually come together, as in the two post-WWII Germanys, whereas societies that are united by ideology or historic circumstances, but divided by culture, as with the U.S.S.R. and Yugoslavia, may disintegrate. He projects into the future that "the most pervasive, important, and dangerous conflicts, will not be between social classes, rich and poor, or other economically defined groups, but between people belonging to different cultural entities. Tribal wars and ethnic conflicts will occur within civilizations."[9] He also notes, citing Yugoslavia and Lebanon, that "people who share ethnicity and language but differ in religion may slaughter each other."[10] Thus, when issues and conflicts exist within a society that embraces a deeply respected value system, the consequences will depend on current circumstances and past experiences, but more significantly on the nature and intensity of hereditary cultural traits. Thus, when culture and change clash, culture usually prevails. The long-standing internal tribal, religious, and cultural violence in the Middle East attests to the fundamental nature and realities of such primitive instincts affecting human behavior.

Individuals are exposed to their society's adopted value system early in life and ultimately adopt, reject, or modify particular elements of the value system. This is based on personal aspirations, opportunities, successes, and disappointments as well as on responses to the influences of family, peers, and other members of their society and the external world. Sorokin recognizes "three supersystems of culture": religion, philosophy, and science, representing respectively the basic distinctive mentalities of faith, reason, and senses.[11] These are also defined in terms of the "Ethics of Principles," the "Ethics of Love," and the "Ethics of Happiness," respectively.[12] The shifting of values as socioeconomic conditions periodically fluctuate is inherent to Sorokin's model of societal maturation, particularly as a mature civilization approaches stagnation and eventual deterioration. Thus, a society's value system is central in the design, implementation, and modification of its socioeconomic system and the associated priorities, ethics, policies, strategies, and tactics. Thereafter, based on established economic systems, corresponding social and political systems are crafted, which may or may not reflect theoretically the essence of the adopted value system. Lawrence James in *The Rise and Fall of the British Empire* captures the typical cultural realities resulting from the interrelationships among economic opportunity and ambition, human nature, and cultural values of the expansionist aspirations of the British Empire:

> **The pursuit of profit remained the most powerful driving force behind Britain's bid for North American colonies. But from the start it was closely linked to a moral imperative founded upon contemporary conceptions of divine providence and the nature of the world and its inhabitants....[13] Commercial advantage, private greed, and a sense of divinely directed historic destiny were intermingled and bound together ... with a high-minded moral cause.[14]**

Authors gravitate to formal definitions of culture that reflect a social order whose organization and functional systems have become more mature and complex following a genesis phase. Consequently, as a result of a sustained period of societal growth, a social order will exhibit characteristics of a more highly integrated socioeconomic and political network. Thus, a *culture* is an organizational unit that is more distinctive and formally organized than a primitive society. Moreover, *the culture* refers to characteristics of a more unified and organized human grouping or society (i.e., more substantive, complex cultural traits). Consequently, societies, cultures, civilizations, and empires exemplify and exhibit specific attitudes, behavior, and skills; distinguishing forms of music, philosophy, literature, and fine arts; and technologies, tools, and ideologies reflective of a particular societal group and its era of existence.

The significance of societal characteristics influencing human behavior and the course of human history resides within a concept of culture that includes a type of social heredity. This constitutes a genetic guidance system for a society that also provides a secure pathway when all else in life fails. Conversely, the natural impulse of this ancestry or genetic-like guidance is also to resist change despite the recognition that life continuously evolves at an accelerated pace. Failure to recognize the need to adapt to socioeconomic change is a common flaw found within civilizations and can be fatal for institutions, commerce, and cultures. Joseph Schumpeter's concept of cultural "creative destruction" envisages established social, economic, and political policies and practices being continuously modified and replaced by progressive growth and transitions.[15] However, such momentum of change is frequently challenged by inherent human resistance to change that is independent of any particular socioeconomic philosophy. Thus, there exists an innate *human nature*, but also differentiating *cultural values and behavior*. All are contributing factors to a society's potential socioeconomic success or failure.

As a *culture* may be viewed as a more highly complex and integrated social order than a primitive society, likewise, a *civilization* may be considered as possessing an even higher order of societal complexity and integration than a culture. Thus, conceptually, the maturation of a society from its genesis phase to a more highly sophisticated, efficient, and effective social order may be viewed as a continuum ultimately leading to a mature civilization. This societal progression may be represented as a multivariable function capable of exhibiting periodic phases of growth and decay. Spengler notes that "every Culture has its own Civilization … the inevitable destiny of the Culture."[16] Carroll Quigley describes Spengler's view of a society's destiny as the "vigorous creativity" of culture leading to "a later stage of weakening moral fiber and devotion to selfish physical comforts that he called 'civilizations.'"[17] Interestingly, Spengler's vivid and extensive description of the deterioration of moral values, creativity, economic productivity, and social unity is strikingly consistent with Sorokin's model of shifting cultural mentalities that accompany increasing civilization maturity. It is noteworthy that despite the diverse definitions, models, and conceptual approaches to describing the various phases of cultural maturation and societal development, civilization studies authors recognize that all recorded civilizations have ultimately experienced common disintegration characteristics and phases.

Many authors have contributed to the comparative analyses of civilizations and have provided important understandings of the nature, identity, and trends of civilizations. Samuel Huntington[18] notes that eighteenth-century French authors intended the concept of civilization to denote the "converse of barbarism." A civilized society, being more urban, commercial, literate, and social, was the antithesis of a primitive society and thus represented progress and enlightenment. As a consequence, a civilization came to be considered as simply a larger and more socially, economically, and politically mature formulation of a culture. Authors have chartered such progressions from the primitive genesis phase undergoing successful socioeconomic growth to achieve the status of a culture and then advancing to a civilization with all of its rewards, accomplishments, failures, flaws, and complexities. Additionally, a culture's expansionism may produce a civilization composed of a collection of individual cultures. Such aspects of societal development will be extensively discussed in later chapters in the context of the mechanistic-thermodynamic paradigm.

To finalize the coverage of organizational terminology, a *nation or nation-state* will be considered as a recognized geographical, governmental, and political entity that engages in interactions beyond its borders. An *empire* will be considered as a large, multinational and multicultural enterprise occupying expansive territories beyond its original borders. An empire maximizes the probability of multicultural stress and conflict.

Human behavioral history, within a multicultural civilization or among civilizations, has demonstrated a fatal lack of mutual tolerance, and more importantly a lack of respect for diverse systems of human thought and cultural values. Unfamiliar characteristics of customs, spirituality, race, and political and religious ideology have been met historically with suspicion, antagonism, intellectual suppression, and violent, inhuman behavior. The prospect that elements of a newly discovered ideology or culture could be viewed as superior to those of one's own culture and readily be adopted is not a typical human reaction. History has witnessed intense disdain and hostility toward differing cultural traits, particularly spiritual ideologies represented by organized religions both within and external to multicultural societies.

> **Religion is the strongest feature of civilization, at the heart of both their present and their past.... In India, for instance, all actions derive their form and their justification from the religious life, not from reasoning.**[19] **Almost all civilizations are pervaded or submerged by religion, by the supernatural, and by magic: They have always been steeped in it, and they draw from it the most powerful motives in their particular psychology.**[20]

Thus, it is not particularly surprising that major civilizations have been closely identified with the world's great religions. Christopher Dawson notes, "The great religions are the foundations on which the great civilizations rest."[21] Max Weber identifies five world religions, of which four—Christianity, Islam, Hinduism, and Confucianism—are associated with major civilizations (the fifth is Buddhism). Authors have repeatedly noted the influence of spirituality and organized religion on the development, and more particularly on the rejuvenation, of civilizations. This includes highlighting the negative effects of sociopolitical activities by organized religions and the major deviation of a population's religious practice from its theological ideology (i.e., the inconsistency of the preaching versus the practice of religion: a reality of human behavior). B. G. Brander quotes Sorokin: "We cannot standardize the ideas of religion, but we can standardize in terms of the Golden Rule. This ethical basis for behavior is what counts." Brander adds: "Religion that is mainly verbal or ritualistic has not been potent, because of a yawning chasm it permits between preaching and practice."[22]

During the 1600s, British imperialism found it convenient to become the agent of divine providence and to confiscate the resources and dominate the lives of non-Christian "savages" living in the New World. Lawrence James describes the British rationalization for their righteous program of conquest:

> The moral question faced by Englishmen was, by what authority could they claim the fertile, untilled lands of North America? A broad and infallible answer was provided by the prevailing view of the divine ordering of the world and man's place in it. "God," wrote John Milton in defence of colonization, "having made the world for use of men ... ordained them to replenish it." The newly revealed American continent was favored with abundant natural resources by a benevolent God, but it was peopled by races who had never recognized nor acted upon their good fortune. Their willful inertia, combined with other moral shortcomings, debarred them from their inheritance which passed to more industrious outsiders.[23]

The breadth of the divergence of the relationships linking spiritual ideology, cultural traits, organized religions, politics, economics, and formal governing processes is an impressive indicator of the vast philosophical gulf that exists among the cultures of the world. Huntington makes this point: "In Islam, God is Caesar, in China and Japan, Caesar is God; in Orthodoxy, God is Caesar's junior partner. The degrees of separation and the recurring clashes between church and state that typify Western civilization have existed in no other Culture."[24] Such fundamental worldwide cultural differences suggest that effective global cooperation and diplomacy should utilize a variety of strategies reflecting respect for the diversity of cultural characteristics and beliefs, particularly in the form of church-state dualism. A lack of such understanding and sensitivity of cultural differences explains the limited success of Western-style methodologies, functions, and organizational structures that have historically been imposed on Middle Eastern and Asian populations. It also raises the issue of the effectiveness, particularly given the complexity and fast pace of the twenty-first century, of religious leaders playing primary, but behind-the-scenes, roles in providing political leadership and strategies directly to the masses as well as to subservient government leaders. Thus, the interrelationships among differing religious ideologies, the re-emergence of non-Christian cultures with their distinctive linkages of religion to governance, a declining Western civilization, and increasing global modernization and related energy requirements are defining the world's twenty-first-century agenda.

The power of a state or group is hence normally estimated by measuring the resources it has as its disposal against those of other states or groups it is trying to influence. The West's share of most, but not all, of the important power resources peaked early in the twentieth century and then began to decline relative to those of other civilizations.[25]

Huntington's acknowledgment that "the power of a state" may be "estimated by measuring the resources it has at its disposal" recognizes that religion, cultural traits, and human nature and behavior do not fully represent the potential variables of the dynamics of societal development. Historically, a society's effective power, influence, and socioeconomic prosperity have been intimately related to its real, available material assets, defined as initially being derived from nature. As will be fully discussed in a subsequent chapter, financial credit, individual wealth, and government currency do not qualify as national wealth. Only wealth-energy capital possessing physical value or energy content is a valid measure of a society's or a nation's wealth.

Joseph Nye[26] distinguishes between *hard power of military and economic strength* that is based on national wealth and *soft power of cultural appeal* as found in the twenty-first-century Middle East. The Islamic resurgence is a spiritually motivated God is Caesar, cultural phenomena while the Asian resurgence is based on a cultural modification of Western civilization's capitalistic economic model. The Islamic effect is based on population expansion and spiritual conversions related to long-term poverty and human deprivation, while the Asian success is a consequence of its growing global industrialized economy. Conventional wisdom is that soft power is inherently ineffective as a long-term strategy. The lack of economic productivity and a respected military establishment has traditionally led to an eventual lack of confidence and support for a society based only on a spiritual foundation.

Of the many diverse influences of human, cultural, and external variables affecting society's creation of successful organizational and functional systems, wealth and energy clearly play a dominant role. One may draw the conclusion that a society's acquisition of a continuous supply of wealth-energy resources is the fundamental imperative for a culture to achieve and maintain socioeconomic growth. Leslie White describes the centrality and context for available wealth-energy capital, thermodynamic principles, and the mechanistic and systems components of societal existence and prosperity:

> **Culture is but a means of carrying on the life process of a particular species, Homo sapiens, ... a mechanism for providing man with subsistence, protection, offense and defense, social regulation, cosmic adjustment, and recreation. But to serve these needs of man energy is required. It becomes the primary function of culture, therefore, to harness and control energy so that it may be put to work in man's service. Culture thus confronts us as an elaborate thermodynamic, mechanical system. By means of technological instruments energy is harnessed and put to work. Social and philosophic systems are both adjuncts and expressions of this technologic process. The functioning of culture as a whole therefore rests upon and is determined by the amount of energy harnessed and by the way in which it is put to work.**[27]

Essentially, White's projection of culture expresses the mechanistic-thermodynamic paradigm: "an elaborate thermodynamic, mechanical system ... harnessed and put to work," where "social and philosophic systems are both adjuncts ... functioning ... as a whole"; therefore, they rests upon and are determined "by the amount of energy harnessed and by the way in which it is put to work."

Wilhelm Friedrich Ostwald states: "The history of civilization becomes the history of man's advancing control over energy."[28] Additionally, George Grant MacCurdy describes the role of energy: "The degree of civilization of any epoch, people, or group of people, is measured by ability to utilize energy for human advancement or needs."[29] The advancement of civilization is synonymous with "man's advancing control over energy ... for human advancement or needs."

While the significance of available wealth-energy resources to a society's basic existence is indisputable, the absence of its inclusion as a major factor in the literature of developing civilizations is almost universal. Yet authors devote much attention to pointing out the failures of other authors "to explain" the historic inevitability of cultural disintegration. Quigley found fault with Spengler's concept of a pattern of transition from a stage of vigorous creativity to a later stage of decay. He comments: "As is usual among writers on this subject, no real explanation was provided for this loss of motion, although the pattern was applied to ten different cultures."[30] His equally critical review of Toynbee's process of change is summarized by "his failure to correlate the *stages* of change with the *process* of change and, above all, in his *failure to explain why* a civilization which has been 'responding' to 'challenges' successfully for centuries gradually ceases to do so, and decays"[31] (italics added for emphasis).

Quigley's comment relative to Toynbee's "failure to explain why" illustrates two major shortcomings of the existing civilizations studies literature. First, authors have universally restricted themselves to a narrow scope of limiting variables in their considerations of major factors affecting societal development. Consequently, they have produced only partially plausible rationales for historically observed events and cultural patterns but have been unable "to correlate the *stages* of change with the *process* of change." Second, authors have universally ignored the necessity for defining and utilizing a science-based economic model that incorporates precise and valid concepts of fundamental economic parameters such as national wealth. Thus, the "real explanation" sought by Quigley may be derived from the thermodynamic principles applied to functional socioeconomic systems, the main focus of this work to be developed in succeeding chapters. This approach provides an explanation for Toynbee's observation that a civilization, successful for centuries, gradually ceases to respond to change and decays.

Comprehending the various stages of societal development requires consideration of a very large and diverse set of variables or factors—most, if not all, being directly or indirectly influenced by the availability of wealth-energy resources and related scientific principles. A society's organizational and functional systems, consisting of a prodigious number of interdependent and complex social, economic, and political subsystems, exist in a constant state of flux. These elements of change are manifest by periodic fluctuations in socioeconomic prosperity or, more quantitatively, in the rate of cultural growth and decay (i.e., productivity). Agreement on the nature, scope, and significance of such variables that adequately and fully represent societal development and the maturation of civilizations is conspicuously absent from the literature.

A volume entitled *Culture Matters: How Values Shape Human Progress*, edited by Lawrence Harrison and Samuel Huntington, "explores how culture ... affects the extent to which and the ways in which societies achieve or fail to achieve progress in economic development and political democratization."[32] The methodology employed is an example of a narrowly conceived analytical perspective of variables affecting human progress, limited by inadequate consideration of a more inclusive and appropriate array of potential interrelated variables that influence socioeconomic maturation. Thus, this approach inherently places severe restrictions on the conclusions drawn from an assessment of such a multifaceted phenomenon as societal development.

In the Foreword, the editors address the well-publicized lack of agreement on the variables influencing societal development: "Increasingly social scientists turned to cultural factors to explain modernization, political democratization, military strategy, the behavior of ethnic groups, and the alignments and antagonism among countries." In addition, they note: "In the scholarly world, the battle has thus been joined by *those who see culture as the major, but not the only, influence* on social, political, and economic behavior and those who adhere to universal explanations, such as devotees of material self-interest among economists, of 'rational choice' among political scientists, and of neorealism among scholars of international relations"[33] (italics added for emphasis).

The editors reject the anthropological definition of culture (i.e., "the entire way of life of a society: its values, practices, symbols, institutions, and human relationships") based on the curious reasoning that "if culture includes everything, it explains nothing." They then define culture "*in purely subjective terms* as the values, attitudes, beliefs, orientations, and underlying assumptions prevalent among people in a society"[34] (italics added for emphasis). This generalized and limited "subjective" number of variables is insufficient to accurately represent societal development. Additionally, the authors leave the impression that somehow a more appropriate expanded array of variables, which may actually be more representative of the totality of societal functions, is either unnecessary or just impossible to contemplate. This is comparable to rejecting the need for pursuing quantum mechanics because it is too difficult mathematically and being content with classical mechanics in the study of atomic structure and explaining spectroscopy.

Additionally, the editors define "human progress ... as movement toward economic development and material well-being, societal equity, and political democracy." However, human progress constitutes a more expansive and diverse view of human existence and is obviously affected by a more significant number of interdependent variables. While it may complicate a full and complete analysis of a more accurate representation of the development of "human progress," the scope of significant variables may actually approach "everything." Regardless, the analysis must be appropriately conducted by a true and full representation of all variables, those that actually represent and affect societies! Thus, it is questionable how such a limited number of variables, representing only a small subset of actual cultural variables, could adequately represent "human progress" and, in particular, adequately represent "economic development and material well-being." How is it possible to address economics and material well-being based only on cultural and behavioral variables while omitting consideration of the availability and consumption of wealth-energy resources?

Reflecting Daniel Patrick Moynihan's chicken-and-egg cultural-political relationship, the editors comment that while culture, not politics, determines societal success, politics may alter culture and save it from itself. "The key issue ... is whether political leadership can substitute for disaster in stimulating cultural change. That political leadership can accomplish this in some circumstances is exemplified in Singapore," which is cited as an example of politics changing a corrupt culture.[35] It is noted that current or potential "economic development and material well-being" have historically been political motivators either to produce change or to maintain the status quo, depending on the particular related human objectives, strategies, and political power structure. Obviously, at some time and place many social, economic, and political factors are capable of individually contributing to, or detracting from, urban corruption and other social ills. However, this narrowly focused Harrison-Huntington approach lacks the rigor of a comprehensive representation of variables constituting major influences on societal development.

It is necessary to fully and accurately recognize and accommodate all variables of the complex systems of societal development in order to adequately comprehend human existence, as well as to maximize reliable predictions of a society's future performance under a variety of conditions and circumstances. The nature and importance of this requirement may be illustrated by the analogy of a *methodology for a systems representation of aircraft flight* to that of civilization development.

In order to provide an accurate and full representation of an aircraft in flight, the aeronautical engineer models all appropriate functions (i.e., mathematically characterizes all variables representing drag, thrust, gravitational, and lift forces, etc. that *may* affect its performance under all known circumstances). All such mechanistic variables must be identified and fully and accurately represented by forces acting on the aircraft so as to adequately explain the nature and potential outcomes of any attempted flight. In addition, a successful flight will depend on the pilot and maintenance crew possessing the necessary knowledge and skill, as well as the good judgment, to professionally perform necessary tasks (i.e., human factors). Thus, a complete understanding and description of the components of a successful aircraft flight must include the appropriate laws of nature, human capabilities and behavior, as well as the totality of the functional and structural systems of the aircraft and their energy requirements. These systems can be characterized utilizing science, engineering technologies and techniques, and operational mechanisms involving energy transformations. Information limited to an aircraft's history of past performances, human values, and the behavioral characteristics of pilots (i.e., the pilot culture) would be an inadequate basis for claiming to fully represent the variables and the nature of the aircraft's flight potential. This modeling methodology also applies to analyses of past successful and unsuccessful flights.

Likewise, culture may depict the historical context and societal climate that influences society's processes, particularly human behavior based on cultural mentalities as described by Sorokin and other sociologists. However, while culture may provide motivation and context for observed human values and behavior affecting socioeconomic and political decisions as well as long-term societal change, it does not represent the complete set of major influences. Clearly, human nature, attitudes, behavior, and the pilot culture represent a very limited array of influences on a successful airplane flight.

Cultural evolution represents a continuing set of outcomes from the dynamic forces of change. Culture is a stalwart guidance system, a genetic framework for the processes of life, whose elements are a motivator for human behavior. However, it is not the primary enabling dynamic that permits a society to survive and prosper economically or to provide a satisfying human existence. Energy is the primary enabler of the processes of nature and society. This deficiency in the modeling of the overall sequencing of cultural development is an aspect of the equation that has essentially been ignored by authors who have performed narrowly focused comparative analyses and explanations of cultural transformations.

The histories of Greece, Rome, Imperial Spain, and the British and American cultures illustrate similar characteristics of socioeconomic and political development that have been described by a number of eminent authors. For example, Toynbee's continuum of four stages from genesis through deterioration involves cultural growth and decay processes and serves as a model for other authors. In addressing social and cultural dynamics, Sorokin, Spengler, Morton Fried, Matthew Melko, and others envision models of progression from a simple society to a culture, to a civilization, and potentially to an empire. Such phases are described using such descriptive terms as crystallized periods, transitional stages, ossified, passive periods, rankings, and stratifications. This literature of the evolutionary process for societal maturation recognizes periodic advancements and setbacks as a society progresses through defined stages, phases, or periods. It will be demonstrated that these concepts are consistent with and enhanced by the mechanistic-thermodynamic paradigm. Likewise, it is consistent with Tainter's view that "as societies increase in complexity they do so on a continuous scale, so that discrete, stable levels will be difficult to define, and indeed may not exist."[36]

As a youthful society develops and evolves into a mature civilization over an extended period of time, an almost infinite number of potential influences and conditions may affect its socioeconomic and political systems and their component processes. These systems all require consumption of wealth-energy capital and produce outcomes, some of which may contribute to the aspirations of society, while others may not. The rates of progression of the almost infinite subsystems and subprocesses of a functioning society will vary with time. For any given time period, the overall rate of observed change may produce a sense of progress for some members of society depending on their status and circumstances. However, the fundamental principles of the energetics and mechanisms of the social, economic, and political processes are the same from a society's birth through its deterioration.

S. N. Eisenstadt addresses the sociological approach to change:

> **One of the major premises of sociological analysis has been that the causes of social change are inherent in the construction of the social order. This premise reflects the specific sociological *Problemstellung*, ... its major focus is the analysis of the conditions and mechanisms of social order and its constituent components, of continuity and change in the social order in general as well as in different types of social orders.**[37]

Eisenstadt's approach to social change focuses on the *transformative properties of social systems,* that is, the central explanations for observed social change are inherent in the construction of the social order and not primarily external or random events. Thus, "the construction of the social order" goes beyond the human elements residing within a society and includes the functional realities of nature and the appropriateness and effectiveness of the created political, economic, and social systems.

Sorokin reflects a similar view:

> **As long as it exists and functions, any sociocultural system incessantly generates consequences which are not the results of the external factors to the system, but the consequences of the existence of the system and of its activities.**[38]

Sorokin rhetorically asks, "Where shall we look for the roots of change of sociocultural phenomena?"[39] His response is that a sociocultural system "bears in itself the seeds of its change" and that change occurs "by virtue of its own forces and properties."[40] His vision of cultural maturation is a "whole integrated culture as a constellation of many cultural subsystems changes and passes from one state to another ... because each of these is a going concern and bears in itself the reason of its change."[41] The *seeds of change* constitute the culture's values, ethics, and integrity as reflected in a composite of its social, emotional, economic, intellectual, political, and physical assets and liabilities. However, the germination of seeds is only possible when a sufficient energy source and other resources are available and appropriately utilized.

> **In a study of a transformation of any sociocultural system, the partisan of the immanent theory of change will look for reasons or factors of the change first of all in the internal properties (actual and potential) of the system itself, and not in merely its external conditions.**[42]

The following chapters will demonstrate that an inclusive understanding and representation of a society's organizational and functional systems relies on the concept of wealth-energy resources and the associated laws of science. The science-based concepts of wealth, energy, and national wealth, as defined by Frederick Soddy, and the principles of thermodynamics provide the missing link to the sought-after full understanding of the inevitable decline of civilizations.

Societal development depends on a human element to manage the acquisition and consumption of wealth-energy capital that fuels societal functions, subject to Mother Nature's laws. However, the enabler of a society's organizational and functional systems of socioeconomic advancement is the acquisition, management, and consumption of wealth-energy resources. Regardless of human wisdom, skills, and behavior, the rate of cultural advancement is limited by the availability of wealth and energy.

Chapter 1
References and Notes

1. Spengler, Oswald. *The Decline of the West*, trans. with notes by Charles Francis Atkinson. New York: Oxford University Press. 1991. 24.
2. *The Encarta Dictionary*, on-line. 2006.
3. Tainter, Joseph A. *The Collapse of Complex Societies*. New York: Cambridge University Press. 2004. 5.
4. Ibid., 3.
5. Ibid., 18.
6. Schumpeter, Joseph A. *Capitalism, Socialism and Democracy*. New York: Harper & Row. 1975. 162.
7. White, Leslie A. *The Concept of Cultural Systems*. New York: Columbia University Press. 1975. 6.
8. Durkheim, Emile. *The Elementary Forms of Religious Life*. Glencoe, Ill.: The Free Press. 424.
9. Huntington, Samuel P. *The Clash of Civilizations and the Remaking of World Order*. New York: Simon and Schuster. 1996. 28.
10. Ibid., 42.
11. Sorokin, Pitirim A. *Social and Cultural Dynamics*. Boston: Porter Sargent. 1970. 20-39, 678-679, and 644.
12. Ibid., 414-429 and 678.
13. James, Lawrence. *The Rise and Fall of the British Empire*. New York: St. Martin's Press. 1995. 11.
14. Ibid., 31.
15. Schumpeter, Joseph A. *Capitalism, Socialism and Democracy*. 81-86.
16. Spengler, Oswald. *The Decline of the West*. 24.
17. Quigley, Carroll. *The Evolution of Civilizations*. Indianapolis: Liberty Fund. 1979. 130.
18. Huntington, Samuel P. *The Clash of Civilizations*. 40.
19. Braudel, Fernand. *A History of Civilizations*. London: Penguin Press. 1992. 22.
20. Ibid., 23.
21. Dawson, Christopher. *Dynamics of World History*. New York: Sheed and Ward. 1956. 128.

22. Brander, B. G. *Staring into Chaos: Explorations in the Decline of Western Civilization*. Dallas: Spence. 1998. 362.
23. James, Lawrence. *The Rise and Fall of the British Empire*. 12.
24. Huntington, Samuel P. *The Clash of Civilizations*. 70.
25. Ibid., 84.
26. Nye, Joseph S. Jr. "The Changing Nature of World Power." *Political Science Quarterly* 105 (Summer 1990): 181-182.
27. White, Leslie A. *The Science of Culture*. Toronto: Doubleday Canada. 1969. 367-368.
28. Ostwald, Wilhelm. "The Modern Theory of Energetics." *The Monist* 17 1907: 511.
29. MacCurdy, George Grant. *Human Origins II*. New York: 1933. 134.
30. Quigley, Carroll. *The Evolution of Civilizations*. 130.
31. Ibid., 131.
32. Harrison, Lawrence H., and Huntington, Samuel P. (Eds.). *Culture Matters: How Values Shape Human Progress*. New York: Basic Books. 2000. xv.
33. Ibid., xiv.
34. Ibid., xv.
35. Ibid.
36. Tainter, Joseph A. *The Collapse of Complex Societies*. 29.
37. Eisenstadt, S. N. *Revolution and the Transformation of Societies*. New York: The Free Press. 1978. 19.
38. Sorokin, Pitirim A. *Social and Cultural Dynamics*. 639.
39. Ibid., 630.
40. Ibid., 633.
41. Ibid., 638-639.
42. Ibid., 633.

Chapter 2
Wealth, Energy, and Science-based Economics

> "Neither individuals nor communities can escape conforming to the laws of matter and energy, however they may apply them to their own ends.... Life works according to, not against, the principles of the physical sciences."[1] ... "Machines are merely imitations of life."
>
> Frederick Soddy[2]

In 1924, Frederick Soddy (1877-1956) published a remarkable manuscript entitled *Cartesian Economics: The Bearing of Physical Science upon State Stewardship*. His expressed intent was "to try to bring the existing knowledge of the physical sciences to bear upon the question 'How do men live?' This question ought to be the first the economist should try to answer.... The modern economist seems to have forgotten that there is such a question, whilst the earlier ones lived at a stage of the development of scientific knowledge when no exact answer was forthcoming."[3]

Soddy, a Nobel laureate in chemistry, proceeds to answer his own question, "How do men live?" by relating the physical sciences to economics, reflecting the thoughts of John Ruskin. Ruskin (1819-1900) is considered by some as "the greatest Victorian bar Victoria" and an accomplished artist, scientist, poet, environmentalist, philosopher, and the pre-eminent art critic of his time. Ruskin provided notable political economic insights in his works: *Unto the Last* (1862), *Essays on Political Economy* (1862), and *Time and Tide* (1867). He developed a comprehensive, but yet incomplete, vision that science (i.e., energy) was an appropriate and necessary foundation for economic theory. His view that economics possesses a scientific nature is revealed in his comment that "the disturbing elements in social problems ... operate chemically."[4] Even more insightful was the physical nature of his definition of wealth as constituting a flow process, "the flowing of streams to the sea is a particular image of the action of wealth.... No human laws can withstand its flow."[5] It is easily appreciated that Soddy, given his scientific knowledge and accomplishments, was able to intellectually relate Ruskin's hypotheses to thermodynamics principles and to economic theory, policy, and practice.

However, in order to rigorously apply the laws of physical science to economic, social, and political systems, it is crucial to demonstrate the appropriateness and validity of the sciences beyond their usual involvement in mechanical and thermal processes and to the more generalized functions of a civilized society. While it has been widely accepted that the physical sciences govern the world of matter, space, and time, authors external to these disciplines have essentially ignored the appropriate inclusion of physical science concepts in economics and comparative historical analyses of civilization development. The universality of physical science principles governing a society's daily existence and the long-term evolution of the animate and inanimate components of the universe have not been transferred and adopted as basic components of societal processes that determine potential survival and prosperity.

Intuitively, one may comprehend that the creation and implementation of activities designed to meet a society's basic needs and aspirations must involve a fuel or energy component, whether such activities pertain to human life, electrical and gasoline engines, or economic activity. Each of the processes that contribute to a society's survival is composed of complex mechanisms that require energy, whether it is an economic, social, governmental, or military function. Whether a society fails, barely survives, or greatly prospers will depend on many factors including its ability to acquire, process, and manage sufficient wealth-energy resources. Therefore, the degree of long-term cultural success, while being influenced by many general factors, is unvaryingly affected by the efficiency and effectiveness of the investment of adequate wealth-energy resources.

The mechanisms of basic cultural functions and their energy transformations that jointly constitute the mechanistic-thermodynamic paradigm are fundamental to adequately and fully representing a society's economic system and its operational existence and consequences. The question exists as to whether economics, as practiced during and since the twentieth century, is based on scientifically valid economic models that represent pragmatic and compelling socioeconomic systems of financial integrity. Surprisingly, *modern economic theory is still based on a mechanistic foundation* comparable to the physics of classical mechanics at a time when the scientific world understands that mechanics alone does not adequately and fully represent the realities of the physical, chemical, and biological world. While mechanics deals satisfactorily with the processes of locomotion (i.e., the when, how, and where of physical events taking place), it does not represent the attendant elements of its energy requirements and thermal consequences.

The late nineteenth century's discovery of thermodynamics provided a more comprehensive and satisfying explanation of life and the physical world. For the first time, it was understood that activities and processes of the universe were not only mechanistic, but also possed a related energy component, managed by Mother Nature's principles of thermodynamics. Thus, the mechanistic-thermodynamic paradigm applies to any process that possesses motion or activity, that is, in thermodynamic terms, performs useful work. As stated by Juan Martinez-Alier: "The starting point of economics should be the first and second laws of thermodynamics.... Energy, in any of its forms, should be the starting point of economics."[6]

> **The fact that a natural law is involved in every aspect of man's behavior is so common that we would not expect the study of the influence of the Entropy Law on man's economic actions to present any unusual complications.... What is more, these avenues lead beyond the boundary not only of economics but of the social sciences as well.**[7]

Friedrich Wilhelm Ostwald (1853-1932), who received the Nobel prize for chemistry in 1909, was one of the first scientists to grasp the intrinsic function of energy in the social sciences. He noted that the development of a civilization was dependent upon an increasing availability of energy, and more specifically on the efficiency of its transformations.[8]

Patrick Geddes (1854-1932) recognized the linkage of historical developments with the consumption of energy and the importance of Ruskin's vision of the role of the physical sciences in the processes of life. Geddes developed an energy accounting methodology that represents net energy and material transformations, illustrating production inefficiencies and the relative value between energy inputs and outputs.[9]

Unfortunately, the social sciences, and specifically economics, have essentially remained in the restrictive world of classical mechanics, ignoring the role that energy plays in the functional processes that define the human existence. Juan Martinez-Alier describes Max Weber, the economics historian and sociologist, as viewing economic theory as:

> a set of deductive propositions derived from the postulates about human conduct and hypotheses about the world which it was not fruitful to discuss, whatever psychologists, historians, and anthropologists could say about real human behaviour and whatever the physicists said about the availability of resources.[10]

Frederick Soddy cites two areas of human confusion and miscalculation that hinder mankind's understanding of cultural development; the first is "the sublimated conception of the mental world inextricably mixed up with the physical."[11] This mixture consists of the human "mental world," managed by free will, and Mother Nature's "physical world," controlled by science and probability. The second blunder is that "life ... is essentially ... a dualism, and any attempt to subordinate either partner is fatal. But the economist is peculiarly liable to mistake for laws of *nature* [for] the laws of *human nature* and to dignify this complex of thermodynamic and social phenomena with the term 'inexorable economic law'"[12] (italics added for emphasis). Soddy concludes:

> We have rather to find the interaction between the commonest forms of matter and energy on the one hand and will and direction on the other.... The principles and ethics of human law and convention must not run counter to those of thermodynamics. For men, no different from any other form of heat engine, the physical problems of life are energy problems.[13]

Clearly, Soddy's "dualism" of human existence has not been the prevailing economic thought during and beyond the twentieth century.

However, it is inconceivable that a globally significant twenty-first-century nation would not embrace a science-based economic model as the foundation for its economic policy.

In the words of Nicholas Georgescu-Roegen,

> **Every subsequent development in thermodynamics has added new proof of the bond between the economic process and thermodynamic principles. Extravagant though this thesis may seem prima facie, thermodynamics is largely a physics of economic value.... Of all physical concepts only those of thermodynamics have their roots in economic value.**[14]

The nature, manner, conditions, and assigned priorities for wealth-energy resource consumption determine a society's *economic effectiveness* (i.e., its conversion of national wealth resources into "economic value" and new wealth-energy capital). The degree of *societal economic effectiveness* reflects an achieved level of technological and engineering capability intended to maximize industrial thermodynamic efficiency as well as a rate and quantity of production. It also reflects the wisdom of investment and of balanced priorities for the production of essential versus nonessential goods and services capable of achieving sufficient *economic value* to attain a high level of cultural productivity. These concepts will be considered more fully elsewhere.

Soddy fully develops the linkages among national economic policy, consumer and public finance, and the laws of thermodynamics. A well-respected chemist, a pioneer in the study of radioactivity, he was awarded the Nobel prize in chemistry in 1921 for his work with Lord Rutherford on the theory of radioactive disintegration. In later life, Soddy diverted his professional attention from chemistry to economics and social issues, writing extensively on the subjects of the nature of wealth, currency, debt, credit, and wealth distribution. This new career apparently resulted from his discouragement with the misuse and nonuse of science and scientific discoveries to resolve economic issues of war and poverty, and flawed and delusive public policy. Soddy was greatly influenced by his exposure to John Ruskin's view that economics required a scientific component.

Ruskin espouses the concept that life and science should be the underpinning of economic thinking and flirts with, but does not fully develop or embrace, the requirement that wealth must be considered as a physical reality. Soddy expands this concept and applies it to national economics and to personal and public finance. Ruskin also notes that his definition of wealth "has often been incidentally given in good Greek by Plato and Xenophon and in good Latin by Cicero and Horace."[15] Most particularly, Ruskin's description of wealth in the physical terms of a dynamic flowing stream led Soddy to apply his knowledge of science and specifically thermodynamics to the creation of science-based concepts of economics, finance, and national productivity. His article, "Economic 'Science' from the Standpoint of Science," illustrates Ruskin's influence: "The economic being of men depends solely upon a continuous flow of energy from the inanimate universe applied by human intelligence and industry to human ends.... This flow ... is the wealth of the world.... Debt is purely human invention and convention."[16]

Soddy came to understand and appreciate the momentous ramifications of financial debt and currency not qualifying as wealth in the sense of not being a real physical quantity possessing energy or economic value capable of fueling economic productivity. Consequently, any non-science-based (i.e., non-energy-based) economic model is inherently and conceptually flawed, and thus inadequate to provide financial integrity for a socioeconomic system. Hence, a most appropriate title was selected for his book, *Wealth, Virtual Wealth and Debt: The Solution of the Economic Paradox*.

In his view, the early twentieth-century economic theories, policies, and practices required revision founded on scientific principles. This revision should recognize the principle that a socioeconomic system must possess a healthy flow of energy and materials capable of generating appreciable *economic value* and that such an energy flow qualifies as a thermodynamic process.

By the end of the twentieth century, economists such as John Kenneth Galbraith understood the necessity and *economic value* of "a steady flow of purchasing power ... the flow of aggregate demand for goods and services." While not basing his view of "the Good Economy" on thermodynamics, Galbraith realizes that the "stabilization of the flow of aggregate demand is the vital factor."

> **The modern economy requires a steady flow of purchasing power—in economic terms, aggregate demand—that is sufficient to utilize the available productive capacity, encourage the requisite expansion therein, and employ all available workers. The flow of aggregate demand for goods and services must keep the economy at or near that limit.... The stabilization of the flow of aggregate demand is the vital factor.... This will lead, in turn, to increased private investment expenditure and more consumer spending.[17]**

Soddy provides a scientific basis for Galbraith's "steady flow of purchasing power—in economic terms" and precisely defines the economic parameters and conditions that permit an accurate determination of economic value capable of maximizing long-term socioeconomic prosperity. In essence, Soddy endeavors to resolve "how men should live" so as to provide society with the prospect for greater economic prosperity, equity, happiness, and world peace which he found lacking in the early twentieth century and which he feared would continue into the future ... which it has.

Soddy's view is that appropriate scientific definitions of economic parameters are necessary in order to represent economic theories as a true science. More explicitly, only if the laws of thermodynamics are applied to national wealth will models of national economics constitute valid and constructive socioeconomic systems. Further, he projects that such a theoretical, scientific foundation for creating national economic policy would permit a more rational understanding, appreciation, and approach to such important cultural issues as unemployment, sound investing, and equitable wealth distribution.

This chapter will explore the work of Soddy and others to develop the thesis that society's prospects for economic survival and prosperity should be viewed from the vantage point of the laws of science. In order to develop this thesis, it is necessary to define and utilize the concepts of energy, wealth, wealth-energy resources, national wealth, and other related economic parameters, all of which must be congruent with the laws of the physical sciences.

At an early age, everyone learns to respect the law of gravity. In contrast to the theory of evolution and economic theories, there is almost universal understanding and respect for the influence of gravity on everyday life. One would expect the law of gravity to operate on all objects free in space regardless of their chemical nature, shape, or density, or the manner in which the object is released to freely fall. Gravity is a good example of a well-known and well-respected physical law of nature that governs the world of matter, space, and time regardless of the nature of the particular physical activity taking place. One may view gravity as a force that is responsible for an object falling when unsupported by an equal and opposing force. The branch of physics referred to as mechanics enables one to calculate and predict the position in space of a freely falling object at any given moment, the time required for it to hit the ground, and its velocity on impact.

If such a massive object falls a significant distance to the ground, it is observed at the point and time of impact that the object and the ground are a bit warmer than the surrounding surface. This observable generation of heat is related to another set of Mother Nature's laws, those of classical thermodynamics that govern energy and heat transfer. This observable thermal phenomena of heat generation is not addressed by the branch of physics referred to as classical mechanics, thus another branch of science, thermodynamics, must be invoked to adequately represent the observed generation and transfer of heat by the impact of the object striking the Earth. Thus, one set of Mother Nature's laws regulates the mechanics or the sequential physical nature and characteristics of the falling object, while another set of laws regulates details of the heat transfer phenomenon. More generally, the overall behavior of physical processes, including the action of motion and the associated energy transfer, can be described, predicted, and accounted for by these two branches of physics.

The integrated knowledge of these two related aspects of the physical world applies equally to the activities and functions of people, machines, and the planet Earth. However, historically, economic theory has dealt with only the mechanics of finance and economics and not with its science-based wealth transfer, even though it has been widely recognized that the activities of man and machine, engaged in life's processes, cannot be fully and adequately represented by mechanics alone. Clearly, a complete and adequate representation of economic theory also requires a science-based energy component. The significance of ignoring thermodynamics in the representation of economic modeling may be compared to the folly of attempting to fully represent the characteristics and properties of an automobile traveling 50 MPH by invoking only the physics of its motion while ignoring its kinetic energy and rate of energy consumption.

> **Every single physical activity that humankind engages in is totally subject to the iron-clad imperatives expressed in the first and second laws of thermodynamics.... The laws of thermodynamics provide the overarching scientific framework for the unfolding of all physical activity in the world.**[18]

Thus, if economics is to be considered as a science-based discipline, the principles of physical science, including the laws of thermodynamics, must be appropriately applied. The physical sciences are an integral part of all life as we know it; Mother Nature shows no favoritism toward when, on whom, or where she enforces her laws. Additionally, these principles apply to society's political and social initiatives and are thus essential to the creation of sound economic models and social policy. Such a science-based economic model provides the appropriate framework upon which to conduct comparative historical analyses of societal development, particularly the search for a satisfactory explanation for the inevitable collapse of mature civilizations. The comprehensive and systematic development of such an economic model and its ramifications for ultimate cultural collapse begin with a consideration of the topics of energy and wealth.

National wealth, based on a scientific definition, is the enabler of a society's economy in an analogous fashion to gasoline being the fuel that permits a car's engine to function. Likewise, the operation of a machine by an individual's physical labor or from electricity requires an energy source. The human source of energy is the nourishment received at mealtime, while electrical energy for machinery may originate from coal, oil, falling water, or nuclear material. In an analogous fashion, a national economy requires a fundamental driving force based on an energy source capable of providing dynamic activity for daily survival and a satisfying quality of life for its members. This economic driving force or economic fuel is derived from a nation's wealth-energy resources.

In Frederick Soddy's world, national wealth and energy are fundamental to the existence of a society's everyday life, with all energy consumption being governed by thermodynamic principles.

Furthermore, the overall impact of national wealth consumption extends beyond its primary function of providing the driving force for an economic system. Inevitably, the consequences of economic activity are multifaceted and affect the fundamental nature and structure of a society, having physical, social, financial, and political ramifications.

Wealth-energy resources are consumed as a society formulates and executes economic decisions within the realities of a given social and political environment. Thus, available material and energy resources combine with human mental capabilities and sensibilities to produce economic strategies, policies, decisions, and outcomes. In the past, and most likely in the future, such outcomes have been good, bad, and indifferent for various segments of a society. Such consequences constitute the economic, social, and political realities for a culture, all of which are mutually interactive in a cause-and-effect relationship. Thus, as energy and wealth are consumed within a given culture attempting to achieve a good life, events within a community inherently produce a variety of conflicting consequences having material, energy, and socioeconomic consequences.

It is appropriate to consider the premise that energy and wealth are parallel concepts. Energy drives the human existence, as we know it, and is the fundamental force of nature affecting the status of all animate and inanimate components of this earthly system. Life is a struggle to continually secure energy resources for animal and machine consumption, thereby enabling a society to survive and hopefully achieve a more prosperous and satisfying existence. The necessity for a continuous consumption of energy resources is basic to all life forms. Energy consumption is defined as the transformation of a useable form of energy, whose heat content is dispersed, into some combination of useful work and waste energy. Such an energy transformation process from one initial stable condition or state to a second stable state is characterized as a flow of energy whose directionality is determined by thermodynamic principles. As will be developed in the next chapter, Mother Nature's laws determine the spontaneous direction of this energy flow phenomenon.

Prior to the nineteenth century, man's use of energy was primarily based on revenues from nature (e.g., sunlight) that were renewable in a short time frame relative to the human life span. These expenditures of the Earth's energy revenues originating from the sun and producing food, fuel, clothing, and shelter materials were replenished in a time frame normally sufficient to meet the fundamental, although minimal, requirements for human survival. From the perspective of the universe, the reservoir of available energy is being continually transformed, with only two possible outcomes: Available energy from the sun is harnessed and harvested by people to create *wealth-energy* resources necessary for conducting societal processes, or is wasted by going unused or converted into heat energy.

With the improvement of James Watt's steam engine during the late1770s, human civilization began an energy migration from the primary use of a renewable energy source (wood) to a stored, nonrenewable energy source (coal). In subsequent centuries, society's energy consumption continued its energy-source migration while consuming increasingly larger amounts of nature's resources. This was necessary in order to meet the needs of an increasing population and society's insatiable appetite for a more rewarding and sophisticated standard of living. Obviously, a transition from consuming only the sun's renewable energy resources as fuel to consuming nonrenewable energy deposits such as coal, oil, and natural gas was a monumental turning point, having significant long-term consequences. This was the beginning of an exponential increase in the Earth's population and in science and technology applications to use, abuse, and deplete the Earth's natural resources at a greater rate than Mother Nature could replace them. In general, the planet's consumption of energy resources, which inherently produces waste materials and dissipates energy, has become increasingly out of balance with nature's ability to create them and to absorb waste materials.

Energy may be precisely quantified and standardized using internationally accepted units directly *associated with a physically measurable quantity*. The joule is a unit of energy that is defined as the *work accomplished* when a unit of force, the Newton, causes a displacement of one meter in the direction of the force. Work is the product of force and distance. The Newton is defined as *the force sufficient to accelerate a mass of one kilogram* one meter per second per second. Newton's law states that force equals the product of mass and acceleration. Alternatively, energy may also be expressed in units of ergs or calories, with all three energy units having interconversion factors. Thus, *energy must be defined in terms of a physical quantity and may be viewed as the potential to accomplish work*, as for example to lift a box or operate a machine. This seemingly overabundance of detail regarding the definition of energy will become significant as *wealth is defined only in terms of a physical mass having either value or heat content, in contrast to Wall St. financial banking practices of assigning "value" to paper documents.*

A given quantity of fuel, such as a gallon of ethanol, possesses a precisely defined energy content (e.g., calories per gallon). If a given quantity of ethanol of known energy content undergoes combustion to form carbon dioxide and water, the laws of thermodynamics and the process conditions imposed will determine the portion of the originally available energy content committed to accomplish some desired useful work or outcome. The remainder of the initial energy content will end up as wasted heat energy. Thus, ethanol possesses useable energy, which establishes its *intrinsic value or wealth;* its energy content is related to its specific, unique properties. Such *properties of value symbolize a wealth-energy commodity* and permit precise, scientifically based, quantifiable accounting methodologies for energy transfer, consumption, and storage. Such an energy balance sheet extrapolated to an entire civilization or nation embodies its financial value and provides insight into the potential well-being and viability of its economic system. This approach is a more useful and precise tool of accountability and guidance for financial and economic matters than the usual currency system and methodologies. In the twenty-first century, the accepted yardstick of measuring economic wealth and success is a currency-based system that inherently exhibits volatility and is subject to uncertainty of value. Such systems generate national and international financial misconceptions, insecurity, and economic instability. The U.S. "bailout" of financial institutions during the 2008 banking crisis was funded by incurring additional national debt (i.e., printing more money), supposedly to re-establish social and economic stability.

Consequently, accountants, bankers, and economists routinely manipulate currency-based wealth, which inherently depends on a *value of faith* rather than on a *value of substance* associated with a specific quantity of physical material. A *value of substance* is directly related to its *energy content* and/or its *economic utility* and is inherently capable of producing new forms of wealth. Such *wealth-energy* material properties provide a precise and valuable coin of the realm for a science-based economic system. It is capable of providing a society with a more precise and valid accounting of its national wealth and a greater sense of economic integrity, value, and security.

> **That realm cannot be rich whose coin is poor or base. William Cecil (1520- 1598), an English statesman referring to reform of coinage by Elizabeth I.**

Historically, as a result of the continuous pursuit of progress, the fuel of choice has evolved sequentially from wood to coal, to oil, to natural gas, and to nuclear energy. Each successive, more advanced energy source is technically more difficult to prepare as an available fuel, thus requiring the expenditure of greater amounts of energy to secure a suitable form for the fuel's ultimate use. Each advancing generation of energy sources becomes more costly to discover, recover, and process than each previous energy type. In a like manner, a clothing evolution progressed through the stages of leather to wool to cotton to synthetic materials, which also requires greater expenditure of energy to produce per unit of clothing than the previous stage. Such social progress has gradually depleted the Earth's energy resources, and since the late eighteenth century, the planet's consumption of nonrenewable resources is unwinding the Earth's energy clock at an accelerating pace. Unfortunately, recapturing such expended energy resources is as impossible as recapturing the time of yesterday!

The daily existence of a society depends fundamentally on its national wealth in the same fashion as human life relies on current food sources. National wealth is, in essence, the fuel upon which a society exists in the same way that oil, natural gas, and electricity are able to provide the necessary energy for the functioning of a factory or a city. Governments and the commercial sector create a national economy based on financial planning and strategies designed to achieve established goals and objectives. The primary economic thrust to maintain a basic and stable societal existence will, over time, create social, economic, and political consequences, which may or may not meet expected or desirable positive outcomes. During the last 4,000 years, many opportunities and self-inflicted problems have escalated the rate of the World's wealth-energy consumption as well as expectations for continuous socioeconomic advancement and individual prosperity.

A sixteenth-century emperor expends national wealth and assumes significant debt to finance New World conquests and secures gold and silver in order to achieve economic and related sociopolitical and military ambitions consistent with global supremacy. Given these priorities, a continuous supply of new wealth becomes necessary as territorial and commercial objectives are implemented and partially realized. Thus, the empire's imperialistic expansion necessitates the unremitting creation of new structures and functions, all of which consume wealth-energy resources in support of expanding commercial interests, military supremacy, and global dominance. Essentially, the empire becomes locked into an endless cycle of wealth and energy consumption in order to continue functioning and pursuing a progressively more costly imperialistic agenda.

National wealth is the driving force of a socioeconomic system, and its productivity will depend on the degree of (a) innovation and discoveries, (b) available wealth-energy capital, (c) knowledge and technology, and (d) commercial markets and skilled labor. These major economic elements provide society with the capability of maximizing the probability of creating an efficient and effective economic engine.

It is important to note that only wealth-energy resources that are in a usable form and available to the pertinent production process can affect an economy; this characterizes a *flow of energy* and constitutes true wealth. Consequently, crude oil deposits are not considered as being available and in a usable form, as they do not comprise a useful flow of energy. If energy sources are unable to directly interface with a society's economic processes and contribute to socioeconomic productivity, they do not represent *national wealth*. Modern Middle Eastern history provides an appropriate illustration of abundant, unavailable energy deposits being irrelevant as a major economic factor and in promoting regional socioeconomic prosperity.

Thus, wealth, defined precisely as a scientifically based parameter, is the foundation for an energy-based economic model that incorporates the principles of thermodynamics. The nature and functions of a society's economic processes, all of which are energy driven, are judged in terms of what is popularly, but often inaccurately, referred to as *wealth*. While it is generally recognized and accepted that something of value is required to buy or pay for the necessities, benefits, and luxuries of life, a rigorous scientific definition of wealth is not the usual component of a society's economic model or its characterization of national wealth. The practical, modern financial methodologies commonly utilize debt, credit, and national systems of paper currency as components of wealth.

However, wealth must have the physical form of matter, as opposed to being an imaginary financial creation such as credit. Additionally, *wealth is produced only as a result, and at the expense, of an energy consumption process.* This transformation of useful and available energy to new wealth-energy forms constitutes a flow of energy that makes possible economic production and desirable societal outcomes. Consequently, *the primary requisite of economics is the flow of energy* whose significance is exemplified by its primary and unique role as the driving force for societal production processes. It follows that, if wealth is defined as physical matter possessing useful and available energy and/or value, wealth is a positive quantity. Based on this premise, debt must be a negative quantity, that is, an imaginary negative number. Soddy describes debt as being "subject to the laws of mathematics rather than physics. Unlike wealth, which is subject to the laws of thermodynamics, debts do not rot with old age and are not consumed in the process of living."[19]

It is instructive to note that coal has economic value and is also a source of usable energy, which may be utilized in production processes to create new forms of wealth. Coal is a wealth-energy resource (i.e., a commodity). However, while steel has value in potentially generating new wealth, it does not possess available, useful energy; its energy content is referred to as being unavailable or, in scientific terms, bound energy. Materials such as steel possess only bound energy and no consumable energy for further production purposes (i.e., to perform useful work). Steel is an example of a *wealth commodity or wealth capital* that may serve as an *agent of production*. Thus, wealth may be placed into two categories, depending on whether its value is considered relatively perishable or permanent. Obviously, no form of wealth is truly permanent and, as a practical matter, durability may vary from clothing to diamond.

Soddy defines "two thermodynamic categories of wealth" production.[20] *Perishable wealth possesses energy content* capable of being consumed at a later time for useful purposes, formally designated and distinguished from other types of wealth by Soddy as "Wealth I" (e.g., food and fuels). *Permanent wealth is the product of a process* in which the energy of transformation is totally consumed in the process of producing it, leaving no usable stored energy in the product to potentially energize additional new-wealth producing processes. He designates such products as "Wealth II" (e.g., steel). The difference signified by the two "Wealth" categories is based on the nature of the wealth-energy content resulting from the commodity's production process. In the production of food, as a wealth-energy commodity, its resulting energy content is in a useable form capable of being transformed into energy for further use in supporting life. Possessing usable energy content defines the value of food commodities. However, the material form of the food and its energy content no longer exists in its initial form after consumption, although some portion of its energy content will be utilized for human processes of life. Conversely, steel is an example of permanent wealth, Wealth II, and while not possessing usable and available energy in the same sense as food, it is capable of being used to generate new wealth as an *agent of production*. Thus, *Wealth I, representing both value and stored energy*, loses both as it is consumed by the transformation or human processing sequence. *Wealth II represents an asset of value* capable of being used in the production and supply of new forms of wealth, but does not possess consumable energy content itself to pass on for any subsequent production process. Accordingly, Wealth II commodities are considered as *wealth-energy capital* that enable and facilitate potential new wealth-energy capital production by virtue of possessing *materialistic value* rather than *energy content value*.

Note that if coal reserves remain in the ground and are never used, they are not considered as a wealth-energy resource until such time as the coal becomes available in a usable form and enters the economy's production flow process. It will be shown that a civilization that is able to maintain a continuously high level of economic effectiveness in its conversion of national wealth-energy resources into new forms of wealth-energy capital increases its probability for continued prosperity and long-term survival. It should also be noted that Wealth I commodities, as a group, constitute required production energy sources and are the primary enabler for the production of Wealth II commodities (i.e., *agents of production*). Thus, the availability of Wealth I, t*he fuel of socioeconomic processes,* is the limiting and determining agent for a civilization's rate of economic development and continuous productivity. That is, regardless of any other factors, the availability of Wealth I is the controlling element of the rate of cultural growth and the primary fuel for society's basic support systems, including transportation and the provision of food, clothing, shelter, and security.

If a civilization is unable to acquire and maintain adequate levels of Wealth I resources, the inevitable consequence is a decline in its Wealth II production of goods and services and, most likely, its economic productivity. Thus far in the history of the world, the great mature civilizations have been unable to avoid this pitfall of inadequate Wealth I capital, albeit as a result of excessive, grandiose aspirations and nonproductive, unwise expenditures of wealth-energy resources, reflecting a self-centered, materialistic society. The oil crisis faced by Western civilization during 2008 when worldwide oil demand outpaced oil production and per-barrel prices ranged from $50 to $150 illustrates the potential impact of inadequate Wealth I capital resources on economic stability.

> **A definition of wealth must be based upon the nature of physical or material wealth, in the sense of the physical requisites which empower and enable human life—that is, which supply human beings with the means to live.... These enabling requisites are derived from and produced by the flow of available energy in Nature.**[21]

For a national economy to "empower and enable human life" requires wealth-energy resources and human time and labor regardless of any issues related to wealth ownership and distribution. Individual members of a society all contribute to the sum total of the life or activities of the community, but the wealth of the community is not as easily represented by individual claims upon current and projected future wealth, as scientifically defined. Thus, Soddy summarizes this importance distinction: "The production of wealth is, in civilized communities, communal rather than individual.... But the use and consumption of wealth, in the enabling and empowering of life, is individual and not communal."[22]

Remarkably, in 1862, Ruskin apparently understood the concept of community or national wealth and that individual wealth and debt are not true wealth:

> **If, in the exchange, one man is able to give what cost him little labour for what has cost the other much, he "acquires" a certain quantity of the produce of the other's labour. And precisely what he acquires, the other loses. In mercantile language, the person who thus acquires is commonly said to have "made a profit"; and I believe that many of our merchants are seriously under the impression that it is possible for everybody, somehow, to make a profit in this manner. Whereas, by the unfortunate constitution of the world we live in, the laws both of matter and motion have quite rigorously forbidden universal acquisition of this kind. Profit, or material gain, is attainable only by construction or by discovery; not by exchange. Whenever material gain follows exchange, for every plus there is a precisely equal minus.... Unhappily for the progress of the science of Political Economy, the plus quantities ... make a very positive and venerable appearance in the world, so that everyone is eager to learn the science which produces results so magnificent; whereas the minuses have, on the other hand, a tendency to retire into back streets ... which renders the algebra of this science peculiar, and difficulty legible; a large number of its negative signs being written by the account-keeper in a kind of red ink, which starvation thins, and makes strangely pale, or even quite invisible ink, for the present.** [23]

In the science-based definition of wealth, "Money, credit, and other legal claims to wealth are debts rather than wealth."[24] Thus, there exists an extremely important fundamental distinction between national wealth and individual wealth, as individual wealth, unlike national wealth, may merely be national debt. As stated by Soddy:

> **Money, credit and other legal claims to wealth are debts, rather than wealth. Labour and inventions are not wealth, though essential factors in its production.... The physical definition of wealth is a form or product of energy or work which enables or empowers life.**[24]

Given the definitions of wealth-energy capital and national wealth that provide the framework for economic models that conform to the requirements of thermodynamics, it becomes appropriate to define other economic and financial terms. Money or currency is a necessary societal device that enables individuals to function in societal relationships by providing *purchasing power* within one's socioeconomic environment (i.e., relating the concept of virtual wealth to the everyday processes of acquiring goods and services). Purchasing power is a device for exchanging wealth and may include, in addition to wealth itself, labor, credit, and money. However, *money is only a claim to wealth*, an acknowledgment of a community's indebtedness to its owner, or a token of wealth.

> **All money, properly so called, is an acknowledgment of debt ... a documentary promise ratified and guaranteed by a nation.**[25]

Many forms of purchasing power exist that theoretically may be exchanged for wealth but, in actuality, represent *ownership transferal devices* without constituting an actual *wealth transfer mechanism* or *exchange process*. Thus, money is a form of national debt, and the amount of money issued by a nation is not normally offset by adequate material reserves of equal value.

> **The quantity of money in a country is the quantity of Debt which there would be if there were no money.**[26]

However, a hypothetical illustration of such a balance or *equal value exchange* involving currency is illustrated by a precisely defined system of the total number of theater tickets available for a given performance, matching the number of available seats. In this case, the money paid per seat has *a measure of value* that has been precisely equated to one seat for one performance. Accordingly, the wealth of an individual expressed in terms of money does not necessarily, and most probably does not, constitute a measure of value or national wealth, as a nation's money supply is essentially national debt or a deficit in real wealth. Thus, modern currency systems and other tokens of wealth do not constitute valid wealth-energy parameters.

Wealth must be a real physical quantity possessing energy content and/or value, not an imaginary or negative entity. A form of debt does not meet this requirement for wealth. If entity A holds a debt for entity B, the debt is not *community wealth* but individual debt. If the debt held by A is a property mortgage for B's debt, the community is responsible for the debt and its value is an unavailable quantity incapable of being invested in the national economy. Thus, transferring wealth and assuming debt may be a natural and necessary societal practicality, but the person holding the debt has not gained wealth in the sense of national wealth value, as transferring ownership does not affect the totality of community or national wealth. If holding debt were considered as wealth, two entities, creditor A and debtor B, would own one property. Alternatively, holding debt means that the creditor A permits the borrower B to use the property. Obviously, accepting debt as a form of wealth would require it to be considered as a positive or negative quantity, which is physically impossible.

In H. D. MacLeod's *Theory of Credit*, the author poses the question: "How is a debt created?" He then proceeds to respond that debt may be created "by the mere consent of two minds ... the mere fiat of the Human Will" and then adds, "When two persons have agreed to create a debt ... it is a valuable product created out of Absolute Nothing, and when it is extinguished it is a valuable product decreated into Nothing by the mere fiat of the Human Will."[27] Obviously if national wealth could be created out of nothing, the laws of physics would be violated.

While credit may be viewed as an informal exchange between individuals or businesses, national credit is quantified as a formalized national debt. Credit is an imaginary financial creation that, for pragmatic reasons of convenience, is sometimes incorrectly considered as capital. Soddy refers to such *imaginary* fiscal innovations of currency, credit, and debt as *virtual wealth*. These are forms of debt *owned* by individuals but *owed* by the community, whose value or purchasing power is not directly determined by wealth. Soddy defines *virtual wealth* as an imaginary negative quantity that "is of psychological origin."

> **The Virtual Wealth of a community is not a physical but an imaginary negative wealth quantity. It does not obey the laws of conservation, but is of psychological origin.**[28]

Thus, *virtual wealth* is an aggregate national debt that includes the value of the currency possessed by members of the community and owed by the community, which is theoretically exchangeable on demand for real wealth. Virtual wealth is not a measure of a true value of wealth. Based on this concept, the virtual wealth or community debt determines the value of money; money does not measure the value of true wealth. Thus, a constant purchasing power requires a balance between virtual wealth (i.e., national debt) and the quantity of money placed into circulation.

> **It is the virtual wealth which measures the value or purchasing power of money, and not money which measures the value of wealth.**[29]

The term *capital* is defined as being *an agent of production*, usually a form of permanent wealth, Wealth II, which may be segmented into *personal possessions* and *organs of production* necessary for the creation of new forms of wealth-energy production. Improvements in economic effectiveness require creativity and capital, as is illustrated by the design of new tools and methods of production capable of providing quantitative and qualitative productivity increases. The human free will permits choices in prioritizing Wealth II production for either personal goods and services or investments capable of generating new wealth-energy capital. Human values and judgments affect these priorities and decisions regarding the consumption of limited amounts of Wealth I resources, a crucial requisite in the production of Wealth II goods and services. Based on these values and priorities, investments of national wealth will be dispersed between sound futuristic investments to regenerate national wealth and nonessential goods and services, determining and limiting a society's long-term socioeconomic success.

> **The distinction between Capital and Not-Capital does not lie in the kind of commodities, but in the mind of the capitalist, in his will to employ them from one purpose rather than another.**[30]

The stewardship of a society's socioeconomic system is essentially the effective management of its wealth-energy resources designed to create a productive, self-sustaining national economy and social stability and satisfaction. This responsibility requires the implementation of strategies, policies, and judgments founded on specific cultural values, attitudes, and traits that guide behavior and reflect potential economic success and the long-term best interests of a society. Such stewardship may be anticipated to produce a broad spectrum of potential socioeconomic outcomes based upon unpredictable variables of free will, the quality of human judgments, and mathematical probabilities associated with human behavior and natural events. These variables operate within a framework of numerous complex circumstances and conditions associated with thermodynamic processes representing wide variations in the efficient and effective utilization of wealth-energy resources. However, society's expectation is that economic productivity, based on sound societal investments, will generate a high profitability that benefits the total community.

It is important to emphasize two distinct categories of human processes and decision-making that are involved in the stewardship of a society's socioeconomic system that have profound long-term ramifications. First, economic priorities are established and implemented that determine what portion of national wealth-energy resources will be devoted to nonessential consumption versus investments in potential wealth-producing endeavors (i.e., the comforts of life vs. potential long-term cultural growth). Second, human capabilities and judgments are involved in the acquisition and utilization of scientific and engineering production processes and techniques and business strategies intended to maximize a society's competitive economic advantage. However, these production capabilities establish only *the potential* of a society's ability to be economically productive and to regenerate its wealth-energy capital. Society's cultural values, ethics, discipline, integrity, and ambitions will influence choices among competing investment priorities and socioeconomic opportunities. The wisdom of such decision-making in a creative environment of invention and discovery, utilizing advanced knowledge and technology, will enable a society to maximize the probability for achieving a high return on its investment of national wealth. Thus, the overall *economic effectiveness* or productivity of a national economy is a function of its technically based efficiency, the effectiveness of its production systems, and the human wisdom and priorities of its wealth-energy investments.

Given that no civilization has ever avoided inevitable socioeconomic deterioration and stagnation, one is motivated to identify inadequacies in the stewardship of national wealth among present and past civilizations. While the factors related to the failure of civilizations are numerous, inadequate wealth-energy capital necessary to meet expanding and unwise ambitions of mature civilization is commonly encountered as a major societal shortcoming. This inadequacy may be historically traced to leaderships' confusion or disregard of the devastating socioeconomic consequences resulting from unwise and uninformed economic choices resulting from inappropriate and ineffective use of national wealth-energy resources. Also, such actions are often the result of unrepentant cultural attitudes of avoidance, leaving known negative and cumulative socioeconomic issues to be resolved by future generations. Regardless of the degree of ignorance, indifference, ambition, or greed, inadequate stewardship of national wealth-energy resources has resulted in mature civilizations being unable to sustain continuous economic growth.

Malthusian socioeconomics views human resource demands as ultimately exceeding the Earth's ability to provide the required necessities for human civilization. The Malthus premise is that the Earth's population would increase according to a geometric progression, while available subsistence would increase only arithmetically. While Malthus probably underestimated society's creativity and productivity to overcome such one-sided competitive odds, his thesis is applicable to the finite limit of the Earth's energy resources, consistent with the principles of thermodynamics. The second law of thermodynamics directs that the energy of the universe will eventually be depleted or be fully utilized and unavailable to provide further dynamic activity for a dramatically increasing population. However, prior to the last gallon of gasoline being available for transportation, there are considerations relative to commercial energy acquisition, production, and distribution that would bring a civilization's socioeconomic system to a standstill.

As society becomes more economically and technologically sophisticated, greater amounts of energy are consumed at a faster rate, resulting in energy resources becoming scarcer and more difficult to acquire and prepare for consumption. Consequently, the processes of producing energy become more energy intensive and more costly. Thus, it is not surprising that in the twenty-first century, as worldwide energy consumption reaches historic levels, the cost of bringing the Earth's remaining energy resources to the consumer drastically escalates.

The debate of how long the world's energy resources will support increasing global energy consumption will persist; however, it is not a question of *if* the Earth's supply of oil, for example, will be depleted, but *when*. Additionally, while the *monetary cost of energy* is the usual consumer concern, the more fundamental and often ignored issue is the increasing amount of energy that is required to obtain and convert more inaccessible raw energy sources into available consumable forms. It is not sufficient just to locate new coal or oil deposits; they must also be transformed into a state suitable for consumption at an affordable price. This entails expending as little energy as possible in the sequential processes of finding and converting raw energy forms into the necessary usable forms and transporting them to consumption sites.

A rationale for assessing the efficiency of the costs of processing raw energy resources (e.g., coal ore and crude oil) into usable and available fuels for consumers has been developed by Cutler Cleveland et al.[31] This *net energy analysis* compares the amount of energy provided for society by a given technology to the total energy required to find, retrieve, process, and distribute the energy. The energy return on investment, EROI, is the ratio of the amount of energy produced from a given raw energy source relative to the energy expended in the required extraction-refining-distribution processes. Such an *assessment of value or return* is not expressed in monetary units, but in energy units. For a coal company, the EROI is the amount of energy contained in a ton of coal in a consumable state relative to the total energy required to produce it from the digging phase through making it available for consumption. As the processes of extracting oil and coal from the Earth into useable forms for consumption become more energy intensive, the EROI declines. As the value of EROI approaches 1, the useful energy made available approximates the amount of energy required to produce the fuel. The EROI for oil, which provides about 40 percent of the world's commercial energy and 95 percent of U.S. transportation energy, has declined for decades. For U.S. production, the ratio has declined from 25 to 15 since the early 1970s. As global oil demand soars, oil is being extracted in more extreme environments that require more energy being invested per unit of energy acquired. Such environments include deep water and distant offshore sites, sand-oil mixtures, and inhospitable, distant Arctic regions. The Alberta tar sands, an identified future prime oil source, has an estimated EROI of about 4 due to the complex, high energy requirements of the extraction process.

Thus, the primary threat to the World's long-range socioeconomic prospects is not measured by the increasing energy cost for a given quantity of fuel, paid in currency, but the inevitable increasing amount of processing energy required to deliver a given quantity of energy to the consumer. The time will arrive for a given commercial energy production process where no *net usable heat content* will be derived in excess of the input energy required by the production process itself. This would correspond to an EROI value approaching 1. EROI values vary greatly for a given technology and among processing techniques applied in varying geographical locations. Current coal-fired power generation values range from 5 to 10, nuclear power values are less than 5, wind is about 18, and hydropower usually exceeds 10. However, few major favorable hydropower sites remain worldwide.

As an energy-based parameter, rather than currency-based, EROI is a valid measure of economic value, essentially defined by the difference between the energy content of output and input for an energy refinement-production process, such as for oil. Effectively, it quantifies a society's energy capital available for supporting the social order. A high EROI indicates a *large energy surplus or net return* for a given energy source using a particular production process. People's wisdom, behavior, and human nature ultimately determine how, and to what degree, such resources will be effectively utilized and constructively affect society. Unfortunately, the history of past civilizations demonstrates that economic self-interest wins the battle of priorities, as materialism traditionally trumps altruism.

The laws of science will ultimately control the functions of a society's socioeconomic system regardless of other factors affecting cultural growth; hence, the utility of a science-based economy would appear to be an essential element of a stable social order. The next chapter will provide greater insights into these scientific concepts.

Chapter 2
References and Notes

1. Soddy, Frederick. *Wealth, Virtual Wealth and Debt: The Solution of the Economic Paradox*. London: George Allen & Unwin. 1983. 24-25.
2. Ibid., 39.
3. Soddy, Frederick. *Cartesian Economics: The Bearing of Physical Science Upon State Stewardship*. London: Hendersons. 1924. 3.
4. Ruskin, John. *Unto this Last*. Dent: Everyman Edition. 1968. 115.
5. Ibid., 147-148.
6. Martinez-Alier, Juan and Schlupman, K. *Ecological Economics: Energy, Environment and Society.* Oxford: Basil Blackwell. 1987. 135.
7. Georgescu-Roegen, Nicholas. *The Entropy Law and the Economic Process*. Cambridge: Harvard University Press. 3-4.
8. Refer to Martinez-Alier, Juan and Schlupman, K. *Ecological Economics*. 183-191.
9. Ibid., 89-98.
10. Ibid., 188.
11. Soddy, Frederick. *Cartesian Economics*. 6.
12. Ibid., 7.
13. Ibid., 8-9.
14. Georgescu-Roegen, Nicholas. *The Entropy Law*. 276-277.
15. Soddy, Frederick. *Wealth, Virtual Wealth and Debt*. 99.
16. Soddy, Frederick. "Economic 'Science' from the Standpoint of Science." *The Guildsman: A Journal of Social and Industrial Freedom*. July 1920. 3-4.
17. Galbraith, John Kenneth. *The Good Society: The Humane Agenda*. New York: Houghton Mifflin. 1998. 34-35.
18. Rifkin, Jeremy. *Entropy*. New York: Viking Press. 1980. 8.
19. Soddy, Frederick. *Wealth, Virtual Wealth and Debt*. 70.
20. Ibid., 115-119.
21. Ibid., 108.
22. Ibid., 130.
23. Ruskin, John. *Unto this Last*. 1862.

24. Soddy, Frederick. *Wealth, Virtual Wealth and Debt.* 118.
25. Ruskin, John. *Unto this Last.* 1862.
26. MacLeod, H. D. *Theory of Credit.* Longmans, Green & Co. 1893.
27. Ibid.
28. Soddy, Frederick. *Wealth, Virtual Wealth and Debt.* 296.
29. Ibid., 140.
30. Mill, John Stuart. *Principles of Political Economy.* 1909.
31. Cleveland, Cutler J. et al. Energy and the U.S. Economy: A Biophysical Perspective. *Science.* August 1984. 890-896.

Chapter 3
The Driving Force of Cultural Complexity and Disorder

And in the beginning Mother Nature spoke: The energy and material content of the universe will remain constant. All energy transformations will flow from a usable to a less usable form, i.e., from an orderly state to one of greater disorder. Let us call these principles the laws of thermodynamics.

Thermodynamics governs the socioeconomic processes and prospects for human survival and prosperity. This is an inescapable consequence of *a force* that, above all other considerations, is fundamental to society's existence and to insuring nature's eternal journey into the future. *The force* is derived from Mother Nature's fuel (i.e., energy), her *master resource* and the enabler of all activities of the universe. Energy enables the motion of atoms and molecules comprising matter, involved in spontaneous and irreversible physical, biological, and chemical mechanisms vital to the subsistence of Earth and its inhabitants. Collectively and at the macro level, these mechanisms constitute the basic processes of existence for a social order.

The spontaneous pathways of the processes of nature and society toward more stable energy states also involves mathematical probabilities that specify what events may occur and under what conditions. Thus, the thermodynamic perspective combines the concepts of the energy content of matter and the mathematical probability of specific and well-defined events occurring. These concepts are precisely and comprehensively captured and enunciated by Mother Nature's laws as found in her principles of science. These principles are the basis for the mechanistic-thermodynamic paradigm utilized in this work to depict society's design and exploitation of socioeconomic systems.

Mother Nature flaunts her strict discipline toward her responsibilities for the management and behavior of the universe. Each of her processes possesses energy requirements and a definitive directionality that is governed by her laws (i.e., the laws of nature). One such fundamental law controls the driving force for all of nature's physical, chemical, and biological processes including restrictions on process reversibility. Photosynthesis is a well-known illustration. Water, carbon dioxide, and sunlight spontaneously produce carbohydrate energy sources for plant growth. The reverse direction for this biochemical process (i.e., the spontaneous decomposition of plant carbohydrates to form carbon dioxide, water, and light during daylight hours), is not the observed outcome for this reaction, thus green vegetation flourishes. Mother Nature's design establishes a spontaneous direction for the growth of green plant life.

Likewise, it would not be anticipated that an abrupt change in nature's behavior would produce a reversal of the established sequencing of the four seasons of the year. Similarly, apples that normally fall to the ground from trees will not spontaneously reverse direction in their long-established pathway and return to the tree branches, form blossoms, and contract to buds. Additionally, it would not be anticipated that people would become biologically younger rather than aging over time.

This directionality of nature's processes toward a more probable event or condition is often depicted as *time's arrow*, or more formally described as the thermodynamic effect of *entropy*. Entropy, despite its menacing and mysterious sounding label, simply expresses a probability that the process as defined (e.g., photosynthesis or human aging) would occur in a particular direction that produces a specific end result. Thus, the expression *entropy is time's arrow* indicates the more probable direction taken by a thermodynamic process. Whether the process actually takes place (i.e., spontaneously occurs), and at what rate, depends on specified conditions and energy considerations of the system.

Thus, entropy may be portrayed as nature's probability parameter that represents a process's directionality, irrespective of the time frame required to achieve the defined end result. It should be emphasized that thermodynamic principles and energy considerations control all processes conducted by nature and society (i.e., determine the potential process spontaneity and directionality). Clearly, the law of entropy insures that a fountain of youth will not exist to reverse the human aging process and that the Earth and it inhabitants will continue to age and evolve over time. Entropy has presided, and will continue to preside, over this horizontal world of time. Time marches on and Mother Nature, via her principles of thermodynamics, will continue to reign in all her glory.

In addition to the directional property of entropy, the entropy principle represents the efficiency of processes conducted by nature and by society. The maximum efficiency of a steam engine, which is theoretically unable to reach 100 percent efficiency, may be calculated by knowing the temperature of the combustion chamber and the temperature of the exhaust gases exiting the engine. Theoretically, in order to obtain 100 percent efficiency, it would be necessary for the exhaust temperature of an engine to be absolute zero, which is not possible. An absolute temperature of zero symbolizes a complete absence of thermal energy, and neither humans nor nature possesses the ability to achieve such an outcome. Therefore, maximum thermodynamic efficiency is always less than 100 percent and cannot be achieved regardless of the highest level of intelligence, inventiveness, and creative machines that humans might conceivably bring to the task. Consequently, all processes of nature and society will result in the degradation of some portion of the original energy content of a fuel, irrespective of society's ingenuity. It then follows that an engine's exhaust will always possess some portion of the fuel's original heat content that will be lost forever as a useful energy source.

Whether considering the combustion of an engine's fuel supply or the overall consumption of a society's wealth-energy resources, energy consumption processes represent the spontaneous journey toward an equilibrium state. This corresponds to the transition from a relatively orderly state to one of greater disorder. During this journey, by-products of material degradation and heat are created, in addition to potentially desirable and productive outcomes. Upon reaching equilibrium, a system achieves a condition of maximum disorder for the given set of process conditions, which corresponds to a state that no longer possesses sufficient energy content to be capable of spontaneously producing additional useful work. Consequently, in order to avoid this unproductive state and maintain a continuously functioning system, an unremitting replenishment of wealth-energy resources is a necessity for a fully functional society as well as for an operational gasoline engine. However, increasing molecular and societal disorder is an inevitable, continuous, and perhaps undesirable by-product of such energy consumption. It follows that a society must continuously replenish its wealth-energy capital in the struggle to survive and prosper, a thermodynamic necessity to prevent the socioeconomic system from reaching an equilibrium condition.

Consequently, a social order must continuously and defensively create sufficient order within a society (i.e., expand social, economic, and political structures and functions), so that its numerous basic subsystems function effectively and become more productive. This requires the continuous production of useful work, which necessitates a continuous available supply of wealth-energy resources; otherwise, the socioeconomic system will approach *cultural equilibrium*. The absence of sufficient national wealth (i.e., financial insolvency) will be shown to be a common symptom of Toynbee's cultural "breakdown" or "stagnation" stage of societal development, which is the prelude to cultural disintegration. Sorokin relates the concept of the equilibrium condition to life's requirement for continuous change:

Life can never be in equilibrium. Complete equilibrium is never attained (by organisms) and would be fatal if it were attained, as it would mean stagnation, atrophy, and death.[1]

Thus, the fundamental socioeconomic processes of human existence necessitate and inevitably promote continuous cultural change, which is inherently accompanied by accumulating societal complexity and disorder. Analogous examples exist of spontaneous physical processes, which increase the disorder of molecular systems. The opening of a perfume bottle produces a fragrance throughout the room. This random diffusion process proceeds spontaneously from a more orderly arrangement of perfume molecules, in a smaller volume within the bottle (initial equilibrium state), to a more disorderly state occupying a larger volume of the whole room (new equilibrium state). The driving force for this diffusion of perfume particles to occupy the total available space is referred to as the *probability of uniform distribution* (i.e., the spontaneous and irreversible tendency toward increased randomness of a system to achieve the equilibrium condition). This provides meaning to the statement that *entropy is a measure of the randomness or the chaos of a system* and to the reference of entropy as the *statistical law*. This is the basis for thermodynamics restricting and controlling a social order's political and socioeconomic processes, producing greater cultural complexity and disorder.

Obviously, random events associated with the diffusion of perfume particles constitute a more simplified model of mathematical probability than a society's struggle to deal with the processes and functions of life. However, the same laws of science that manage perfume particle diffusion will also control the activities representative of a society's processes. Thermodynamic principles restrict and control the processes of nature and society that spontaneously and irreversibly generate socioeconomic disorder. Probability is a driving force.

A more extensive and rigorous treatment defining the role of thermodynamic principles in social, economic, and political processes affecting civilization growth, stagnation, and decline can be found in Appendix A.

Human interactions may alter the character and outcomes of nature's established and normally predictable processes while society continuously creates and alters cultural processes to support survival and the search for social satisfaction and economic prosperity. A society's organizational and functional activities are inherently subjected to unpredictable and irrational forces of change that are unalterably regulated by probability. Thus, regardless of society's success in achieving high levels of technological efficiencies and economic effectiveness, "disorder, arbitrariness, and randomness in social life and organization" are normal outcomes of the human existence.

> **"Organization not only fails itself to solve the problems of potential social disorder but also in a sense exacerbates them by transposing these problems from organizational givens into foci of conscious concern that are formulated in symbolic terms and emphasizing even more the possibilities of disorder, arbitrariness, and randomness in social life and organization."[2]**

The generation of disorder within a society's socioeconomic systems most often contributes to further physical, social, emotional, and economic consequences that impact long-term cultural development.

It is noted that increasing disorder within a molecular system is commonly referred to as *increasing the degrees of freedom of the system.* For the perfume illustration, this signifies achieving greater freedom via increasing the volume available for expansion, enabling the particles to occupy the whole room as opposed to the bottle. This also defines the new equilibrium condition, or maximum entropy state. The concept of *increasing disorder or societal entropy production* within a social order is analogous to, but significantly more complex than, the system of diffusing perfume particles. In the perfume illustration, all members of the set are identical and engage in identical physical behavior under an identical set of conditions. On the other hand, in the cultural analogy, all members of the set are dissimilar, and each member's more complex potential behavior is independent of other members.

The high level of human capabilities to independently think, analyze, and reason stimulates creativity, shapes communal development, and provides a large number of potential degrees of freedom or directions affecting future societal events. That is, the diversity of the human population's thought processes, opportunities, values, and capabilities represents a large number of potential societal options or variables, which may be translated into a diverse set of probabilities, potentially affecting future societal outcomes. Such a large number of potential outcomes, expressed in terms of probability, may or may not contribute to a productive and satisfying civilization, but thermodynamics determines that societal complexity and disorder will be inevitable outcomes.

To summarize, probability, as a driving force of society's socioeconomic processes, defines the spontaneous direction of wealth-energy consumption processes, which is associated with an increase in a system's degrees of freedom. The system's ultimate outcome will represent a state of greater probability, more degrees of freedom, and increased disorder. The diversity of human judgments and behaviors inherently influences socioeconomic and political outcomes generally via unpredictable and complex mechanisms. Additionally, the endeavors of a culture may, and often do, produce perturbations of nature's normal and probable pathways, which are governed by chance. The Earth's environmental changes during the last century offer many examples of society's degradation of natural resources (air, water, and creature) and the alteration of nature's processes. Equally, the potential pathways and anticipated outcomes of society's initiatives may be inadvertently altered by chance due to the diversity of human behavior. Cultural processes utilize Mother Nature's resources, human intelligence and creativity, and the knowledge and tools of science and technology, but in the last analysis, they are subject to the mathematical principles of probability and the limitations of associated thermodynamic principles.

Processes involved in people's pursuit of personal and societal advancement produce outcomes that create cultural complexity and disorder while perhaps benefiting the community by accomplishing desirable goals. The expanding requirements of a viable society necessitate a continuous and sequential series of interrelated social, economic, and political processes. The number, complexity, and technical sophistication of these societal processes expand exponentially over time, resulting in dramatic and significant consequences, most importantly for mature civilizations. Three major factors contribute to, and intensify, societal complexity and disarray for a significant portion of a mature civilization's membership that lead to a subtle, but inevitable, declining rate of cultural development.

First, an established self-initiated cycle of accumulating societal disorder is a reality of a culture's attempts to provide and maintain a productive economy, useful private and public services, efficient governmental operations, and security for all members of society. Achieving such desirable outcomes requires the continuous flow of wealth-energy resources intended to create and maintain more efficient and effective products, services, organizations, and functional systems to meet an expanded menu of human needs, desires, and expectations. Thus, a society's social, economic, and political *work* may be considered as utilizing Mother Nature's resources and society's capabilities to create systems of orderliness, inherent in the processes of providing the necessities of food, clothing, shelter, security, transportation, and governance for its members. It is noted that even as a civilization undergoes successful economic advancement, the rapid pace of change, involving an escalating number of increasingly complicated societal mechanisms, produces an increase in the diversity and rate of accumulating societal disorder. This mounting cultural disorder creates subtle, but long-term, negative socioeconomic consequences.

It is emphasized that in order to produce goods and services, thermodynamics requires society to create orderly structural and functional systems. However, such system transformations to more orderly states have been shown to be thermodynamically improbable (i.e., nonspontaneous). Thus, in order to survive and prosper, a society must energize processes that create greater order within a system that are not thermodynamically spontaneous or energetic favorable. This feat of creating more highly organized systems and thereby opposing the spontaneous direction of Mother Nature is accomplished only at the expense of a considerable energy requirement and also the generation of greater disorder in the surroundings. This concept is more fully developed in Appendix A.

Accordingly, *one of the fundamental generalities of the entropy concept* is that whenever a society creates order to accomplish useful purposes, an even larger disorder is generated in the surrounding environment. This concept extends beyond the simple examples of exhaust and waste material produced by a gasoline engine and the smoke and airborne particulate matter from an industrial plant. Global warming and acid rain from society's atmospheric contamination is a prime example, but the concept applies to all of society's processes of existence. Achieving desirable and worthwhile socioeconomic goals requires activities that create order within social, economic, and political systems involving the consumption of wealth, energy, and other resources. However, such processes create greater disorder externally in the surroundings than the degree of desirable order achieved within the system. Thus, pockets of order may be created but, when all related factors are considered, the world suffers a net increase in disorder (i.e., increased entropy production).

The Industrial Revolution and modern manufacturing have produced spectacular socioeconomic successes for certain segments of society, but undesirable, significantly negative social, environmental, and economic consequences have also been created. Elements of societal disorder within a population include financial, physical, and psychological anxiety, stress, and trauma. Such societal entropy ramifications affect human attitudes and behavior and the probability of future outcomes, even from potentially constructive socioeconomic activity.

As increasing cultural dysfunction is inevitably created, enhanced efforts become necessary to counteract the continuously increasing accumulation of cultural anxiety, complexity, and new and unresolved issues. These defensive initiatives include creating additional organizations, processes, regulations, and greater government oversight to implement and manage additional subsystems thought capable of overcoming the negative consequences of societal progress. Thus, an expanding, self-perpetuating cycle of societal entropy production and human counteractions is established. Regardless of the effectiveness and efficiency of a society's socioeconomic system, each *phase of cultural disorder production will normally stimulate a counteractive response (i.e., the creation of additional order),* thus producing a continuous cycle of accumulating cultural dysfunction. Pockets of systemic orderliness and socioeconomic progress are established concurrently with the creation of additional complexity and disorder.

Business, industry, and government are required to consume large portions of national wealth in order to combat increasing societal disorder during times of prosperity as well as during periods of economic decline. Such solutions create more orderly systems, intending to improve economic and social stability while protecting commercial profits and political interests. However, initiatives to relieve the pain of societal entropy production, while possibly providing some relief, involve further resource consumption and the generation of additional societal dysfunction. The worldwide post-9/11 antiterrorism security initiatives have required the creation of enormous bureaucracies and a proliferation of policies, procedures, and laws, all at great costs to governments and businesses. While deemed necessary and prudent, this worldwide security initiative has created enormous bureaucracies and expenditures, which reduce the time, creativity, and financial resources available for the investment in regenerating national wealth.

The second major factor contributing to greater complexity, disorder, and a declining rate of cultural growth is the financial and sociological price paid for the *cycle of continuously accumulating societal entropy.* The consequences of the human and financial costs in meeting a culture's expanding agenda of goals and objectives place an unsustainable burden on society. Ultimately, the pragmatic complexities and psychological stress of a rapidly maturing social order and the associated wealth-energy capital required to perpetuate socioeconomic growth generates significant hardships for the population. At some point, this negative cultural environment becomes counterproductive to establishing and maintaining social and economic stability.

Recall that a society's economic effectiveness is due to the combination of thermodynamic efficiencies (engineering and technology capabilities and techniques) and the impact on productivity of human values, wisdom, and behavior that ultimately affect the priorities and decisions for the investment of wealth-energy capital. The level of societal complexity, economic ineffectiveness, and social instability may become a major deterrent in achieving and maintaining a positive rate of cultural growth. The sociological perspective of the population, particularly the labor force, is a major element or variable affecting economic productivity and social stability.

Anson Rabinbach addresses the nineteenth-century literature of *social energy* as related to the labor force, which was inspired by the machine-oriented thermodynamic principles developed by Carnot and Helmholtz. Citing Carl Neumann and others, Rabinbach suggests that the economists of that era had come to recognize:

> **that the devastating effects of entropy could not be ignored: The gradual diminution of forces over time presaged inevitable economic and social decline. For the proponents of energy conservation in the national economy, both economic intensity and energy efficiency were constantly thwarted by the negative effects of entropy: All commodities are directed to maximize their utility through exchange. The human being must constantly increase his intensity or economic energy, since in the course of nature all intensity diminishes.[3]**

The focus of his work, the "social implications of thermodynamics," was that "all economic life could be expressed by an energy exchange" and that social energy was to be "considered as a commodity." Thus, the labor force was considered a source of social energy, prompting him to question "whether the productivity of the nation was being optimally used and whether the regulation or even reduction of the overall expenditure of energy might not increase the yield if the nation's energy supply were better conserved."[4] This focus on the labor force as social energy to be wisely managed and conserved as a component of national wealth is of major significance. The labor sector of a society not only represents a valuable wealth-energy commodity, but also is able to create and inflict, or absorb and endure, actions that may contribute to or detract from economic and social stability. Such outcomes emanate from a culture's implementation of society's functional processes that are administered by human behavior but subject to human nature, which has historically tended to generate social instability.

Thus, the third major contributing factor exhibited during the inevitable final stage of declining socioeconomic growth is the unsettling extent of cultural dysfunction within the population. At some point in time, the rate of cultural decline begins to dramatically and negatively affect the population's psychological mind-set. The stress of increasing emotional, social, and financial issues may reach the level of initiating a chain reaction of a massive loss of confidence in the existing social order and its leadership. The population senses that despite past socioeconomic progress, their culture has entered a phase of prolonged economic decline and sociopolitical decay that threatens their survival. With time, and out of necessity, the civilization's leadership typically employs strategies that only exacerbate the deteriorating situation and contribute to cultural collapse. At this point, the impossibility of acquiring the resources and leadership necessary to avoid socioeconomic stagnation and cultural deterioration becomes widely evident. This situation necessitates a frantic reshuffling of the leadership's strategies and tactics, which historically has produced drastic disproportionate and unacceptable hardships on those of less wealth and with little political power. Consequently, the basic mechanisms and functions of a society degenerate further, reducing economic effectiveness and cultural stability. Finally, the general population formally rejects the failed governance structure and its authority, mission, and leadership, depriving it of revenue and manpower, which quickly leads to financial insolvency and cultural collapse. America in the twenty-first-century finds itself on this path to cultural stagnation as will be discussed in chapter twelve.

History has recorded momentous negative humanitarian abuses that accompany this sequence of cultural growth to decay, a consequence of people's ambitions and unwise utilization of Mother Nature's resources. During the nineteenth- and twentieth-century periods of economy success, various forms of slave, illegal, and underage labor were extensively utilized. The general abuse of labor forces, widespread poverty, and inequitable distribution of wealth were common economic characteristics, even during eras of general prosperity.

S. N. Eisenstadt captures important elements of human nature and behavior, historically adopted to manipulate societal functions and organizations, thereby creating societal disorder, instability, and distrust, particularly within the labor force, arguably the most significant segment of society:

> **These disruptive possibilities of the organizational aspects of the social division of labor are rooted in the fact that in any given setting of social interaction, a combination of conflict and cooperation develops among different groups and actors over the production, distribution, and use of resources.... The mechanisms of the social division of labor gives rise to attempts by different actors to monopolize access to positions and resources and to promulgate rules to support and perpetuate such arrangements. But while these rules produce stability in interaction, they are usually perceived as arbitrary, coercive, and unjust. They may generate a perception of ambiguity and disorder among actors. As a result, they cannot provide the basis of trust among participants in these relationships; instead, they build instability into the social relations they structure. The potential for instability and disorder and the likelihood that actors will perceive the social divisions of labor as arbitrary are increased by the fact that these fundamental indeterminacies are systematically related to the basic organizational features of social interaction—the structuring of collectivities, institutions, and macrosocietal orders.[5]**

John Ruskin condenses and expresses this economic reality of human nature and behavior; a society inherently produces winners and losers:

> **The art of making yourself rich, in the ordinary mercantile economist's sense, is therefore equally and necessarily the art of keeping your neighbour poor.[6]**

In spite of the inherent negative consequences of societal entropy production, a society's increasing intellectual maturity and sophistication in matters of finance, marketing, science, engineering, and technology continue to produce substantial socioeconomic advancements and humanitarian accomplishments. However, as civilizations mature, each new set of cultural issues is more difficult to unravel than the previous ones. As each challenge is overcome, inherent complexity and societal disorder multiply and continue to accumulate. A sociopolitical accomplishment or technological achievement in one decade may ultimately evolve into a more threatening, long-term issue. Successive technological improvements or phases are referred to as *transformers,* and the faster transformers appear and multiply, the faster additional disorder is generated and the more rapidly additional problems proliferate. Also, each transformer requires more advanced and expensive systems and related technology.

It is in the nature of the [social] system that its productive process or its products can have harmful social effects. And there can be long-run effects that are currently invisible or enthusiastically, even righteously, ignored but that are potentially disastrous.[7]

Ultimately, opportunities, challenges, and issues appear more rapidly within a societal system than creative solutions can be devised and implemented. The treadmill of societal life begins to accelerate more rapidly than many members of the civilization can intellectually and emotional accommodate. The minor annoyances of low-level societal complexity and disorder subtly increase and evolve into unsettling sociological issues for a significant portion of the population.

These cumulative negative outcomes intensify, eventually overshadowing and diminishing limited positive cultural outcomes. A limit is attained where the accumulated social chaos, economic inadequacies, and individual and societal complexities of life inhibit the continuation of a formerly flourishing socioeconomic system. From a financial perspective, the point is reached where the economic system is unable to generate sufficient productivity or new wealth-energy capital to sustain financial solvency for an increasingly more expensive and advanced civilization. This phase of cultural instability marks the beginning of the civilization's demise to a less prominent position in the world order.

Complex societies are unstable, not just in certain kinds of environments, but inherently.[8]

There is a difference between a society merely becoming complacent or greedy, or losing the inspiration to maintain successful cultural growth, and its ability to maintain socioeconomic stability in an environment of continually escalating cultural change and disorder. In the latter situation, a large segment of the population eventually reaches the limit of its tolerance for coping with rapid and disruptive cultural change, an entropy effect. In addition to the increasing complexity and disorder of life's daily routine, major problems, issues, and future directions become innately more complicated, arbitrary, disruptive, and expensive. The mature civilization, having created and accumulated numerous unresolved or partially resolved social issues and dealing with economic instability, ultimately becomes a dysfunctional, out-of-control society. This situation constitutes a cultural Peter Principle: Cultural systems tend to develop and evolve to the limit of their adaptive competence and capabilities and then undergo cultural degradation.

Historians and economists have long contemplated the nature of the fundamental forces that contribute to the affluent years of great civilizations that appear unable to sustain continuous cultural growth. Joseph Schumpeter, the Austrian economist, who wrote extensively on the process of capitalism and successful economies, applied the term *"creative destruction"* to explain the phenomena of great societies losing their momentum and declining from the world stage.[9] Schumpeter describes the tendency of successful societies to self-destruct as being due to their failure to be progressive and abandon outmoded economic methods and strategies. A healthy economy requires a continuous infusion of new goods, technologies, and forms of communication, management, and transportation, as well as new services capable of altering, when necessary, a culture's socioeconomic structure. Thus, he views societies as being subject to forces that require continuous change in order to remain competitive in the race for economic dominance and to achieve cultural survival. Historically, maintaining such economic momentum for a major, mature civilization has required not only substantial, escalating quantities of wealth-energy resources, but also the management of intricate, logistical support for the imperialistic and militaristic ambitions of building empires.

A refinement of the *creative destruction* concept is that *continuous, progressive change* also requires the *continuous renewal* of expended wealth-energy capital through financially sound investments and the utilization of the most modern technology. Driving out the old economic practices and creating new products with appropriate new investments requires that a society maintain a high marginal return on investment or a high level of economic productivity. The great civilizations have ultimately failed to achieve this goal.

Essentially, Schumpeter recognizes the necessity for a successful economy to continuously create new generations of *creative* production capable of maintaining the momentum of a growing economy. Likewise, he recognizes that such a growing, creative economy inherently generates *destruction,* that is, societal entropy that eventually will be society's downfall. Schumpeter has identified, albeit unknowingly, the thermodynamic consequences of functioning socioeconomic systems and has prescribed the appropriate strategies to minimize the elements of cultural deterioration.

The Prologue to Part II of Joseph Schumpeter's historic volume *Capitalism, Socialism and Democracy* is entitled "Can Capitalism Survive?" He poses this question in the first sentence of the Prologue and then quickly responds, "No. I don't think it can" and then proceeds to state:

> **The thesis I shall endeavor to establish is that the actual and prospective performance of the capitalist system is such as to negative the idea of its breaking down under the weight of economic failure, but that its very success undermines the social institutions which protect it, and "inevitably" creates conditions in which it will not be able to live and which strongly point to socialism as the heir apparent.**[10]

The current view of the twenty-first century is that while global capitalism is healthy, the more mature and established Western economies are in the process of losing their historic global leadership positions as more successful economies emerge from historically less-developed cultures. This is consistent with Schumpeter's thesis and his portrayal of economic mechanisms: "Success undermines the social institutions which protect it, and 'inevitably' creates conditions in which it will not be able to live." The full quotation above reflects the nature and characteristics of the thermodynamic concept of societal entropy production.

Schumpeter's description of impending self-destruction of the capitalist economic system is strikingly congruent with the socioeconomic outcomes expected from the entropic perspective presented in this chapter and Appendix A. Hence, the thermodynamic-based concept of *societal entropy* resonates with Schumpeter's economic foundation of *creative destruction,* as both provide a fundamental rationale for the observed disorder and ultimate deterioration of civilizations. Societal disorder and chaos are unfortunate but assured by-products of economic progress that will accumulate as inevitable socioeconomic outcomes of thermodynamics. The priorities, judgments, and actions of a culture can only affect the intensity of cultural entropy production or disorder ranging from a theoretical minimum to a maximum of chaos.

Interestingly, Schumpeter separates the culture and ethics of economics from philosophy and politics, in that he believes that one need not embrace the socialist culture to accept his economic thesis that capitalism will self-destruct:

> **Nor need one accept this conclusion [self-destruction] in order to qualify as a socialist. One may love socialism and ardently believe in its economic cultural and ethical superiority but nevertheless believe at the same time capitalist society does not harbor any tendency toward self-destruction.**[11]

However, Schumpeter's prediction that socialism is the "heir apparent" to capitalism incorrectly assumes that a totally different economic theory will be more successful in achieving long-term economic success, rather than some evolutionary combination of various elements of capitalism and socialism. It is clear in the twenty-first century that developing nations have evolved and developed their own versions of social organizations and economic systems that blend various elements of capitalism, socialism, and democracy.

A significant aspect of Schumpeter's characterization of the capitalist economic system is its consistency with the observations and predictions that would be expected from maturing economic systems based on the laws of thermodynamics. In the Introduction to Schumpeter's 1974 printing of his original work, Tom Bottomore remarks that "capitalism, in Schumpeter's view, would be killed by its economic successes, not by its failures, because these successes create an unfavorable social and political climate, or in his words, an 'atmosphere of almost universal hostility to its own social order.'"[12] Bottomore then proceeds to highlight three processes that he finds representative of Schumpeter's "anti-capitalist outlook" of creative destruction:[13]

(1) "The development of the capitalist economy itself undermines the entrepreneurial or innovative function, which Schumpeter regards as the essential feature of capitalism because technological progress and the bureaucratic administration of large enterprises tend to make innovation itself a routine matter and to substitute the activities of committees and teams of experts for individual initiatives."
(2) "Capitalism erodes its own institutional framework by destroying the protective strata—the gentry, small businessmen, farmers, and others—which had survived from an earlier form of society and by weakening individual proprietorship in favor of a more diffuse kind of ownership in the modern corporation."
(3) "Capitalism encourages a rational and critical attitude which is essentially turned against its own social system, and this process is greatly assisted by the creation of a large stratum of intellectuals who have a vested interest in social unrest."

As a scientist, Frederick Soddy would agree with Schumpeter's three major characterizations as described by Bottomore, but would attribute them to the inherent and inevitable self-induced societal disorder, conflict, and chaos resulting from the scientific principles related to a society's functional processes. It should be noted that Bottomore's characterization of Schumpeter's "outlook" includes the use of "committees and teams of experts," "a more diffuse kind of ownership," and "a rational and critical attitude which is essentially turned against its own social system." Such elements, described as constituting an "atmosphere of almost universal hostility to its own social order," exemplifies societal complexity, disorder, and bureaucracy that is associated with the concept of societal entropy production. Human behavior, influenced by human nature, attempts to deal with self-induced complexity and disorder, interjected into socioeconomic systems. This sociology of civilization development as provided by Sorokin and others will be extensively considered in later chapters.

Schumpeter provides an accurate portrayal of the inherent negative elements or outcomes of capitalism that are simply the inevitable consequences of the thermodynamics of energy consumption, but are independent of any particular economic model. Consequently, as a culture's socioeconomic system matures, business entrepreneurs and innovators find it increasingly difficult to survive and prosper in a more competitive world of business that involves greater complexities, investments of wealth, and legalities. Such entropic consequences of cultural advancement are the *inevitable price* to be paid for limited societal success regardless of the philosophical nature of the economics and politics being practiced. Hence, socialism, or any other economic or social model, will not alter the ramifications of the laws of thermodynamics. Economic practices emerge over time as a result of human actions within societies that are usually dictated by cultural realities and experiences derived from the lessons of bare survival, prosperity, and failure. However, entropy considerations, as a driving force for socioeconomic systems, insure that over time the system will increasingly produce a higher level of disorder, turmoil, cultural dysfunction, and hostility.

Economic progress, in capitalist society, means turmoil.[14]

Thus, Schumpeter's conclusion that "the capitalist order tends to destroy itself" correctly acknowledges the disorderly and destructive nature of a society in its normal processes of consuming wealth and energy, as described by Soddy. However, Soddy also understood that social issues could not be resolved simply by adopting an alternative economic model, but by appreciating and utilizing a science-based model for national economic theories, policies, and practices. In this way, it would be possible to minimize the entropy effect of societal disorder and maximize the use of national wealth for economically profitable and socially useful and just purposes, including a greater reliance on renewable energy sources.

Often, economists and historians interpret the observed characteristics of societal disorder as resulting *solely* from society's resistance to change. Thus, a frequent assumption of authors addressing cultural deterioration is that a civilization is unable to comprehend or unwilling to accept, promote, and invest in constructive cultural change that would benefit a broad segment of society. While this resistance to change is an observed human characteristic, other common elements of cultural aging are also conspicuously present in the literature of the historic disintegrations of ancient Greece and Roman, Imperial Spain, and the British and American empires. These exhausted social orders all have a history of increasingly embracing materialistic, self-serving values and ultimately suffering from an acute shortage of resources and related social and economic instability. This continuum of empire failures, each of which rose to great heights only to ultimately deteriorate, illustrates the effect of *creative destruction* (i.e., how "success undermines the social institutions which protect it, and 'inevitably' creates conditions in which it will not be able to live"). Schumpeter's "creative destruction" concept parallels Eisenstadt's analysis of social disorder: "Problems of potential social order from organizational givens ... are formulated in symbolic terms and emphasizing even more the possibilities of disorder, arbitrariness, and randomness in social life and organization."

The continuum of empire failures from ancient Greece through America exemplifies cultural transformations that exhibit common societal entropic characteristics. Hans-Friedrich Mueller's 1957 critical foreword to Gibbon's volume *The Decline and Fall of the Roman Empire* states that given "the military supremacy of the United States of America over the entire planet," one "may well indeed wonder what our New World Order might hold." His response is, "The story of the Roman Empire offers instructive lessons: oppressive taxes for the sake of a despotic military establishment, tyrannical government, religious bigotry, endless warfare, and finally, collapse."[15] This was a 1957 point of view most applicable to twenty-first-century America.

During the latter phases of maturation, civilizations since ancient Greece have systematically accumulated and exhibited common entropic characteristics of cultural complexity, disorder, and decay that reflect an escalating momentum of cultural degradation. This roller-coaster ride represents the spontaneous and irreversible directionality of civilization development in the same sense as the spontaneity and directionality of nature's photosynthesis process. A chaotic runaway society becomes mired down in disorder and faces the consequences of ever-increasing, inevitable complexity and its associated financial insolvency and materialistic cultural values. The magnitude and intensity of societal entropy accumulation eventually overcomes a society's capabilities to cope with and counteract the impact of a perplexing, structurally complex, and functionally convoluted culture. Remarkably, the creation of these distinctive negative cultural elements is the inevitable companion of socioeconomic success and prosperity. Nevertheless, the enormous broad-based human achievements of the Graeco-Roman Western civilizations during the last 4,000 years are testimony to the ever-impressive capabilities of humans to survive and prosper.

Historian J. H. Elliot documents the unique accomplishment of Imperial Spain during its developmental years, particularly the impressive creation of a governance system capable of managing the global affairs of the Crown. In Elliot's words:

> **Perhaps the most remarkable of all Spain's achievements— the ability to maintain its control over vast areas of widely scattered territories at a time when governmental techniques had scarcely advanced beyond the stage of household administration, and when the slowness of communication would seem at first sight to have made long-distance government impossible.[16]**

Obviously, a great deal of initiative, creativity, and investment was required over an extended period of time to establish this expanding empire, whose objective was to increase its wealth, power, and prestige. The empire accomplished many great things, as Elliot states:

> **For nearly two centuries Spain had sustained a remarkable creative effort which added immeasurably to the common stock of European civilization.[17]**

However, it is equally evident from Elliot's analysis that during the sixteenth century, the task of Spain adapting its medieval political, social, and economic processes to a vastly expanding empire had taxed to the limit the capabilities of the people, as well as the empire's infrastructure and operational systems.

While insufficient revenues, a lack of strong and wise leadership, competing nations, and acts of God could be cited as contributing factors for Spain's decline, societal entropy production is also an obvious fundamental factor. Elliott captures the essence of socioeconomic decay and the people's loss of spirit and capability to cope with the accumulated disorder that eroded the empire's ability to deal with the escalation of cultural complexities that led to final collapse:

> **The men of the 17th century belonged to a society which had lost the strength that comes from dissent and they lacked the breadth of vision and the strength of character to break with a past that could no longer serve as a reliable guide to the future. Heirs to a society which had over-invested in empire, and surrounded by the increasingly shabby remnants of a dwindling inheritance, they could not bring themselves at the moment of crisis to surrender their memories and alter the antique pattern of their lives. At a time when the face of Europe was altering more rapidly than ever before, the country that had once been its leading power proved to be lacking the essential ingredient for survival—the willingness to change.**[18]

It is appropriate to analyze the mind-set of the population of this great civilization at a time when its societal disorder was at the point of being a significant negative influence on maintaining its global dominance. Elliott characterizes the prevailing attitude as "lacking the essential ingredient for survival—the willingness to change." Alternatively, his descriptions of life during that period could also be represented as a prevailing realization that the treadmill of existence was moving too fast for the coping skills and abilities of the general population. Thus, the people sensed a disintegration of the social order and a common loss of spirit and purpose. As a result, moral, financial, and military support of the empire were absent during the latter stages of Imperial Spain's era. J. H. Elliott references the work of Gonzalez de Cellorigo:

> **The Castile of Gonzalez de Cellorigo was thus a society in which both money and labour were misapplied; an unbalanced, top-heavy society, in which, according to Gonzalez, there were thirty parasites for every one man who did an honest day's work; a society with a false sense of values, which mistook the shadow for substance, and substance for the shadow.**[19]

An uninspired and disillusioned population did not possess the confidence, morale, or willingness to financially, emotionally, or physically support the common cause of the empire's survival. This negative sociological environment was obviously linked to the inability of the empire to acquire and maintain a sufficient stream of wealth-energy capital into the socioeconomic system to maintain a stable society while also pursuing an imperialistic agenda. People at all levels became disengaged from society and were merely consumers of wealth, leaving its production and other constructive societal responsibilities to others (i.e., to no one).

During the empire's initial rise to power, the accumulating baggage of societal or entropic disorder did not retard overall progress, as rapid economic success outweighed these annoying but relatively minor accumulating negative by-products of advancement. Fortunately, abundant new sources of wealth were acquired from the New World to deal with the expenses of setbacks or new issues. However, eventually the escalating expansion and complexity of the empire became an appreciable drag on the socioeconomic system, a true cultural wake effect. This presented the threat of financial insolvency and created considerable social disorder.

Imperial Spain is a good illustration of a successful civilization's collapse that is often characterized as resulting from an unwillingness to accept the need for cultural change. However, rather than simply an *unwillingness to adapt* based on a failure of leadership, an inappropriate economic model, or a lack of cultural initiative, the fundamental reason for cultural decline was the inherent accumulation of unmanageable societal disorder and complexity at all levels of the empire. As Imperial Spain matured, the number, nature, and complexity of cultural issues increased exponentially and consumed more time, resources, creative thought, and initiative. Over time, sociological effects of human stress resulted in increasing stagnation of its essential operational functions. The ramifications undermined the overall quality of life and affected governmental administration and support services, the military, commerce, and key public services. Importantly, this continuously escalating cultural complexity required the consumption of an increasingly large proportion of the empire's dwindling wealth-energy resources. Consequently, severe socioeconomic constraints placed increasing limitations on its economic productivity, contributing to the demise of the overextended empire.

To summarize, the consequences of complexity and disorder created by the maturation of a civilization's socioeconomic and political processes have negative financial and sociological effects on some segment of the population. In addition to the human element, a society's pursuit of survival and prosperity consumes increasing quantities of wealth-energy resources and produces proportional quantities of waste heat and materials. These by-products include air, water, soil, noise, and visual pollution and a history of strip mining; denuding large virgin forests; global warming; and excessive hunting, trapping, and fishing of various species to extinction. This progressive degradation of the planet and the creation of higher levels of social, economic, and political disorder, dysfunction, and human distress are stark illustrations of the entropic effect on human civilization.

> **The relationship between the economic process and the Entropy Law is only an aspect of a more general fact, namely, that this law is the basis of the economy of life at all levels.**[20]

Chapter 3
References and Notes

1. Sorokin, Pitirim A. *Social and Cultural Dynamics*. Boston: Porter Sargent. 1970. 635.
2. Eisenstadt, S. N. *Revolution and the Transformation of Societies*. New York: The Free Press. 1978. 20.
3. Rabinbach, Anson. *The Human Motor: Energy, Fatigue, and the Origins of Modernity*. Berkeley: University of California Press. 1990. 71.
4. Ibid., 70-83.
5. Eisenstadt, S. N. *Revolution and the Transformation of Societies*. 20.
6. John Ruskin, *Unto this Last*. Dent: Everyman Edition. 1968.
7. Galbraith, John Kenneth. *The Good Society: The Humane Agenda*. New York: Houghton Mifflin. 1998. 80.
8. Tainter, Joseph A. *The Collapse of Complex Societies*. New York: Cambridge University Press. 2004. 55.
9. Schumpeter, Joseph A. *Capitalism, Socialism and Democracy*. New York: Harper & Row. 1975. 81-86.
10. Ibid., 61.
11. Ibid., 62.
12. Ibid., ix.
13. Ibid., ix-x.
14. Ibid., 32.
15. Gibbon, Edward. *The Decline and Fall of the Roman Empire*. New York: Random House. 2003. xxxvi.
16. Elliot, J. H. *Imperial Spain*. London: Penguin Books. 1990. 383.
17. Ibid., 382.
18. Ibid.
19. Ibid., 318.
20. Georgescu-Roegen, Nicholas. *The Entropy Law and the Economic Process*. Cambridge: Harvard University Press. 1999. 4.

CHAPTER 4
Socioeconomics: A Wealth-Energy Based Perspective

> "All economic history is the slow heart-beat of the social organism, a vast systole and diastole of naturally concentrating wealth and naturally explosive revolution."
> Will Durant[1]

Scientifically defined wealth-energy economic and financial principles complement and supplement traditional economics by expanding the discipline from its historic, narrowly conceived mechanistic perspective to a more precise and accurate model of cultural subsistence. This more intellectually inclusive and rigorous view of economics, an integrated energy-driven, mechanistic perspective, is an indispensable foundation for socioeconomic principles, policies, and practices. It will be demonstrated that a wealth-energy based economic model also provides additional insight into, and support for, well-respected traditional concepts of sociology, political economics, and comparative historical analyses of past civilizations. This model possesses a unifying property that embraces economics and the social and physical sciences to more appropriately express the underlying principles and rationale for the cultural growth and degradation experienced by civilizations since the time of ancient Greece.

> Only an analysis of the intimate relationship between the Entropy Law and the economic process can bring to the surface those decisively qualitative aspects of this process for which the mechanical analogue of modern economics has no room.[2]

The character, extent, and quality of human behavior represent numerous, unpredictable factors affecting a civilization's socioeconomic potential success or failure. However, the laws of nature (e.g., thermodynamics) also restrict cultural processes of existence and a civilization's pursuit of prosperity by predictably influencing the outcome of wealth-energy consumption processes. Thus, human behavior and available wealth-energy resources constitute major variables of cultural development. Given this reality of human existence, it is remarkable that economists ignore wealth-energy based economic models. Accordingly, it is appropriate to validate the significance of the integration of mechanistic-thermodynamic principles with the work of eminent historians, sociologists, and economists. For example, is the understanding and significance of Toynbee's sequence of cultural growth, stagnation, and disintegration stages of cultural development enriched by the scientific principles of energy consumption? Are these energy principles also consistent with, and supportive of, the economic principles contained in the work of John Kenneth Galbraith?

The first step in seeking appropriate responses to these inquiries is to identify the mechanistic economic steps that constitute the required societal processes of human existence. Specifically, do substantive disagreements and conflicts exist between the principles of a wealth-energy based economic model and accepted social processes and economic concepts? Consequently, the objective is to determine if the interjection of the mechanistic-thermodynamic paradigm is consistent with and supportive of accepted views of the civilization studies literature.

Since it is not the mission of this work to extensively survey economic models proposed during the modern age, no attempt is made to represent the continuum of such economic views. The work of John Kenneth Galbraith has been adopted as a respected economic framework upon which the mechanistic-thermodynamic paradigm will be applied so as to validate the applicability of the science-based perspective to economics. Consequently, a major objective of this chapter is the review of the general process of socioeconomic maturation in order to assess the congruency of Galbraith's economic thinking with the mechanistic-thermodynamic paradigm. While various sociological aspects and implications of economic principles will be noted in this chapter, later chapters will provide a more in-depth, integrated scientific context for sociological concepts applied to cultural development.

From a macro view, the first law of thermodynamics (see Appendix A for details) requires that a society must provide a continuous flow of energy into an economic system in order to perform its necessary useful work in producing a continuous production of goods and services essential to human existence. The second law of thermodynamics mandates that only some portion of the consumed energy will be successfully captured for intended production outcomes (i.e., useful work). The remaining energy and material by-products will be dissipated, thereby contributing to the accumulating material and energy residues of the universe as prescribed by the entropy principle. A society's skill, creativity, knowledge, and technology will determine the extent of success in achieving "useful work" (i.e., a highly efficiency production capability and economic effectiveness and profitability). Recall that *wealth* is scientifically defined as *a flow of energy* and may be invested, depending on society's priorities, for superfluous outcomes or for regenerating national wealth.

The socioeconomics of society's processes inherently produces cultural complexity, disorder, and dysfunction. This continuous accrual of societal entropy and the related future increases of communal wealth-energy capital obligations represent an ongoing challenge for a society attempting to maintain a positive rate of cultural growth. A society constantly struggles to maintain a positive rate of growth in order to prevent its socioeconomic system from approaching an equilibrium condition corresponding to cultural stagnation. This sequence of a society slipping from a positive to a negative rate of cultural growth (i.e., cultural deterioration) can result from a prolonged period of inadequate, and/or unwise, investment of national wealth. The corresponding socioeconomic principles of societal development will be explored in this chapter from the perspective of a scientific model of economics.

> **Economics like other social life does not conform to a simple and coherent pattern. On the contrary, it often seems incoherent, inchoate, and intellectually frustrating.**[3]

History illustrates the character and intensity of the human instinct to survive as well as to pursue the potential rewards and pleasures of life. From the most primitive to the most mature economic system, the primary objective of a society's controlling economic forces is to produce consumer goods and services capable of providing socioeconomic satisfaction for the population, as well as financial benefits for those responsible for a successful economic system. Thus, economic productivity is fundamental to a surviving society and establishes the foundation and ground rules for sociopolitical structures and functions. In order to appreciate this history, and more specifically human behavior, it is instructive to examine how the economics of survival and prosperity becomes the prime motivator within sociopolitical systems and plays a central role in promoting cultural change as well as cultural deterioration.

Societies place their highest priority on providing food, shelter, clothing, orderliness, and security for its members. In primitive societies, the pursuit of these basic human needs consumes most, if not all, of its available time, labor, and mental efforts. Characteristically, the initially simple techniques, tools, and forms of raw materials and energy resources become incrementally more sophisticated and more effectively utilized with the passage of time. Ultimately, the initial simplicity of societal processes supporting life becomes appreciably more complex and, if economically successful, capable of production that exceeds the basic needs for survival. Consequently, a society eventually achieves the ability to provide adequate necessities for survival as well as for desirable (but nonessential) products. Generally, a society's expectations for survival, wealth, and the frivolities of life become the primary objectives of economic policies and practices, thus shaping the goals and objectives of sociopolitical systems.

Will Durant offers a comprehensive review of the economic elements of early civilizations, beginning with the following statement:

> **The moment man begins to take thought of the morrow he passes out of the Garden of Eden into the vale of anxiety; the pale cast of worry settles down upon him, greed is sharpened, property begins, and the good cheer of the "thoughtless" native disappears.**[4]

The transition from a preoccupation with today's survival to more thoughtful plans for the requirements of next week's existence signifies a defining step in the human pathway toward a more highly organized society. This pathway increasingly intensifies the cultural necessity for combining political and economic strategies for common survival outcomes, all of which will inherently and irreversibly alter the social order's organizational structures and functions. Durant cites Polynesian fishing practices that use sophisticated netting techniques requiring hundreds of men as a primitive example of sociopolitical organization resulting from economic necessity.[5] Moreover, the primitive development of more advanced and effective hunting and fishing techniques not only achieved economic and sociopolitical advances but also became an inspirational source for early civilizations' literature, philosophy, heritage, and the arts. Durant comments:

> **In the last analysis civilization is based upon the food supply. The cathedral and the capitol, the museum and the concert chamber, the library and the university are the façade; in the rear are the shambles.**[6]

It could be added that the affordability and quality of the "façade" depend on available national wealth generated by the culture's economic production, which also affects the level of poverty within the "shambles."

Historically, the key role of animals within the social order expanded from that of food to becoming beasts of burden and, in some cases, to be part of the family. The soil was recognized as an economic asset, and a farming economy prospered and matured with the invention of simple tools for animal husbandry and crop planting techniques that also eased the labor burden, all of which catalyzed a prosperous agricultural industry. People learned to preserve food and thereby accelerated the development of agricultural techniques, tools, and knowledge associated with a broader distribution and an increased longevity of food materials. Agricultural advancement stimulated structural and functional changes in primitive life; for example, women assumed a new and important role in directly producing the crops, which had associated ramifications within the social order.

The discovery and use of fire resulted in metal tools and weapons being forged. Creativity and ingenuity resulted in more advanced mechanical techniques and tools being crafted from natural materials, which improved and broadened society's array of economic products. These included food, clothing, building materials, home use products, and decorative items, all of which became more widely available. Some products, upon becoming widely accepted as possessing economic value, were equated to wealth and contributed to the success and expansion of the communal marketplace concept. These developments led to more sophisticated techniques and processes for the transportation, distribution, and exchange of goods beyond the tribe and local region, thus facilitating wider geographical trading zones. The central, unique role of wealth-energy resources in such cultural advancement is obvious, as are the social and political ramifications.

Economic advancement has required continuous modification of social and political systems in order to become more responsive to the changing needs and expectations of a more aggregated population. Until the concept of trade became a reality for primitive societies, the concepts of property ownership, money, and profit were nonexistent. The community owned the land, and food was generally a communal commodity regardless of who was responsible for its acquisition. However, such philosophical equality, inherent to these *communal* concepts, soon came to be considered an economic liability, a deterrent to progress, and thus gradually disappeared. In the more *progressive society*, the communal philosophy was viewed as lacking desirable incentives and rewards for the individual creativity, initiative, and hard work necessary for economic advancement. Interestingly, the primitive society as a rule does not distinguish between slaves and free persons, or between a chief and followers, and does not recognize serfdom and castes. Durant comments:

Darwin thought that the perfect equality among the Fuegians was fatal to any hope of their becoming civilized.[7]

Communal ownership of food and property appears to exist only in the primitive societies that share a common fear of starvation, God, or violent death. In economically and physically secure times of plenty, when starvation and security are no longer concerns, individualism predominates, cohesion lessens, and self-interest increases. The communal concept is gradually displaced by self-oriented priorities, ruthlessness, competitiveness, and the pursuit of materialistic pleasures and rewards. Clearly, the degree of economic prosperity alters human behavior, which becomes evident in the design and implementation of continuously evolving socioeconomic systems. As a civilization continues to evolve, inequalities in wealth, power, and status increase, ultimately becoming sufficiently severe to destabilize the social and economic system. In Durant's words:

> **Individualism brought wealth, but it brought, also, insecurity and slavery; it stimulated the latent powers of superior men, but it intensified the competition of life, and made men feel bitterly a poverty which, when all shared it alike, had seemed to oppress none.**[8]

This mounting, insidious economic inequity and sense of cultural insecurity, symptoms of a generalized social and economic instability, may be traced to the evolution of more competitive, intricate societal processes, which inescapably increase societal complexity and disorder. Aspirations of wealth, the chief human motivator, become more intense and culturally dominant, creating a more competitive economic and social environment. Emerging materialistic priorities increasingly place less emphasis on altruistic attitudes, ethics, and behavior. Consequently, the shifting of cultural values toward materialism, combined with economic opportunity and prosperity, gradually create a more inequitable distribution of wealth and a more chaotic social order.

Galbraith approaches the issue of social and economic instability in terms of the potential for "human satisfaction" and "social achievement" with the question, "Just what should the good society be?" His response is that "the good society fails when democracy fails." Further, Galbraith expresses the primary objective of the *good society*:

> **In all the industrial countries, there is the firm commitment to the consumer economy—to consumer goods and services—as the primary source of human satisfaction and enjoyment and as the most visible source of social achievement.**[9]

He views those denied the opportunity to participate in the benefits of a healthy economy as being denied a basic liberty and the freedom of opportunity. Those denied the prospect of acquiring the necessities of life and beyond are relegated to a secondary social status. However, history has demonstrated that it is not possible for all people to benefit equally from economic opportunity and productivity. The reality is that some level of productive consumption of national wealth will benefit some people while providing less desirable financial and social outcomes for others, relatively disadvantaged, as society struggles to survive and prosper to the next level of success.

The more pointed message that emerges from thermodynamic-based economics is that the more successful societies religiously conform to two key socioeconomic principles: First, economic wealth-energy based investments must be diversified and balanced over time among production categories that continually generate *sufficient new wealth-energy capital* to maintain a positive rate of cultural growth. This entails wisely and creatively utilizing the most appropriate science, engineering, technology, and a general knowledge base to create and implement processes of the highest possible economic effectiveness. That is, a sufficiently large portion of consumed wealth-energy resources must be restricted to the regeneration of national wealth.

Second, and equally important to a society's continuous net generation of new wealth, is the equitable and wise utilization of economic productivity to create and maintain a relatively strong and satisfied labor force and a content general population. More generally, a social order that is economically and sociopolitically stable is generally content and constitutes a *good society*. Sociological outcomes that promote common, constructive cultural purposes, approaches, and agendas increase the probability of creating a more altruistic cultural environment among the population, which will promote positive contributions toward cultural growth and longevity. This includes the wise and effective use of wealth-energy resources, including an equitable distribution of wealth.

Galbraith and other modern economists have understandably focused attention on the circumstances and fates of those disadvantaged by economic and political systems as well as on strategies and tactics of investing, marketing, and negotiating for profit and success. Less thought has been provided to long-term socioeconomics that deals with critical issues of global energy availability, more efficient use of diverse energy sources, minimizing waste heat and materials from fuel combustion, and adopting science-based economic policies. Instead, economists have emphasized the mechanics of the financial and political aspects of economic policies and practices. Such topics as income distribution, income tax strategies, tariffs, anti-trust issues, and trade unions have been major mechanistic-oriented themes, influenced more by politics than by the science and economics of energy and true wealth. The main focus of modern economics is achieving unabated increases in corporate profitability benefiting a narrow segment of society. However, such private enterprise success must also provide a society with such elements of social equality and stability as Social Security, unemployment insurance, and old age and survivor's pensions and health insurance. In *The Affluent Society*, Galbraith cites his purpose:

> **The purpose of these chapters ... is to see what economics assumed in the origins about the ordinary individual and his fate.**[10]

While the above microeconomic topics deserve the attention they receive, the absence of a rigorous science-based economic context for these socioeconomic issues, addressing the priorities and effectiveness of national wealth consumption, is an unfortunate and serious omission. This exclusion of energy-based economics from this conversation during the twentieth century permitted economists to rely on philosophy, political ideology, and speculation supported by empirical data as the basis for economic principles, policies, and practices. Nevertheless, it is appropriate that Galbraith and others, in considering the general benefits to society, place great significance on productivity and the equitable distribution of its rewards. However, this constitutes only a theoretical, mechanical (non-wealth-energy) approach to minimize unemployment and poverty, thereby achieving a more equitable and inclusive income distribution. His vision is a healthy economy of increasing productivity, capable of benefiting a broad segment of society rather than just the affluent minority. That is, economic productivity can be a pathway to an improved social order and vice versa.

> **Our preoccupation with production is, in fact, the culminating consequence of powerful historical and psychological forces—forces which only by an act of will we can hope to escape. Productivity, as we have seen, has enabled us to avoid or finesse the tensions anciently associated with inequality and its inconvenient remedies.**[11]

Galbraith views increasing production (i.e., increasing aggregate output) as a methodology to potentially achieve greater equity of income and wealth distribution, thus reducing serious historic societal tensions. However, in actuality, it is "the increase in output in recent decades, not the redistribution of income, which has brought the great material increase in the well-being of the average person."[12] Thus, it would appear that the consumer's *acquisition of interesting products would satisfy modern human desires in lieu of improvements in the broader distribution of wealth*. From an egalitarian perspective, this view represents significant progress from the prevailing thinking of the time by such reputable economists as Richardo, Malthus, and Adam Smith and from the accepted view that the "iron law" of wages is intrinsic to a successful economy. The principle of the iron law, held by the most respected economists in the early 1900s, is that for most workers, a minimally survivable wage is the most that a worker could ever hope to achieve. The generally accepted ideology was "Men will forever live on the verge of starvation."

At this point, thermodynamic-based economics will be related to specific conventional twentieth-century economic thoughts and principles expressed by Galbraith. The discussion will center on the following three questions: First, did he envision an accumulation of social complexity and disorder as being *inherently* generated by economic systems as societies matured economically and politically? Second, did he comprehend the necessity for an economic system to consume an uninterrupted flow of wealth-energy resources, which must be wisely invested in order to regenerate national wealth? Third, did his work reveal economic characteristics and properties that could be described or interpreted in thermodynamic terms? These issues will be the foci of this review to test some accepted twentieth-century economic concepts as being consistent with wealth-energy-based economics.

Galbraith's volume, *The Good Society: The Humane Agenda*, depicts modern economics as advancing initially from an era of conflict between capital and labor to a system whereby the economically successful and affluent segment of society acquires sufficient economic and political power to dominate. Interestingly, his journey of socioeconomic evolution treats history (i.e., time and change) in a manner consistent with thermodynamic principles:

> **To ignore a far deeper truth—to fail to appreciate the more fundamental thrust of history.... In reality, it is history that is in control.... History, the truly relevant source of change, will not be reversed.**[13]

Galbraith's description of change, "it is history that is in control," "the more fundamental thrust of history," conveys a similar message as entropy's descriptor of *time's arrow*. The daily processes of nature and society, all of which require wealth-energy resources, function as directed by Mother Nature's laws and create history as affected by human influences. His statement refers to the directional character of the progression of a social order: "History, the truly relevant source of change, will not be reversed." This characterizes the spontaneous, irreversible nature of a civilization's progression from a more orderly state to one of greater disorder, *independent of time*. More specifically, Galbraith traces the era from rural America of the Great Depression, when half of the employed worked in agriculture, to the dominant modern urban industrial environment. This exemplifies the irreversible, spontaneous directionality of a socioeconomic system, considered as thermodynamic processes, from a more orderly society to one of greater disorder. He notes that societal evolution derives from economic change that creates complexities, arising from economic success, and that societal complexity and disorder are among the elements associated with productivity. These concepts as expressed by Galbraith are equivalent to those of the mechanistic-thermodynamic paradigm and are applied to wealth-energy based processes that constitute essential cultural processes of existence. Additionally, his assertion that history is "the truly relevant source of change" is consistent with sociologists' concepts of change.

As noted in a previous chapter, Sorokin rhetorically asks, "Where shall we look for the roots of change of sociocultural phenomena?" His response is that a sociocultural system "bears in itself the seeds of its change" and that change occurs "by virtue of its own forces and properties."[14] Galbraith describes history as "the more fundamental thrust, ... the truly relevant source of change." This is consistent with Sorokin's view that a culture itself contains the "seeds of change, ... its own forces and properties" of change. Sorokin's sociological characteristics of cultural change are based on human nature, values, and behavior (i.e., a *mentality to do the right things* for all humanity). Sorokin views *the right things* as an adherence to, and the implementation of, altruistic values and behavior.

On the other hand, Galbraith views *doing things right economically* as primarily the effectiveness of the private sector's production and profitability and secondarily the degree to which cultural priorities utilize profits for the public good. This cultural objective relies on wise management of economic processes to generate a positive rate of cultural growth. Thus, *doing things right economically* entails efficiently, effectively, and wisely utilizing science-based economic principles, policies, and practices. Both authors embrace fundamental factors that influence socioeconomic systems (i.e., human nature and behavior and economic management capabilities), consistent with the mechanistic-thermodynamic paradigm. However, both authors fail to incorporate the one limiting and essential factor of cultural existence and change: Mother Nature's *master resource* (i.e., energy), the enabler of all human processes.

While reviewing Galbraith's economic journey through the twentieth century, it is instructive to reflect on shifting human values and behavior with increasing socioeconomic prosperity. Although the main focus of this chapter is economic change and resulting sociopolitical consequences, it is imperative to address the underlining foundation of human nature and behavior that drives cultural transformations.

Inevitably, as a culture overcomes one set of socioeconomic challenges with creative responses, a new set of even greater challenges appears with each cycle, creating its share of new and accumulating cultural complexity and issues. In the dominant agricultural economy of the early twentieth century, life expectancy was less than that at the end of the century, and one generation cared for the aging population of previous generations. There was less need for creating a Social Security-type program until the emergence of the urban industrial economy and related new employment concepts in the latter part of the twentieth century. Additionally, societal advancements including medical and health sciences, the technologies, home and workplace safety, and nutrition systematically increased life expectancy in developed countries.

Modern industrialization has produced regional and periodic unemployment, which was unknown in the predominantly agricultural economy, and thus, the concept of unemployment insurance was created. Ultimately, this was followed by new government worker initiatives including old age and survivor's pensions and health insurance. Also, with the expanding and increasing complexity of socioeconomic systems, a need emerged for societal protection in banking, employment, housing, sanitation, and food safety. In America, poor rural black and white populations were discovered to possess little education and inadequate civil rights, ultimately resulting in social legislation with the related costs shared by the private and public sectors. As the nation became a global economic and political player, national commerce, military forces, and foreign diplomacy assumed a larger, new dimension of government activity. Thus, as a result of economic success and societal progress, more intrusive government increasingly became a new reality in people's lives, providing much-needed services, but at considerable expense.

In Galbraith's description of the economic evolution from a predominantly agricultural economy to a modern industrial economy, he documents the increase in complexity and disorder that accompanies a society's maturation process. Additionally, the economic and sociopolitical processes are described in terms of directionality and irreversibility. Galbraith appears to accept that a civilization's socioeconomic activity inherently generates and integrates by-products of complexity and disorder into the fabric of mature societies. He recognizes the necessity to devote resources and to adopt policies that will counteract these negative social consequences of cultural success (i.e., societal entropy production is to be expected but is costly to the national economy).

Galbraith appears to share Soddy's view that sufficient national wealth must be invested in high-priority production areas capable of continuously creating new wealthy-energy capital, as opposed to excessive investment for outcomes that do not possess the capability to generate new wealth. Galbraith recognizes that there are categories of "production" that are "most frivolous" and others that are "the most significant and civilizing services":

> **There is another respect in which our concern for production is traditional and irrational. We are curiously unreasonable in the distinctions we make between different kinds of goods and services … some of the most frivolous goods with pride … some of the most significant and civilizing services with regret.**[15]

It is clear that Galbraith understands that unwise and unbalanced investments of society's wealth will eventually cause a society to "stagger and fall." Societal stability depends on available resources being able to support both the private sector's production mission and the public sector's social services mission.

> Economists in calculating the total output of the economy—in arriving at the now familiar gross domestic product—add together the value of all goods and services.... No distinction is made between public and private produced services.... It is privately produced production that is important.... This adds to national well-being. Its increase measures the increase in national wealth. Public services, by comparison, are an incubus. They are necessary, and they may be necessary in considerable volume. But they are a burden which must, in effect, be carried by the private production. If that burden is too great, private production will stagger and fall.[16]

Galbraith and other economists express similar views regarding the importance of private production: "It is privately produced production that is important.... This adds to national well-being," while the public services "are a incubus ... a burden which must, in effect, be carried by the private production." However, economics does not provide a formal, quantitative model or equation that identifies and allots the appropriate minimum sustaining level for a society's investment in public services that will also permit fiscal integrity and economic productivity. The approach to the fundamental issue of financial responsibility for public services within a financially sound economic model is vague and qualitative. On the other hand, the thermodynamic-based economic model reflects quantifiable wealth-energy resource parameters that provide for a socioeconomic system of financial integrity. This approach is also capable of identifying a sound balance of wealth-energy investment between essential and nonessential goods and services.

Galbraith recognizes qualitatively that for a society to prosper financially while maintaining social stability, *national wealth must be intelligently and proportionately invested* in areas capable of creating new national wealth. However, this necessitates an economic balance capable of paying for the continuously increasing costs of public services that proliferate as a socioeconomic system matures.

> Failure to keep public services in minimal relation to private production and use of goods is a cause of social disorder or impairs economic performance.[17]

Clearly, he also recognizes that these "public services" address the complexities and disorders of a social system that result from economic progress and, if not appropriately addressed, are capable of destroying a prosperous, stable society.

> **It is in the nature of the system that its productive process or its products can have harmful social effects. And there can be long-run effects that are currently invisible or enthusiastically, even righteously, ignored but that are potentially disastrous.**[18]

Poverty and unmanageable personal and national debt are modern examples of "harmful social effects" that are "potentially disastrous." Galbraith credits Thorstein Veblen for his view that "poverty, or more accurately both the moral and material debasement of man, was part of the system and would become worse with progress.... The economic costs of progress are, however, even less severe than its cultural consequences."[19]

> **"With economic advance and accompanying social responsibility, the problems facing government increase in both complexity and diversity, perhaps not arithmetically but geometrically."**[20]

The mere existence of a social order necessitates processes that inevitably result in some degree of poverty, and even if these processes are generally successful, such "cultural consequences" are capable of ultimately destroying a socioeconomic system. This is the essence of the societal entropy concept as well as the consequences of Galbraith's theories of social balance[21] and investment balance.[22]

An abundant collection of interrelated, self-perpetuating *cultural consequences* inevitably emerges from "economic advancement and accompanying social responsibility," constituting a major source of societal complexity and disorder. What has previously been referred to as a typical element of societal entropy production is more formally defined in economics as the *bureaucratic syndrome* or *bureaucratic stasis,* characteristic of both private and public organizations. This results from the proliferation of large organizations and the expansion of managerial and related support staffs, a primary source of increasing institutional complexity, confusion, and stress. This phenomenon produces costly cultural expansion and integration of institutional structures for communications, planning, and general administration as well as for the implementation of emerging technologies, marketing, and public relations functions. Significantly, as a society matures, such organizational expenditure increases have a limited capability to generate incremental profitability and national wealth. Wealth that is unproductively spent is wealth unavailable for generating new wealth-energy capital to support long-range societal prosperity. Galbraith states:

> **It is the wonderfully perverse tendency of economic behavior that results are frequently more visible than cause.... Tendency to nonfunctional proliferation must be accepted. Organization is organization wherever it exists and under whatever auspices.... The answer lies in accepting that bureaucratic stasis and unnecessary personnel proliferation are the basic flaws of all organization.**[23]

The principles of energy-based economics would necessitate an addendum to the end of the above quote: *"and the inherent by-products of economic wealth-energy resource consumption."*

Thus, Galbraith conceptually portrays, in his own words and without scientific terminology, the societal entropy phenomenon that inherently and inevitably accompanies the consumption of national wealth. His economic principles and rationales are consistent with the concepts of an energy-based economic model. Interestingly, his description of economic outcomes is similar to the social consequences of the capitalism economy expressed by Joseph Schumpeter, that the very success of capitalism "undermines the social institutions which protect it, and inevitably creates conditions in which it will not be able to live."[24] This also depicts the societal entropy production of culture disorder and complexity created by economic processes.

The element of *stasis or imperfect societal efficiency* and *effectiveness* is a by-product of the management of social, economic, and political life within a society. It is a property of inevitable societal entropy production for a society implementing its socioeconomic agenda. The economists, historians, sociologists, and political economists all see the same societal characteristics of cultural progress, but label and describe them differently depending on their discipline's frame of reference. However, the social and political complexities and disorder emanating from a given economic system complicate society's achievement of its aspirations, but are consistent with the expected outcomes of an energy-based economic model and its wealth and energy consumption. A society can only take steps that will influence and attempt to minimize the degree of societal complexity and entropy production, but it is theoretically impossible to completely avoid it. This is inherent in the second law of thermodynamics.

As has been emphasized numerous times in this work, a fundamental principle of the energy-based economic model is that a sufficient portion of national wealth, appropriately defined, must be wisely and effectively invested in the economy in order to generate sufficient new wealth-energy capital to continuously grow the economy. Tainter refers to this challenge as combating a "seemingly inexorable trend toward declining marginal productivity in hierarchical specialization."[25] Historically, the correlation of marginal productivity reduction with the increasing maturity of civilizations is also reflected in Galbraith's discussion of the need for *social balance* of goods and services produced and consumed. He notes that societies have a problem with investment decisions and choices of what to produce, "an implacable tendency to provide an opulent supply of some things and a niggardly yield of others."[26]

He continues:

> **The same forces ... distort the distribution of investment as between ordinary material capital and what we may denote as the personal capital of the country. This distortion has far-reaching effects [and] ... impairs the production of private goods.[27] The inherent tendency will always be for public services to fall behind private production. We have here the first of the causes of social imbalance.... Failure to keep public services in minimal relation to private production and use of goods is a cause of social disorder or impairs economic performance.[28]**

Galbraith emphasizes the need for "social balance" between the investments of "personal capital" for the "production of private goods" and the economic financial burden of investing in "public services." "Failure to keep public services in minimal relation to private production and use of goods is a cause of social disorder." It is recalled that Soddy defines *capital* as an *agent of production,* usually a form of permanent wealth, Wealth II. He divides capital into *personal possessions* and *organs of production* necessary for creating new forms of national wealth-energy capital.[29] Conceptual consistencies among the economic, sociological, and scientific perspective of Sorokin, Galbraith, Tainter, and Soddy relative to the investment of national wealth are noteworthy.

Even though wealth-energy resources are the fuel of a society's existence, the late twentieth-century economists did not position science-based economics as the cornerstone of national socioeconomic models, policies, and practices. Rather, the major emphasis and investment priorities of society have been, and continue to be, the creation and satisfaction of nonessential consumer desires. Marketing-oriented commercial strategies promote production, profit, and short-term business interests, but at the expense of long-term societal prosperity, stability, and longevity. Modern economic practice deals with the mechanistic aspects of the discipline while ignoring the science of energetics. Increasingly, consumer economics emphasizes increasing production output to meet synthesized consumer needs by cultivating and transforming created desire into sociological "must have goods" through advertising and salesmanship of new products one never knew were necessary. In addition, financial institutions continually create financial gimmicks and ethically questionable tactics to enable the consumer to incur debt in order to participate fully in modern consumerism.

> **Today's market economy, which so competently supplies consumer goods and services, does so in pursuit of relatively short-run return; that is its measurement of success. It does not invest readily, sometimes not at all, for long-run advantage. Nor does it invest to prevent adverse social effects from its production or its products.[30]**

The increasing sophistication of societal advancement has stimulated human desires to create and pursue an expanding menu of new goods and services, a socioeconomic evolution from basic *needs* required for survival toward *wants* that satisfy psychological desire. Satisfaction is never fully achieved, but grows insatiably with increasing economic prosperity. Galbraith refers to this phenomenon as the "dependence effect"; higher-level production results in a higher level of *want creation*, which in turn necessitates a higher level of *want satisfaction*, ad infinitum. The record levels of personal and national debt of the twenty-first-century American culture of extreme consumerism, greed, and unethical practices of major financial and corporate institutions illustrate Galbraith's view.

> **Plainly, the theory of consumer demand is a peculiarly treacherous friend of the present goals of economics.... The economist does not enter into the dubious moral arguments about the importance or virtue of the wants to be satisfied. He doesn't pretend to compare mental states of the same or different people at different times and to suggest that one is less urgent than another.... He sets about in a workmanlike way to satisfy desire, and accordingly, he sets the proper store by the production that does.**[31]

Obviously, if the private sector's highest priority for investment of national wealth is profit generated by providing goods of "marginal utility" and "slight consequence" in order to satisfy consumer desires, the long-term economic health of society will suffer. Such priorities will result in fewer resources being allotted for sound investments in opportunities capable of regenerating national wealth. This restricts a responsible and effective flow of national wealth-energy resources into the economy.

> **While our productive energies are used to make things of great urgency—things for which demand must be synthesized at elaborate cost lest they not be wanted—the process of production continues to be of nearly undiminished urgency as a source of income. The income men derive from producing things of slight consequence is of great consequence to them. The production reflects the low marginal utility of goods to society. The income reflects the high total utility of a livelihood to a person.... Income and employment rather than goods have become our basic economic concern.**[32]

Thus, an important task is to identify the responsible agent for formulating and responding to the "dubious moral arguments" associated with managing socioeconomic processes and cultural progress. History teaches that civilizations that successfully avoid difficult socioeconomic moral dilemmas, while striving for short-term economic gains without regard for long-term socioeconomic consequences, are on the pathway to cultural deterioration. Unwise investments and the abuse of wealth-energy resources based on poor moral choices and uncivilized behavior have historically resulted in cultural degradation and collapse. Historians, sociologists, anthropologists, and political economists have extensively documented the growth and deterioration of these civilizations, which will be reviewed in later chapters. However, a brief commentary will be presented on the question of how a culture arrives at the point of accumulating such chaos as to lose control of the social order.

The moral dilemmas that ultimately produce self-serving, materialistic, and ruthless outcomes, characterized by Sorokin as "sensate values," are obviously contributing to cultural decay. Sorokin describes such cultural characteristics as "the seeds of its change." Additionally, that societies typically make poor moral and ethical choices supports Schumpeter's contention that the very success of capitalism "undermines the social institutions which protect it, and inevitably creates conditions in which it will not be able to live."[33] Sorokin, Galbraith, and Schumpeter describe the same negative societal characteristics as by-products of a successful civilization pursuing self-serving socioeconomic aspirations. These properties of societal entropy production associated with energy-based economic models illustrate the nature of Schumpeter's reference to "conditions in which it [*society*] will not be able to live." As previously described, this is consistent with the mechanistic-thermodynamic paradigm, the basis for a science-based economic model as formulated by Soddy.

Additionally, the societal entropy effect provides more precise insights into Karl Marx's description of *self-created alienation and class conflict* generated by the dynamics of society's environment, resulting from economic-related attitudes and behavior associated with capitalism. Specifically, Marx notes that the rate of profit declines as capital accumulates, thus leading to more frequent and damaging periods of economic decline.[34] Essentially, this is analogous to the economic argument of Tainter, referred to as a declining marginal return on the investment in cultural complexity (i.e., "the best explanation of collapse").

In the latter stages of a capitalist or market-oriented economy, a society's once-successful financial and social incentive-reward system effectively and systematically becomes a corrupting influence on society's sense of values, self-discipline, and integrity. Society's initial, more primitive focus on rigorous, moral, and cooperative economic standards of behavior supporting societal goals is replaced by individual self-interest and materialistic priorities. The philosophy of doing the right things for society yields to a human instinct of doing the right thing for the individual in a competitive, ruthless environment that offers abundant financial opportunities and rewards. The turbulent 2008 American financial and economic crisis created by the Wall Street investment bank scandals and the $50 billion fraudulent global Ponzi scheme by Bernard L. Madoff Investment Securities established a record for brazen, unethical, and callous human greed. The unprecedented financial losses for millions of people produced life-altering ramifications for families, foundations, and businesses. One of victims of the Madoff scam was Nobel laureate Elie Wiesel's Foundation of Humanity.

Sociologists have proposed models to explain such behavior that usually emphasize competing forces within an individual that represent conflicting altruistic and self-serving values. An individual is torn between achieving materialistic, self-centered goals versus following an innate responsibility to support, as a good citizen, important objectives leading to societal-oriented success. History teaches that self-serving values and the attractiveness of potential wealth, status, and power will often win the battle of competing human values.

In the volume entitled *The Good Society: The Humane Agenda,* Galbraith asks the question, "Just what should the good society be?" He responds: "Before knowing what is right, one must know what is wrong."[35] It is clear that one major element of "what is wrong" is the highly attractive nature of potential economic success and its related rewards that inherently create incentives capable of motivating people to make poor choices. Human behavior, derived from self-serving values, and often ignorance, is responsible for creating and managing societal processes that inherently generate a high level of social complexity and disorder. Thus, the identity of "what is wrong" with a society's socioeconomic system may be illustrated by the nature and characteristics of its generated societal entropy.

This list of cultural elements that have typically gone wrong for mature civilizations is associated with large-scale economic systems and a corresponding sociopolitical environment intended to expand wealth, territory, and power. The *wrongs* reflect such cultural characteristics as inequitable wealth distribution, racial and religious hostility, wars, financial insolvency, environmental pollution, and widespread government and business corruption. These consequences expose a value system that accepts promoting the well-being of a few at the expense of many. These are by-products ... *the things gone wrong* ... created by a civilization's otherwise successful economic, social, and political processes. Ironically, the level of societal entropy production is proportional to the level and intensity of a society's economic investment and the adopted value system and priorities. Increasing socioeconomic advancement also increases undesirable waste material and energy residues and examples of abhorrent human behavior.

The theories and models of social transformations found in the literature place varying importance on individual, group, and societal motivations, attitudes, and behavior. Such models seek to capture cultural characteristics that codify current societal issues and opportunities, values, and economic conditions that appear to explain a particular chapter of human social history. Given this work's emphasis on energy-based economics, it is noteworthy that many of these sociological models recognize that societies often undergo rapid and significant socioeconomic advancement during periods of extreme challenge, stress, and chaos. Such swift, cultural advancement inherently requires enhanced wealth and energy consumption to support the expansion and integration of a society's structural and functional systems. However, such modeling has not provided a suitable and full rationale for the historically observed, sequential stages of cultural development (i.e., the sociocultural transitions from cultural growth to stagnation to deterioration). Finally, it is noted that more than at any time in the history of the world, the first decade of the twenty-first century demonstrates that wealth-energy resources represent a primary controlling variable of global socioeconomics, whether applied to the need for and cost of oil and coal or to the rejuvenation of the Asian civilization.

Chapter 4
References and Notes

1. Durant, Will. *The Story of Civilization: Part I: Our Oriental Heritage.* New York: Simon and Schuster. 1954. 19.
2. Georgescu-Roegen, Nicholas. *The Entropy Law and the Economic Process.* Cambridge: Harvard University Press. 1999. 3.
3. Galbraith, John Kenneth. *The Affluent Society.* New York: Houghton Mifflin. 1998. 6.
4. Durant, Will. *Our Oriental Heritage.* New York: Simon and Schuster. 1954. 6.
5. Ibid.
6. Ibid., 7.
7. Ibid. 18.
8. Ibid.
9. Galbraith, John Kenneth. *The Good Society: The Humane Agenda.* New York: Houghton Mifflin. 1998. 3.
10. Galbraith, John Kenneth. *The Affluent Society.* 28.
11. Ibid., 101.
12. Ibid., 79.
13. Galbraith, John Kenneth. *The Good Society.* 9.
14. Sorokin, Pitirim A. *Social and Cultural Dynamics.* Boston: Porter Sargent. 1970. 633.
15. Galbraith, John Kenneth. *The Affluent Society.* 108.
16. Ibid., 108-109.
17. Ibid., 193.
18. Galbraith, John Kenneth. *The Good Society.* 80.
19. Galbraith, John Kenneth. *The Affluent Society.* 45.
20. Galbraith, John Kenneth. *The Good Society.* 71.
21. Ibid., 186-199.
22. Ibid., 200-208.
23. Ibid., 108-109.
24. Schumpeter, Joseph A. *Capitalism, Socialism and Democracy.* New York: Harper & Row. 1975. 61.
25. Tainter, Joseph A. *The Collapse of Complex Societies.* New York: Cambridge University Press. 2004. 106.
26. Galbraith, John Kenneth. *The Affluent Society.* 186.

27. Ibid., 200.
28. Ibid., 193-195.
29. Soddy, Frederick. *Wealth, Virtual Wealth and Debt: The Solution of the Economic Paradox.* London: George Allen & Unwin. 1983. 122-123.
30. Galbraith, John Kenneth. *The Good Society.* 20.
31. Galbraith, John Kenneth. *The Affluent Society.* 130.
32. Ibid., 217.
33. Schumpeter, Joseph A. *Capitalism, Socialism and Democracy.* 61.
34. Robinson, Joan. <u>An Essay on Marxian Economics</u>. London: St. Martin's Press. 1960. 42.
35. Galbraith, John Kenneth. *The Good Society.* 1.

Chapter 5
Socioeconomic Transformations: Labor Power and Social Energy

"When the population rapidly increased, more and more of its members were able to enjoy a certain collective civilization. No doubt the social cost of this transformation—unconscious, admittedly—was very heavy.... The great problem for tomorrow, as for today, is to create a mass civilization of high quality. To do so is very costly. It is unthinkable without large surpluses devoted to the service of society.... The problem is more complex in the world as a whole. Much of the world's population is what one essayist has called 'the foreign proletariat,' better know as the Third World.... Unless humanity makes the effort to redress these vast inequalities, they could bring civilizations—and civilization—to an end." Fernand Braudel[1]

The nineteenth and twentieth centuries produced an unprecedented global socioeconomic transformation based on advancements in science, engineering, and technology and on the availability of abundant, inexpensive energy sources. Initiative, creativity, and perseverance utilizing newly found knowledge and developed skills initiated a new era of the Industrial Revolution and capitalistic economics, which gave birth to the rapid development of industrialization and commercialism. The resulting prosperity of this *new economy* catalyzed momentous cultural transitions that have revolutionized the social order of the previously dominant agriculture society.

In a chapter entitled, "The Sociological Approach to Change," S. N. Eisenstadt states, "One of the major premises of sociological analysis has been that the causes of social change are inherent in the construction of the social order."[2] He identifies the need to analyze the conditions and processes of the social order in order to comprehend the dynamics of a civilization's evolving social history. This sociological viewpoint of cultural change focuses on the fundamental nature of continuous socioeconomic interactions among members of a society. The membership relies on the established infrastructure, values, laws, and other formal and informal cultural understandings, all of which may eventually become subject to modification or obsolescence.

Sociological analysis reveals two opposing tensions always prevalent within a social order, one being cooperation in the hope of mutually beneficial outcomes, while the opposing tension calls for achieving outcomes that meet the individual's objectives and self-interest. At any particular time, a society must contend with both constructive and destructive forces, simultaneously spawned by a variety of conflicting altruistic and sensate cultural mentalities and behavior. Such conflicting motivations lead to competitive priorities and agendas, generating some degree of social instability. Such conflict within organizations and sociopolitical entities regarding the acquisition and disposition of communal wealth and related equity and power issues are common elements of controversy, stress, and dissension. As cultural complexity increases with societal maturation, a society increasingly approaches an extreme level of socioeconomic disorder, and a corresponding political instability generates a general sense of a society being out of control.

Consequently, an objective of sociological analysis is the identification of a society's organizational and functional properties and characteristics that might reveal details of how and why a population is able or unable to create and maintain a high level of social and economic stability. Understanding these basic dynamics is the first step toward identifying variables that adequately represent successful socioeconomic development. The next step is to apply these findings to major societal transitions that represent cultural phases of rapidly changing economic growth and social progress *as empirically identified and interpreted from historical events*. Such historical information and sociological analyses may provide a better understanding of human behavior associated with major cultural change, which is a different issue than identifying specific attributes, skills, and resources that inherently will lead to a successful socioeconomic transition. For example, the success of the Industrial Revolution required human creativity and inventiveness, but was only possible when specific scientific and technological principles were discovered and ultimately applied to people's vision of possible societal advancements. Specific resources and conditions were also necessary to achieving rapid technological and societal progress and ultimate economic productivity.

Sociological analysis also identifies the proliferation of socioeconomic bureaucracy, societal disorder, and other examples of cultural deterioration, the intensity of which appears to be in direct proportion to the rate of cultural progress. There is general agreement that societies continually increase their degree of complexity and dysfunction as cultures inevitable evolve. As stated by Eisenstadt, "Any social system tends to change totally in the general directions of increasing complexity and differentiation."[3]

In their pursuit of the root causes for the demise of past great civilizations, authors have provided an abundant literature of societal transition analyses. Historians, sociologists, political economists, and anthropologists have pondered recorded history, searching for signs and symptoms of societal decay described as cultural decline, collapse, and failure. Theories are advanced as to how and why great civilizations are transformed, not just into economic or military disappointments but, in addition, how people are subjected to suffering as successful cultures decline and, in some cases, become extinct. Based on late eighteenth- and nineteenth-century scientific discoveries, it became more evident how and why a "social system tends to change ... in the general directions of increasing complexity and differentiation," and in proportion to the rate of socioeconomic activity.

Many of the literature's causal linkages of history to societal impediments and ultimate decline are directly related to a lack of wealth-energy capital during critical periods to meet mounting resource requirements of an increasingly expanded, complicated, and aggressive society. Other linkages are often by-products of unwise investments of national wealth, particularly related to opportunistic commercial, military, and territorial expansions as well as to pursuing religious and political ideological conversions. Also, elements of societal deterioration sometimes ultimately appear as unintentional consequences of a social agenda that was considered desirable at an earlier time as an appropriate solution to a pressing problem, issue, or common need. Societal complications, characteristic of a failing social order, include unmanageable sociopolitical policies and practices, disintegration of cultural values, unresponsive large bureaucracies, abuses of technology, breakdown of central authority and infrastructure, financial insolvency, dispirited populations, lawlessness, and inadequate responses to emerging issues. These properties and characteristics of cultural deterioration, examples of societal entropy production, have been observed in the declining years of the great civilizations from ancient Greece to America.

Such consequences result from long-term sequential transitions from an era of positive cultural growth to an era of prevailing socioeconomic decline. The downward spiral of cultural deterioration from initial success to a state of disintegration corresponds to the gradual accumulation of societal entropy from unremitting and escalating wealth and energy consumption. The initial stages of cultural development may produce significant, useful socioeconomic advancement but, at the same time, are accompanied by a limited degree of accumulating societal dysfunction. While each subsequent cycle may produce a measure of societal advancement, each also creates additional issues that require further attention and resources. This *wealth-energy consumption-societal entropy production cycle is the inherent input-output relationship* of a society's socioeconomic activity, in accordance with the laws of thermodynamics. Consequently, the cultural burden and consequences of accumulating societal entropy are managed to the extent that human abilities and resources will permit.

At some point, this accumulating cultural degradation evolves into a chronic and insidious socioeconomic deterioration stage approaching cultural stagnation. Subsequently, the combination of psychological stress, ineffective public services, inadequate leadership, and insufficient financial resources to support basic societal functions produces a critical point of unmanageable crisis and chaos. This severe social instability represents a critical fork in the road for a society that leads either to a constructive cultural transformation and socioeconomic revival or to continued deterioration and potential cultural collapse. The pathway of cultural rejuvenation is possible only if significant changes occur in human values and behavior, resulting in significant reformulation of the socioeconomic system, which historically has also required overcoming inappropriate financial priorities and inadequate material resources.

Previous chapters have provided the foundation for this inherent social disorder being a natural consequence of the fundamental laws of nature that govern wealth and energy consumption during the pursuit of opportunity and social advancement.

> **In the application of the laws of thermodynamics to cultural systems we have one of the most illuminating and profound interpretations of cultural systems that is currently available to us. The extent to which the laws of thermodynamics have been applied to cultural systems is so far very limited and not well understood.**[4]

Appropriately, it is the objective of this chapter to trace Western civilization's nineteenth- and twentieth-century socioeconomic transformation with emphasis on "the application of the laws of thermodynamics to cultural systems, … one of the most illuminating and profound interpretations of cultural systems." Such interpretations reveal the historic, fundamental importance of the role and consequences of societal entropy production on the inevitable deterioration of mature civilizations.

Civilizations experienced major cultural transformations as a result of advances in the sciences, applied technologies, and engineering, which catalyzed major socioeconomic changes including the Industrial Revolution and the capitalist economic system. These scientific advances not only identify Mother Nature's controlling principles of energy consumption but also provide the knowledge and technology for the commercial production and distribution of a broader range of goods. Appropriately, the evolution of modern social history may be traced by following nineteenth-century economic development from a scientific perspective. Importantly, modern scientific understandings of fundamental energy exchange principles enabled the development of more progressive, science-based economic concepts and applications that influenced sociopolitical thought which, in turn, revealed the limiting potential of civilization progress.

By the late nineteenth century, the continuing development of classical thermodynamics by scientists such as Hermann von Helmholtz, Rudolf Clausius, and Sadi Carnot caused economists to examine the potential relevance of these new energy principles to the nature of human labor and ultimate socioeconomic consequences. Such investigations led to science-based definitions of *labor power* and *social energy*. Labor power relates the mechanics of the human body and its energy consumption to the labor process (i.e., physical work). This energy consumption is essential to the performance of useful and necessary work, whether it benefits only the individual or a specific group; thus, the appropriate label of *social energy*.

Helmholtz's historic work of 1847, "On the Conservation of Force," was significant for the disciplines of physiology as well as physics.[5] Thiss was the first fundamental scientific statement that permitted physiologists to use material and energy relationships that had been considered as the sole domain of physics. For the physical sciences, this work was one of the most precise and useful statements to date of the principle of the conservation of energy. Helmholtz's thermodynamic principles enabled the calculation of the work, defined in the physicists' formal sense of energy, performed by humans as a result of the activities or motions of the body. *This approach constituted a unified mechanistic and thermodynamic view of human labor based on scientifically defined parameters*. Thus, labor power and social energy were based on people assuming the role of nature's agent in the performance of useful work on behalf of a society. The work performed by the human body, fueled by nature's energy resources, also had ramifications for the disciplines of economics and social philosophy. That is, Mother Nature empowers the forces of societal activity and processes; the human body is her machine and agent! The authors of the period seriously considered not only the nature and effects of the emerging energy concepts of Helmholtz on the physical world and the human machine but also the impact on interrelated political and socioeconomic systems.

It was ultimately appreciated that human pursuit of survival and prosperity was dependent on wealth-energy resources for the creation and functioning of a society's organizational systems, which was also influenced by cultural principles, values, and behavior. Such factors define a culture's social, economic, and political systems and reflect distinctive culture traits and history.

The influence of Helmholtz's work from the late nineteenth century and beyond has been extensive. For example, the work of Charles Darwin was based solely on mechanistic concepts and principles pertaining to the evolution of natural history but devoid of energy considerations. It was not until Helmholtz's concepts of thermodynamics were related to human energy and labor power that a complete view of mechanistic-thermodynamic principles was applied to the evolution of the animal world. That is, the Darwin literature is deficient by the absence of the energy and social components of evolutionary theory.

> **As Darwin's writings left open the social implications of evolutionary doctrine ... thermodynamics permitted an equally broad range of interpretations from progressive social reform to the more apocalyptic conclusions of Nietzsche's image of history.[6]**

The thermodynamic principles emerging from that era provided both conservative and progressive social thinkers with new and broader concepts to ponder that were capable of providing an enhanced understanding of social change. Thus, the fundamental concepts contained within the first two laws of thermodynamics, the conservation of energy and the entropy concept, were being thought out, applied, and debated by such prominent people as Karl Marx, Hermann Henri Gossen, Adam Smith, and Carl Neumann. As a result, it was accepted that thermodynamic principles, as professed by Helmholtz, had significant implications for economic and sociopolitical thought. Specifically, the concept of the conservation of energy was identified as potentially providing significant insights into possible long-term economic benefits for society that could be obtained from the wise use of both nature's energy resources and human labor. This is reflected by Anson Rabinbach's review:

> **By the 1850s, the implications of energy conservation were becoming clear to political economists: The nation that most efficiently used and conserved the existing supply of the world's energy—including both labor power and technology—would also win the race for industrial supremacy.[7]**

Hermann Henri Gossen notes the wisdom of conserving social energy and illustrates his thinking on the *social energy* and *economic energy* concepts by his statement: The "totality of commodities over which a person disposes constitutes his economic energy, his wealth, which devolves to the benefit of the social whole."[8] Gossen emphasizes the necessity within a society to balance individual self-interest with the best interests of society by providing appropriate and satisfying cultural conditions for all members as a quid pro quo for the individual's personal contributions to society's labor power. Additionally, Gossen identifies the lack of capital as a major impediment to a society maximizing its energy for *useful purposes*. That is, additional capital investment is required for the necessary facilities, technology, management, and staff to conserve sufficient energy to improve long-term economic prosperity as well as to enhance society's general working and living conditions. However, he notes that such financial investments usually detract from short-term profits and are contrary to observed human behavior (i.e., the self-interest of investors for quick profits usually prevails). Gossen recognizes the necessity for balanced self-interest with societal-interest, for what Galbraith refers to as the creation of "the good society" and describes in his theories of investment balance and social balance.[9] Furthermore, such *balances* illustrate how inherent human values and behavioral characteristics, albeit possibly possessed by only a minority of the population, are capable of contributing to social, political, and economic deterioration that accumulates over time.

The entropy concept, which governs the efficiency of converting energy into useful work, implies that a society's functional processes are capable of minimizing its long-term energy wastefulness and environmental pollution while positively affecting economic efficiency and productivity. Additionally, the conditions that people impose on the essential processes of cultural existence via values, priorities, and decision-making can result in a broad range of positive as well as negative socioeconomic outcomes. Thus, wealth, energy, and human values and behavior are a society's primary variables of potential success and failure.

By the late nineteenth century, as a result of the conceptual considerations of social energy and labor power, it was generally accepted that energy conservation was related to long-term economic health, and the notion of a planned national economy emerged. Carl Neumann, who characterized economic life as principally involving energy exchanges, expressed such logic as "the transfer of energy of physical bodies, … the energy of the self as the capital of the body."[10] Thus, humans consume energy via food materials, become a commodity, and are utilized as labor power (i.e., an energy transfer to meet economic outcomes as useful work). It should be noted that an individual's or a given society's *perception of economic need* or *useful work* may or may not be classified as such by other individuals or groups. Such diversity of societal viewpoints of *perceived needs or wants versus actual survival needs* is illustrated by investments in production that regenerates national wealth in contrast to those providing luxury goods and services incapable of creating new wealth-energy capital. Over the course of history, this issue of balancing human "wants" or "must haves" with society's basic needs, useful work, and purposes has posed a serious challenge to a social order prioritizing initiative, creativity, and resources necessary to stimulate constructive cultural advancement.

Thus, the revolutionary socioeconomic implications derived from the new thermodynamic principles that emerged from early nineteenth-century thinkers eventually led to thermodynamic-based concepts of national wealth, energy-based economic models, and a national economy as more precisely and completely defined by Frederick Soddy in the early twentieth century.[11] These views include energy possessing a controlling influence on cultural processes and, as noted by Rabinbach, entropy being a negative social force: "The devastating effects of entropy could not be ignored: The gradual diminution of force over time presaged inevitable economic and social decline."[12]

Gossen's view is expressed in more substantive scientific language by Rabinbach's observation that incremental capital investment is continuously required for a society to prevail over its continuous accumulation of the "devastating effects of entropy." This is, a society's accumulation of entropy production becomes increasingly expensive over time, requiring a continuously increasing rate of economic growth to maintain the momentum of advancement. Thus, the achievement of long-term economic prosperity is dependent on minimizing societal entropy, which translates into minimizing nonessential production, maximizing productivity, and adopting altruistic cultural values. However, the human drive to maximize short-term profits, regardless of ultimate long-term societal consequences, has historically ruled human behavior. Self-serving sensate values and ambitions are stimulated and thrive during periods of economic prosperity, thereby maximizing societal entropy production.

Many constructive examples exist of financial investments intending to achieve long-term wealth-energy capital production efficiencies that simultaneously benefit the individual and society. On a micro scale, homeowners may elect to invest in certain construction costs and adopt certain environmental attitudes and practices that, as a homeowner, will provide positive financial returns. This can result in long-term saving for the homeowner, as well as reduced societal energy consumption and environmental pollutants. Achieving a theoretically desirable win-win scenario of a balance between the individual's self-interest and society's benefit is hypothetically possible for an entire social order. However, historically, this objective has been unattainable on a large scale for mature civilizations. Human nature prevails, and entropy continues to accumulate and take its social and economic toll.

By the late nineteenth century, a new appreciation of the potential negative impacts of entropy production on socioeconomic systems created concerns for the long-term well-being of society. This apprehension was based on an evolving view of socioeconomic relationships involving such concepts as labor power, human energy, national energy consumption, and the newly formulated thermodynamic principles. As a result, formidable social thinkers suggested that productivity could be improved while reducing energy consumption and requiring fewer working hours from the labor force. This would be accomplished by utilizing processes and techniques that minimize wasting wealth-energy capital. Equally surprising, this perspective was not the exclusive domain of conservative economic and social thinkers, as even the socialists shared many of the same views, most notably Karl Marx:

> **The more the productivity of labor increases, the more the working day can be shortened, and the more the working day is shortened, the more the intensity of labor can increase. From the point of view of society the productivity of labor also grows.... This implies not only economizing on the means of production, but also avoiding all useless labor.**[13]

Thus, Marx comprehends the social and economic benefits to humanity of minimizing working hours through the use of technology and conserving natural resources, thereby minimizing "useless labor." Additionally, he sees the necessity for "human freedom," which he relates to the worker being spared the time and physical effort of useless labor that results from production practices and motivations that unnecessarily waste time and energy.

According to Marx, production is best achieved "with the least expenditure of energy and under conditions most favorable to, and worthy of, their human nature."[14] He captures important cultural properties necessary for long-term societal prosperity, including energy conservation and a satisfied, well-treated, and appreciated labor force. Idealistically, Marx correctly views the conservation of human labor and natural resources as thermodynamically, economically, and environmentally desirable (i.e., "economizing on the means of production" and "avoiding all useless labor,"), while providing the production necessary to satisfy human needs.

However, Marx incorrectly assumes that people's perception of the sophistication of *perceived needs* would remain relative constant over time at a survival level and would not escalate toward greater materialism with increasing economic prosperity, as was later recognized by such sociologists as Sorokin. Additionally, Galbraith addresses more modern strategies of marketing and salesmanship that increasingly create new and more elaborate *perceived consumer needs*.[15] That is, continuing economic success creates and promotes an appetite for greater consumer spending for nonessential goods and services and provides the financial mechanisms to pay later, creating a culture of debtors.

Engels, expanding Marx's thinking, considers labor power as a commodity with a unique potential for creating value, "a value creating force, the source of value.... When properly treated, the source of more value than it possesses itself."[16] Rabinbach states that "labor power is ... both social and physiological, historically specific and at the same time a form of universal energy.... [It] represents a quantitative aspect of labor under capitalism.... Both a social and a physiological magnitude, it is a measure of value and a measure of energy."[17] Marx expresses the same thought:

> **On the one hand, all labor is an expenditure of human labor power in the physiological sense, and it is in this quality of being equal, or abstract, human labor that it forms the value of commodities. On the other hand, labor is an expenditure of human labor power in a particular form and with a definite aim.**[18]

Thus, *labor power* represents a quantity corresponding to a culture's necessary labor force, a force associated with a specific bodily energy equivalent. Marx also made an important distinction between *laboring activity* and *labor power,* as in simply expending energy versus actually completing useful work. Labor power has a standardized value as a unit of production and is a productive force in the sense of providing energy capital corresponding to a thermodynamic quantity. Also, Marx recognizes that energy must be renewed and continuously flow into a production system. These interrelationships among labor power, economic value, wealth, and energy also correspond to those of Soddy's scientific-based definitions of wealth and related economic and financial parameters found in earlier chapters.

Possessing labor power, considered a social objective, has a magnitude, consequences, and an energy equivalent associated with the work of the human body. *Greater human freedom* is an important component of a socioeconomic system that involves fewer working hours and thus constitutes a methodology to reduce energy consumption and entropy production associated with a society's economy. While the creation of less societal disorder and achieving human freedom or labor-free time to enjoy life is theoretically sensible and personally rewarding, human nature drives people to disregard this sensibility and to pursue the sensate values of materialism and self-interest to an extreme. Consider American cultural changes during the last half century when, increasingly, a larger proportion of families had both spouses employed, with some having more than one job, not necessarily to survive but often to acquire nonessential consumer products. The modern drive to acquire and maintain a socially acceptable and respected financial and social status and to afford the abundance of nonessential consumer goods leads to "the competitive waste of capital" by individuals and to commercial "profitability" that "yields proportionately less value for the capitalist.... The rate of profit declines" as large proportions of national wealth are consumed and not regenerated.

Rabinbach captures Marx's appreciation for the importance of the entropy effect on socioeconomic practice:

> **Marx discovered the principle of entropy at work in capitalism: The unidirectional time flow of history produces the inevitable tendency of capitalism to decline as the productivity of labor power increases.... But because of the well-known tendency of capital to require greater and greater intensity to maintain or increase the profitability of labor in a competitive atmosphere, the conversion process yields proportionately less value for the capitalist, or as Marx expressed it, the rate of profit declines.... History exercises its sovereign "cunning" through the competitive "waste" of capital.**[19]

It is noteworthy that the phraseology "the conversion process yields proportionately less value, ... the rate of profit declines" is the essence of the second law of thermodynamics and the accumulative effect over time of societal entropy production reducing a society's marginal return of investment in its future. Tainter refers to this as a return on the investment in cultural complexity.

Rabinbach presents the argument that, based on the revolutionary thermodynamic perceptions that surfaced late in Marx's life; he held the view that "the productive potential of nature and technology is the organizing principle of society."[20] Also, he says that Marx "held to a 'paradigm of production' in which the distinction between the natural forces of production and the productive forces of society is no longer decisive."[21] That is, economic production, whether through the direct efforts of people or nature's processes, is the basic force for the existence of a social order.

Engels is credited with utilizing Helmholtz's scientific work, specifically the law of the conservation of energy, and extending the Marxian philosophy to a more scientific formulation, thus making the transition from a social philosophy to scientific-based materialism. Engels, as reported by Rabinbach, "believed that all events in history and nature were manifestations of the fundamental character of matter."[22] Thus, the energy concept was expanded to include the social implications of energy consumption by machines and industry, as well as for human labor involved in the societal processes of living. Thus, Engels placed political economics within the historic energy-dependent disciplines of physiology and the physical sciences.

Accordingly, by the beginning of the twentieth century, a more sophisticated and comprehensive view of energy evolved; specifically, the effect of thermodynamics and the "Helmholtzian cosmos" was firmly established as a factor, if not a primary societal driving force, for a culture's economic, cultural, and political activities. This view was later corroborated and extended through the efforts of Etienne-Jules Marey, whose study of human body motion as just another machine—"the body as a social instrument"—projected new insights into a thermodynamic-based concept of human labor power.[23] With great precision, Marey applied the same laws of mechanics and motion to the "animal machine" as would be applied to man-made machines. His view was that the "animal organism is no different from our machines except for their greater efficiency."[24]

As a result, based on the conservation of energy, the mythical concept of "vitalism" as the essence of human life was replaced by energy and thermodynamic principles. Marey makes the comparison of an engine's combustion chamber producing heat by burning a fossil fuel with the human body's heat generation relying on consumed food as fuel. The mechanical engine and the human body are both capable of producing forces that may accomplish useful work for society's goals, as well as just generating activity and wasting energy. Marey considers the essence of life to be explained by the principles of chemistry and physics and analyzes the human body and its functions utilizing mechanics, optics, and thermodynamics. Thermodynamics is viewed as "the most remarkable theory of modern times,"[25] as it is the foundation for achieving motion and useful work, which is central to society's existence.

Hopefully, at this stage in the presentation, the interrelationships of the energy consumption required to fuel society's purposeful processes that permit an organized social order to exist and evolve have been successfully linked to thermodynamic principles, specifically to the entropy concept. While such energy consumption may produce beneficial outcomes for some members of a society, "the devastating effects of entropy," as expressed by Rabinbach, will inevitably be generated by society's social, cultural, economic, and political courses of action, all of which require some form of human thought and continual correction or improvement. Therefore, from a historical perspective, it is not surprising that the emerging energy-related social thinking of the late nineteenth century became interwoven with the fast-paced, embryonic, and scientific-based Industrial Revolution. The principles of the "Helmholtzian cosmos" and the concepts of human labor and labor power as represented by Marey's work became components of the social transformation to industrialism. The industrial or modern society produced profound socioeconomic, cultural, and political change, all of which consumed increasingly larger amounts of wealth-energy resources in each stage of its evolution toward the more advanced society of the twenty-first century. It is clear that the industrial society has encompassed all segments of a society, extending beyond simply the considerations of the economy, scientific discoveries, technology, and engineering. The social order has been completely transformed.

This phase of social history is referred to as possessing *modernism* or *cultural modernity,* a society transformed by a radically new environment of rapidly changing socioeconomic characteristics emanating from the processes and consequences of industrialization and capitalism. Such modern and novel developments as more diverse human values and art forms, advancements in science and technology, unique economic opportunities, and progressive social philosophies blend into fresh, novel cultural attitudes, behavior, and characteristics. New cultural elements gradually replace more traditional ways of life and reflect a progressive view of existence in a new age. However, this *modernization* also produces increasing levels of familiar human behavior and discontent, stemming from new sources of complexity, unmet expectations, and societal stress, as was previously created in less intensity under former, more traditional and primitive societal conditions. Thus, *crises of modernity* appear and new issues accumulate at a rapid pace, resulting in significant increases in societal disorder. It is recognized that a modern industrialized society creates high-quality products and services that provide the necessary ingredients to stimulate the capitalist economic system and economic prosperity. However, unfortunately, it also generates a continuous accumulation of societal complications, discontent, and incoherence within cultural operating systems. This is simply the consequence of an era of more intense societal entropy production generated by one of the most significant and rapid cultural transformations in the history of the world.

Wealth, Energy, and Human Values

The sequence of socioeconomic transitions stimulated by continuous discovery and creative use of new knowledge, materials, and methods during the past 500,000 years of human social history provides insights into the nature of twenty-first-century modernity. The assignment of exact dates that a given civilization enters a new period of socioeconomic growth or decay, undergoes a significant cultural transformation, and ultimately enters a state of prolonged cultural deterioration is exceedingly imprecise. However, the primary variables affecting cultural change and the mechanisms representing transitions of the social order can be identified.

The evolution of social history, particularly significant socioeconomic transitions, parallels the progression of human knowledge and its application to survival and the continuous improvement in the quality of life. The baseline for this continuum is the primitive society's capabilities as hunters and gatherers and the social order's reliance on small group loyalties for security and survival as opposed to the certain failure of embracing values of individualism. About 8000 B.C., a significant economic transition occurred, as people began successfully raising animal stock and cultivating crops. The knowledge base of this era expanded to include the important economic technologies of the plow, irrigation, and the stirrup, which affected and modified the social order, resulting in the adoption of new cultural traits.

This transition provided the motivation and incentive to expand family and clan associations into less familiar but potentially economically rewarding communal groups. Such successful utilization of new tools and knowledge for agricultural products evolved to a degree whereby food surpluses were used as a source of wealth. This transition expanded the concept of labor beyond individuals providing their own food, clothing, and fuel, which led to the appearance of merchants, clergy, and various other new service providers. This significant economic advancement substantially altered the social structure of the culture. Characteristically, socioeconomic advances became associated with the variety and quality of commodities including foods, fuels, and materials and with the creation of new economic opportunities. *Wealth, energy, and human values and behavior are a society's driving forces of change.*

Villages became cities, and about 400 B.C., elaborate trading systems, markets, laws, and armies joined the realities of human existence.

During the sixteenth to eighteenth centuries, the Industrial Revolution rapidly expanded, creating a sharp increase in the rate of change of Western civilization. Human aspirations created new opportunities as inventions, discoveries, and creativity became commercially profitable, further altering the nature of the society. The continuation of this evolutionary social progression to more advanced levels was realized by the continuous achievement of new knowledge, greater production capacity, and increasing profitability. Such productivity increases are required, in part, simply to compensate for the physical and financial costs of the inherently accumulating, usually unproductive, socioeconomic complexity.

This social evolutionary process included transitions in clothing from animal skins and furs to wool to cotton, and ultimately to synthetic materials, with each transition in this sequence requiring an increase in energy consumption per capita. Each stage also produced more complex production and distribution systems inherently, which affects and expands existing sociopolitical systems, an example of *expanded cultural integration and stratification*. Human fuel consumption underwent transitions from wood to coal to oil and natural gas and ultimately to nuclear material, with each of these transitions requiring more extensive labor, technology, and energy consumption. Each transition requires increases in human and energy resources, but also results in higher levels of societal complexity, disorder, and material waste. Significant increases in the consumption of energy also contribute to the declining health of oceans, rivers, land, and animal populations.

An appreciation of this escalating pace of socioeconomic evolution and its consequences may be realized by reflecting on twenty-first-century intricacies, rules, laws, policies, tactics, and strategies of dealing with everyday life. These include the common complexities of urban living: tax forms, motor vehicle bureaus, monthly cell phone bills, health insurance legislation and regulations, and corporate and governmental customer service procedures. This raises the questions of people's physical, intellectual, and emotional tolerance to adjusting to the accelerating pace of a mature civilization and shifting cultural values.

These aspects of human behavior and other variables that constitute major influences on and consequences of societal transitions will be discussed in a later chapter. However, one specific set of cultural traits appears to have had a major influence on the development of the unprecedented socioeconomic transformation to industrialism. Thus, it is instructive to pursue the often-asked question of why the Industrial Revolution evolved initially from Western Europe, England, and the Netherlands as opposed to other more technically and culturally advanced civilizations of the sixteenth and seventeenth century.

A plausible response to this question is found in the similarities between the unique cultural traits of a major population segment within this geographical area and the optimal cultural characteristics that have been identified as maximizing positive cultural change and economic effectiveness. These idealized behavioral characteristics are those associated with the Protestant Reformation of the sixteenth century, as attributed by to Max Weber by Michael Lessnoff in *The Spirit of Capitalism and the Protestant Ethic*.[26] His insights into the work ethic and the spiritual and cultural foundation and characteristics of Puritan Protestantism constitute radically different sociological dynamics compared to the civilizations dominated by Eastern religions. The latter ideologies are motivated by, and focused on, the life hereafter, while the center of attention of Puritan Protestantism is on the here and now of a *worldly religion*. Weber offers the view of Protestantism as a culture with an intense work ethic whereby self-fulfillment of one's earthly destiny is an absolute religious imperative defined in terms of dedication and accomplishment in a profitable economic role in a manner that is pleasing to God.

It is noteworthy that religious philosophy and duty are linked to a work ethic and economic success as a requirement to avoiding eternal damnation (i.e., God expects the individual to be associated with a society's economic profitability and good works). In addition, one's behavior must avoid worldly pleasures, frivolous self-indulgences, and displays of self-aggrandizement. Thus, regardless of one's social or economic status, working diligently at some honorable and profitable job is a religious, divine expectation, not simply an optional economic opportunity of self-interest to become financially successful. Weber's Protestant ethic is represented by Lessnoff as:

> a powerful motivator of action not just for a few people but for large numbers, ... a religious doctrine accepted by many in an age of faith, ... explicit connection made by the Protestant ethic between a certain way of behaving in the world, and the fate of the individual's eternal soul—eternal bliss versus eternal torment.... Weber's thesis is as much (or more) a thesis about human psychology as about the filiation of ideas.[26]

The Protestant work ethic was clearly a societal motivator that provided an ideal environment for scientific and technological discovery, innovation, and commercialization. As such, this philosophy provided the necessary creative environment to nurture the Industrial Revolution, which ultimately led to a thriving new worldwide economy. Lessnoff argues that the Protestant ethic and the spirit of capitalism, while possessing subtle differences, share the same pathway but not the same ultimate objective, hence, the "human psychology " motivator. Protestantism offers salvation, but only if God views positively one's general behavior and one has met economic expectations and responsibilities of service to one's neighbors. Thus, the individual's objective is to serve the glory of God, avoid temptations, and achieve the goal of eternal salvation while contributing to a financially successful, comfortable life for all members of the community. Conversely, for the capitalist the *profit ethic* is primarily for one's own pleasures and not associated with any objective of salvation. However, regardless of the motivations, the Protestant work ethic was a major contributing factor to the early development and modern success of capitalism. It provided a nurturing, protective socioeconomic environment but also an aggressive and creative pursuit of profit maximization, a disciplined and inspired labor force, continuous accountability to God for one's general behavior, and stable sociopolitical structures and functions.

Given our primary thesis of relating thermodynamic principles to society's potential economic success, it is emphasized that these values, traits, and behavioral characteristics of the Protestant Puritan work ethic symbolize a cultural environment favorable to socioeconomic growth. This cultural environment creates a high probability for societal advancement and longevity. The Protestant religious ideology is consistent with crucial economic conditions and sociopolitical characteristics that create a high level of communal cohesion and cooperation even during periods of significant economic prosperity, a unique, major accomplishment. The Puritan cultural traits include prevailing attitudes of personal discipline and thriftiness, community as the first priority, an intense work ethic approaching religious zeal, strict adherence to one's religious ideology, and resource conservation. As will be discussed in a later chapter, the usual social counterproductive behavior that normally accompanies socioeconomic maturity and prosperity is not nearly as altruistic or intense as found in the Puritan culture.

Thus, these constructive, atypical cultural traits of the Puritan work ethic equate to an extremely profit-oriented economic priority that maximizes natural resources while promoting an unassuming way of life. Failure to achieve these religiously influenced communal and economic expectations has the most severe consequence of eternal damnation. In such an environment, societal wealth is scrupulously utilized for potential economic profitability while being consumed sparingly for nonessential purposes. Thus, such atypical cultural characteristics and strict priorities consistently applied are uniquely suited to produce a high yield of new wealth-energy capital from a society's investment in progress. This contrasts significantly with the usual high priority placed on luxury goods and services found within mature civilizations. The highly constructive and motivating Puritan environment is conducive to overcoming individual and group challenges and conflicts by virtue of its inherently creative, entrepreneurial, and altruistic attitudes that inspire responsibility and discipline toward communal interests.

The characteristics of Puritan behavior may be summarized as including perseverance, creativity, motivation, and energy efficiency, which are congruent with favorable thermodynamic requirements for maximizing useful work from the consumption of a given quantity of wealth-energy capital. The methodology for conducting societal processes that minimize the wasteful use of wealth-energy resources minimizes entropy production. That is, processes are conducted in a manner that most fully utilizes the heat content of fuels and, in the thermodynamic sense, maximizes useful work in order to maximize productivity and profitability. This approach is conducive to social and economic stability by placing a high priority on economic success while providing for the needs of all society members rather than just the self-interests of an elite class. Such attitudes and methodologies result in less environmental pollution and minimal political self-serving policies and discord.

An argument may be made that a high probability of socioeconomic success and long-term cultural viability exists when a society's general population adopts an ideology and practices corresponding to the Protestant work ethic. These human characteristics translate into a highly efficient conversion of wealth-energy capital for useful societal purposes. Such cultural traits also provide an environment that is highly conducive to the continuous search for new knowledge; maintaining creativity, motivation, and altruistic values; and achieving sensible and progressive intellectual, economic, cultural, and technological progress.

The principles and practices of the Protestant work ethic combined with the mechanistic-thermodynamic paradigm provide a plausible explanation for the leadership role played by Europe, England, the United States, and the Netherlands during the early developmental stages of the Industrial Revolution and the related capitalist economy. These Protestant behavioral characteristics produce low levels of social disorder due to fewer unresolved and contested issues that are normally based on individual and group materialistic self-interest, dissatisfaction, and social unrest. Thus, a higher level of communal satisfaction is achieved, which is a major attribute toward social stability by promoting social orderliness, mutual cooperation, civility, and a stable economy.

While the Industrial Revolution dealt with new science, innovation, and technology that stimulated a new economy, new sociopolitical thinking was being vigorously debated. The French and American revolutions gave birth to political reforms that, in conjunction with rapid industrial and economic transformations associated with the Industrial Revolution, provided the elements for the modern industrial age and the new economy. Thus, Western civilization underwent rapid changes toward a *modernism* linking science, technology, and philosophy of social thought that defined a new era. Such modernism, promoting a radically different and progressive economic way of life, produced a continuous shift of cultural values and traits while altering sociopolitical structures and functions. Over many decades, an evolving global economy, providing abundant rewards, has eroded the original

Puritan-type characteristics and cultural traits initially responsible for the initial success of modern Western prosperity.

Systematic societal advances during the late twentieth century produced a proliferation of organizational structures and centralizing functions related to the rapid expansion of economic production and distribution, financial services, communication, modes of transportation, and human services. Such socioeconomic progress invariably requires greater governmental and political involvement and expense to establish and maintain security, public finance, laws, health policies, and justice. The rate and extent of such societal integration and stratification is directly proportional to that of socioeconomic activity and the perception of the need for more societal orderliness and oversight.

Thus, there exists a dual nature to becoming more modern, advanced, and progressive or more civilized (i.e., a more organized, efficient, and affluent social order). On the one hand, improved social conditions, a prosperous economy, quality cultural arts, and more consumer luxuries are achieved. On the other hand, the same society produces increasing levels of undesirable by-products such as massive centralized governmental and corporate entities, political apathy, urban disorder and decay, and environmental pollution, all of which contribute to societal deterioration. In addition to this increasing complexity of socioeconomic life, the widening gap between the nature of an individual's labor and the associated financial rewards of the workplace irreversibly erodes the unity and integrity characteristically found in the more primitive phases of a society. Self-centered individualism and the pursuit of personal self-interest and the pleasures of life gradually replace traditional values and the unselfish concern associated with the culture of clans and tribes.

Secularization is a term used to describe a characteristic of the modern industrial society that is representative of a particular scientific approach (i.e., *rationalization*) being applied to some segment of a society's many operational functions. Max Weber describes the ramifications of creating extreme levels of orderliness within a rational system of administering the modern society.[28] He defines "rationalization" as the process of ordering and unifying systems of ideas and interests derived from cultural traditions and motivated by an underlining tendency toward order in human thought. This requires the specification, clarification, and systematization of ideas and normative controls and sanctions applied to a hierarchical social order. Consequently, a more mature society experiences a continuously increasing higher order of societal complexity as a normal by-product of cultural change, thereby reducing the usefulness of operational functions to the point of becoming ineffective. Specifically, Weber views administrative bureaucracies as being created by the application of rational principles such as impersonal and impartial enforcement of appropriate laws and policies. However, the bureaucracy ultimately takes on a life of its own, becoming increasingly hostile, irrational, ineffective, and inefficient in the practical applications of the "rationalization" principle by virtue of size, complexity, inflexibility, and authoritative mind-set. The abused participants of such a system ultimately develop a negative and noncompliant attitude toward such complex, inefficient, dysfunctional, and overbearing controlling practices that no longer achieve their originally intended purposes. Rabinbach, relying on Max Weber's "concept of rationalization," reflects this loss of societal integrity, efficiency, trust, and coherence:

> **In this view, the formal rigidity and rule-bound modern bureaucracy; the division of labor and the rise of complex managerial strategies; and the impact of commercial culture reduced traditional ways of living and interacting to quantifiable relationships, subverting ethics, values, and universal norms.[29]**

The consequences of "rationalization" become signs and symptoms of a society's illness, not its root cause. These symptoms of institutional and cultural degradation have come to be expected from an industrialized society's continuing success and expansion as exemplified by the history of General Motors and America's financial institutions during the past fifty years.

The legal but unethical 2008 American Wall St. and home mortgage scandals illustrate Max Weber's organizational psychology-based concept of the ultimate ineffectiveness of bureaucratic institutions and their "subverting ethics, values, and universal norms." Additionally, this perspective is consistent with *Tainter's financial perspective* of the negative consequences of excessive cultural complexity and *Sorokin's sociological model* of materialistic values dominating and subverting mature civilizations. American financial institutions, in the first decade of the twenty-first century, illustrate extreme organizational complexity, the misuse of national wealth, and the lack of ethical values and integrity that unfortunately permeate governments and corporations of Western civilization.

Finally, the literature cited in this chapter focuses attention on the social and political impact of inherent societal dysfunction resulting from a society pursuing a life style that maximizes economic productivity while inefficiently maximizing its wealth and energy consumption and its labor force. This is characteristic of an expanding and successful civilization already under stress, accumulating mounting complexity and disorder at a rapid pace on the pathway of cultural deterioration.

> **Continued investment in sociopolitical complexity reaches a point where the benefits for such investments begin to decline, at first gradually, then with accelerated force. Thus, not only must a population allocate greater and greater amounts of resources to maintaining an evolving society, but after a certain point, higher amounts of this investment will yield smaller increments of return, ... a recurrent aspect of sociopolitical evolution, and of investment in complexity.**[30]

This "continued investment in sociopolitical complexity" is, to a great extent, inherent in the erroneous philosophy that expansion of corporate and governmental functions is, in itself, evidence of progress and that society is better served by the creation of larger and more highly organized and specialized socioeconomic and political institutions. Consequently, centers of corporate and governmental power proliferate and inherently increase their influence and control over societal functions. This creates enhanced infrastructure and corresponding regulations to manage and control an ever-increasing number of issues of greater intricacy. However, attempts to reduce the rate of intensifying and accumulating disorder within a culture and to resolve the growing number of unresolved issues only increases bureaucracy and identifies additional issues and costly resolutions. New crises and problems proliferate faster than solutions can be created and implemented, while turmoil continues to accumulate.

Such systems, represented by large and continuously integrating and stratifying expensive bureaucracies, become increasingly less cost effective and less responsive to people attempting to cope with inherently increasing complexities of everyday life. These properties and characteristics of a failing, mature Western civilization exemplify Joseph Schumpeter's theory of "creative destruction."[31]

Schumpeter states that economic success requires continual and endless change, an ability to create the new that is able to drive out the old. In Schumpeter's words:

> **The fundamental impulse that sets and keeps the capitalist engine in motion comes from new consumer's goods, the new methods of production or transportation, the new markets, and the new forms of industrial organization that capitalist enterprise creates.[32]**

Accomplishing successful change and producing "new consumer's goods, the new methods of production or transportation, the new markets, and the new forms of industrial organization" inherently necessitates bureaucracies. However, successful change is not necessarily accomplished by simply creating larger, more complex bureaucracies, and simply creating the "new" is not always profitable. Bureaucratic roadblocks tend to undermine an entrepreneurial and innovative organizational spirit and stifle attempts to create necessary changes and remain competitive in the marketplace.

> **The bureaucratic method of transacting business and the moral atmosphere it spreads doubtless often exert a depressing influence on the most active minds. Mainly, this is due to the difficulty, inherent in the bureaucratic machine, of reconciling individual initiatives with the mechanics of its working.[33]**

Thus, an *appropriate, adequate rate of economic change* is a requirement for socioeconomic progress, which also generates undesirable by-products that may ultimately contribute to cultural deterioration. In the modern, global industrial society, major corporations are susceptible to the same realities of socioeconomic transformations and suffer the same consequences as developing societies. General Motors is an appropriate modern illustration of Schumpeter's concept of "creative destruction," as this former American icon of successful capitalism enters the twenty-first century clearly on the pathway of economic deterioration.[34] GM reached its zenith in the 1960s when it represented about 600,000 workers and accounted for about 50 percent of the American automobile production. By 2006, GM products represented about a 25 percent share of the American market while employing about 150,000 persons and posting an annual loss of $4 billion. The often-cited reasons for GM's decades of decline include worldwide competition from global technology transfer and cheaper labor costs, GM's stifling bureaucracy, and the leverage of powerful labor unions during the early, more profitable, and less competitive years.

The long-term economic effectiveness of GM's consumption of capital was inadequate to sustain long-term economic profitability. Over many decades, investment in technological innovation was too little and too late to prepare for the inevitable global technology migration and resulting foreign competition. Expenditures were excessive in areas such as employee wages, benefits, and services that were significantly beyond national norms. The consequences of GM's inefficient use of its resources over time resulted in excessive wasted capital, which was not available and invested in innovation and profitable products.

As a result, GM enters the twenty-first century saddled with a cost disadvantage with Asian automobile manufacturers of over $2,500 per car that has accumulated over a period of four decades by providing unnecessary and unrealistic employee benefits. These costs include health care insurance for over 1.2 million employees, retirees, and their families equivalent to $1,500 per car produced, equal to the cost of the steel used to fabricate one car. Historically, workers have not contributed to their health insurance, and retirees pay only a small portion. GM's cost for its retirement benefits is equivalent to $1,000 per car produced, with retirement being achieved at full pay after thirty years of service, regardless of age. Additional costs result from laid-off workers being eligible for a program that essentially provides continuing income indefinitely for performing charity work or attending school. Historically, GM's hourly wage for high school graduates without technical skills has been significantly higher than the national marketplace rate.

Consequently, the GM bureaucracy's inability to evolve and maintain its historic production competitiveness and to sensibly manage ever-changing technical and socioeconomics factors resulted in unrealistic and costly priorities, policies, wages, and benefits. Early profits were squandered. The history of GM's institutional attitude and behavior is a microcosm of modern society. It exemplifies the deterioration of a functioning organization that results from ignoring the imperatives of technological evolution, the realities of science-based economics, and the dominant consequences of human self-interest in the rewards of the modern industrial age.

Schumpeter's concept of "creative destruction" may be appropriately applied to General Motors' transition from impressive financial success to organizational deterioration. Obviously, such a major, complex socioeconomic entity could not sustain its initial economic prosperity in the realities of the late twentieth-century global economy. It will continue to exhibit the same characteristics of a deteriorating cultural order as it approaches economic stagnation.

There are striking similarities between the characteristics and life cycle of a society and its microcosm, the major corporation. In both environments, the overall human and material logistics and management of an increasingly complex and expanding socioeconomic order become difficult, if not impossible, to control. Ultimately, the expanding large bureaucracy evolves into an initial stage of collapse from the burdens of accumulating complexity, materialistic values, financial insolvency, and escalating sociopolitical instability. Eventually, either an aspiring competitor conquers the weakened victim or tumultuous internal circumstances produce a transition to a less advanced and less complex social order in an attempt at organizational resuscitation. Such events may reduce complexity to a simplified, organized, and functioning social order, thereby creating hope and opportunity for a more secure but less grandiose future. However, such organizational rejuvenation requires drastic alteration of human values, behavior, and resource priorities.

An appropriate illustration of the consequences of the thermodynamic-based economic perspective applied on a broader scope than the corporate example of General Motors is the budgetary priorities of a highly successful nation currently on the pathway of socioeconomic deterioration. The global recognition of an empire's declining greatness is a major symptom of cultural deterioration that becomes discernible as a nation's social instability, political ineffectiveness, and financial insolvency accumulate in the eyes of other major economic and military powers. America, in the first decade of the twenty-first century, is a sterling illustration of cultural degradation from decades of unwise investment of national wealth and an addiction to materialistic values. Instead of wise investments designed primarily to achieve and maintain a stable economy and social order, resources are devoted to nonessential personal and national priorities of dubious social value and long-term benefit.

In an energy-based economic model, the optimal economic strategy that will minimize the rate of societal decline is to maximize national wealth in order to regenerate wealth-energy capital for further equally wise investments beneficial to the whole society. Thus, it is appropriate to identify national budgetary priorities in order to determine the extent of budget expenditures that are incapable of generating new national wealth. The proposed 2006 U.S. budget plan of $2.57 trillion included $850 billion, or 33 percent of the total expenditure, for defense, the war efforts, homeland security, veterans' affairs, and the portion of the debt payment (80 percent) that is attributable to the military.[35]

The overwhelming portion of these military and defense-related expenditures results from the American philosophy that, as the number one world power, it should assume a missionary role to implement worldwide democracy, peace, and American values while protecting economic interests. Meanwhile, the proposed combined federal spending for education, health, human services, housing, and urban assistance totals $726 billion, or 28 percent of the total proposed budget. The proposed allocations to energy and the Environmental Protection Agency are $30 billion, with both agencies having projected budget reductions. The national debt of over $9 trillion is more than twice the annual federal revenue, and the annual interest paid on the national debt is over 10 percent of the annual budget.

The high U.S. priority on military expenditures to fight two elective foreign wars while incurring a record national debt and providing inadequate resources for infrastructure maintenance and repair, education, health care, poverty, and basic research is characteristic of a society in decline. This is reminiscent of misplaced values and priorities of war, debt, and inadequate public assistance exhibited by past great civilizations from ancient Greece through the British Empire. In the twenty-first century, the history of General Motors, representing a major successful worldwide corporation, and the American culture both illustrate the completion of the sequential transitions of "cultures" from an initial creative, genesis phase to a prosperous economic growth phase to a deteriorating phase. On a larger scale, Western civilization, which underwent a historic global transformation based on the Industrial Revolution and capitalism, has made similar transitions from rapid-rate growth and prosperity to an emerging stagnant condition.

Tainter's statement, "Continued investment in sociopolitical complexity reaches a point where the benefits for such investments begin to decline, at first gradually, then with accelerated force," applies to twenty-first-century General Motors, and more generally to the American culture and Western civilization. The evolution of a social order and its socioeconomic entities does not permit continuous, unabated cultural growth and prosperity, even for long-standing successful civilizations or their component corporations and governments. The seeds of deterioration are inherently sown within social orders in their early growth stages and, upon germination, produce a continuous proliferation of sensate values and behavior. Consequently, national wealth is unwisely, if not recklessly, consumed, ignoring the restrictions and wisdom of Mother Nature's principles and the causes of social change inherent in the construction of the social order. The energy and mechanistic principles and processes that gave rise to the socioeconomic Western transformation represent the same variables of wealth, energy, and human values and behavior that are responsible for the twenty-first-century Western civilization era of cultural degradation. The *minor fluctuations of cultural growth and decay*, the more intense *cultural transitions*, and the explosive global *transformations* are based on the relative success and failure of society's attempts to utilize wealth and energy resources to manipulate a culture's socioeconomic system to achieved human aspirations.

Chapter 5
References and Notes

1. Braudel, Fernand. *A History of Civilizations*. London: Penguin Press. 1992. 21.
2. Eisenstadt, S. N. *Revolution and the Transformation of Societies*. New York: The Free Press. 1978. 19.
3. Ibid., 26.
4. White, Leslie A. and Dillingham, B. *The Concept of Culture*. Minneapolis: Burgess. 1973. 57.
5. Helmholtz's historic work of 1847, "On the Conservation of Force," a series of lectures delivered at Carlsruhe in the winter of 1862–1863.
6. Rabinbach, Anson. *The Human Motor: Energy, Fatigue, and the Origins of Modernity*. Berkeley: University of California Press. 1990. 69.
7. Ibid., 70.
8. Ibid.
9. Galbraith, John Kenneth. *The Good Society: The Humane Agenda*. New York: Houghton Mifflin. 1998. 186-208.
10. Rabinbach, Anson. *The Human Motor*. 71.
11. Soddy, Frederick. *Wealth, Virtual Wealth and Debt: The Solution of the Economic Paradox*. London: George Allen & Unwin Ltd. 1983. 70.
12. Rabinbach, Anson. *The Human Motor*. 71.
13. Ibid., 73.
14. Ibid., 74.
15. Galbraith, John Kenneth. *The Good Society*. 20.
16. Rabinbach, Anson. *The Human Motor*. 74.
17. Ibid., 74-75.
18. Ibid., 76.
19. Ibid., 80.
20. Ibid., 81.
21. Ibid.
22. Ibid., 82.
23. Ibid., 83.
24. Ibid., 90.
25. Ibid., 92.

26. Lessnoff, Michael. *The Spirit of Capitalism and the Protestant Ethic*. Hants, England: Edward Elgar. 1994. 1-15.
27. Ibid., 14.
28. Weber, Max. *The Sociology or Religion*. Boston. Becon Press. 1993. xiv-xvii. Also, Weber, Max. The Theory of Social and Economic Organization. New York. The Free Press. 1947. 8-86.
29. Rabinbach, Anson. *The Human Motor*. 85.
30. Tainter, Joseph A. *The Collapse of Complex Societies*. New York: Cambridge University Press. 2004. 92.
31. Schumpeter, Joseph A. *Capitalism, Socialism and Democracy*. New York: Harper & Row. 1975. 81-86.
32. Ibid., 83.
33. Ibid., 207.
34. See for example, George Will, *The Washington Post,* 13 April 2006, and Jeffrey McCracken, *The Wall Street Journal*, 5 April 2006.
35. These figures are intended only to reflect federal priorities and spending trends contained in government documents during recent times.

Chapter 6
The Mechanisms and Energetics of the Human Social Order

"Man is an animal. But he is not 'just another animal.' He is unique. Man alone, among all species, has an ability which, for want of a better term, we call the ability to symbol. The ability to symbol is the ability freely and arbitrarily to originate, determine, and bestow meaning upon things and events in the external world, and the ability to comprehend such meanings. These meanings cannot be grasped and appreciated with the senses. For example, holy water is not the same thing as ordinary water. It has a value that distinguishes it from ordinary water and this value is meaningful and significant to millions of people."
                                    Leslie A. White and B. Dillingham[1]

The successful evolution of a social order from its primitive form to becoming a mature civilization relies on periods of positive socioeconomic progress when favorable elements of leadership, human values, natural resources, societal common cause, and human creativity and self-determination combine with the unpredictable forces of probability and nature. The extent to which such variables or factors of cultural development constructively or destructively coalesce defines the progress and success of cultural evolution on the continuum from cultural failure to cultural prosperity.

The maturation of the human social order from the birth of primitive societies to decay of the more mature civilizations has occupied the minds of some of the most recognized and respective historians, social scientists, cultural anthropologists, and political economists. In particular, Oswald Spengler, Arnold Toynbee, Pitirim Sorokin, and A. L. Kroeber are among those recognized as having produced the most useful works dealing with the characteristics and developmental stages of societies, cultures, and civilizations. In more recent times, the thoughtful work of such authors as Matthew Melko, Morton Fried, Leslie White, Ruston Coulborn, Samuel Huntington, and Carroll Quigley, while recognizing the impressive contributions of past pioneers, has called for more academically diverse approaches to civilization studies. These authors often recognize the need for a more highly developed approach to a more comprehensive understanding of the primary driving forces, mechanisms, conditions, and properties that are common to the continuum of societal development (i.e., the phases of cultural maturation).

It would appear that the early authors, dealing with the creation as well as the deterioration of cultures, have focused narrowly on arbitrary developmental concepts and properties. Considerable effort has been devoted to debating peers regarding definitional wording, speculative rationales, arguable consequences of historical events, and the imprecision of arbitrarily assigned dates of cultural transitions. Additionally, comparative historians have not appropriately considered the random nature of human history and chance happenings relative to attempts to achieve personal and societal objectives. While an analysis of historical events may identify successful and unsuccessful reoccurring social, economic, and political patterns resulting in significant societal change, a social order's implementation of particular strategies and tactics and their resulting effectiveness is always subject to the restrictions and limitations of probability and human wisdom and capabilities.

Modern authors have utilized more comprehensive approaches to create new paradigms of civilization development, which provide more in-depth understandings of the continuum of human history. Such paradigms must accurately represent major operational variables, mechanisms, properties, and conditions that accurately portray cultural dynamics fundamental to the social, political, and economic functions of the social order. Models should accurately represent short-term fluctuations in the rate of socioeconomic growth and the more intense sociocultural transitions, as well as major global transformations. Cultural outcomes evolve over time from concurrent constructive and destructive elements affecting societal development.

Three diverse methodologies are used to analyze past civilizations: The *evolutionists* seek to understand how and why societies form, prosper, and decline; the *historians* catalog personages, events, cultural characteristics, and circumstances; and the *structure-functionalists* utilize a science-oriented mechanistic approach to cultural change. Obviously, all three categories of analyses are capable of contributing useful insights into societal development.

Prior to considering the cultural dynamics and long-term sociocultural transitions related to the evolution of human social history, it is appropriate to address the fundamentals of the human creation of a social order. This includes the nature and maturation of the functional systems of human existence (i.e., the mechanics and energetics of the social order involved in cultural change). Consequently, a precise definition and discussion of human nature and its relation to behavior is an appropriate starting point for conducting analyses of the historic creation and development of the human social order.

Since the time of the ancient Greeks, sociologists have documented that as major successful cultures have matured and prospered, human behavior and more specifically cultural mentalities have systematically shifted from intense communal values toward more self-centered, materialistic values. This gradual appearance and proliferation of individualism is also consistent with, and has contributed to, the cultural deterioration and stagnation of prominent civilizations. Negative consequences of human nature and behavior are among the often-cited contributing factors to cultural degradation. However, such empirical evidence is unable to identify the full range of fundamental variables affecting the implementation of actions and to fully explain the universally observed phenomenon of ultimate civilization deterioration and stagnation. Thus, an important civilization studies issue is how and why human nature and behavior inevitably and negatively modify initial altruistic cultural values of primitive societies, transforming a prosperous socioeconomic era into one of cultural deterioration.

It is noted that while human behavior has been extensively portrayed as the product of culture, the concept of innate human nature originating from a noncultural, biological source existing at birth (i.e., behavioral genetics) appears to be lacking in many such discussions. The twenty-first-century knowledge base and issues related to sociobiology and the perceived genetic predisposition at birth of medical abnormalities and diseases, gender orientation, creativity, introverted and extroverted personalities, and the behavioral studies of identical twins separated at birth would appear to provide important insights into the *genetic role of human nature*. That is, are there hardwired and software components of human nature?

Thus, individuals reacting to cultural stimuli, particularly during times of socioeconomic stress, fear, and opportunity, are continually engaged in a personal battle of behavioral dominance between culturally driven, revered constructive values and more animalistic, genetically engineered instincts of human nature. This conflict creates emotional competition between a *genetically based human nature* and a *culturally modified or acquired, learned human nature*. The issue emerges as to *genetic human nature* being a significant variable affecting positive and negative elements of societal progress and a potentially troublesome hurdle for individuals, as well as collectively for a maturing society. During an era of economic prosperity that offers potential alluring materialistic opportunities and rewards, human nature, defined in terms of both the culturally learned and the genetically endowed influences, may provide a compelling motivation for the emergence, domination, and victory for sensate values winning the battle of individual self-interest over altruistic values and mutual societal benefits.

Consequently, the *nature of human behavior* will be considered as a multifunctional, continuously evolving phenomenon influenced by genetic inheritance (i.e., innate human nature), the cultural environment, and the various social learning mechanisms, as for example outlined by Morton Fried. *Human nature*, as reflected by values, attitudes, and both endowed and learned behavior, continuously evolves over time based on cultural influences and experiences, most notably socioeconomic factors.

White addresses the influence of *cultural context* on the intellectual translation of "beliefs" into behavior:

> **Culture is the name of a distinct order, or class, of phenomena, namely, the things and events that are dependent upon the exercise of a mental ability, peculiar to the human species.... Culture consists of material objects—acts, beliefs, and attitudes that function in contexts.**[2]

According to White, the "contexts" for culture and the human "exercise of a mental ability" in dealing with societal change and survival are "freely and arbitrarily to originate, determine, and bestow meaning" and value to its "acts, beliefs, and attitudes." For example, "holy water is not the same thing as ordinary water. It has a value," that is, for some people. Such an intellectual and cultural context (i.e., faith based) is consistent with the innate altruistic value system that serves as a behavioral standard for more primitive societies but which is gradually modified and ultimately transformed over many generations into a dominant sensate context (i.e., a more rational science-based cultural mentality). During this evolution, the *cultural context* is subject to the forces of change, whether economic, political, or social, thus influencing human behavior. The instincts of human nature that create and motivate primitive societies and define perceived needs and values necessary for survival gradually evolve into more opportunistic, self-serving, and self-indulgent values and beliefs that redefine *survival instincts* and *perceived needs*.

Anthropologists tend to view societies in terms of the explicit values, attitudes, and traditions exhibited by thinking and feelings and by attitudes and behavior (i.e., cultural traits). The origin of a society begins at the more primitive level when a specific group of people discovers common needs, purposes, and ambitions that permit the group to pursue an agreed-upon agenda in an atmosphere of trust that most importantly promises to provide the basic elements of survival. This primitive *kinship system* is the *cultural glue* of the social order that epitomizes altruistic values and provides a constructive social element that is unique to the less mature cultures compared to that of mature civilizations. While the social relationships of kinship stress an individual's duty and responsibility to the group, the quid pro quo is society must provide for individual human needs and group survival. Such a social order is defined by its agreed-upon and respected societal understandings and conditions that constitute the framework and guiding principles for the functioning of social, economic, and political systems. In the more mature civilizations, people acquire more aggressive, selfish, and materialistic attitudes, behavior, and expectations, generally at the expense of other persons, while attempting to satisfy a grandiose vision of perceived individual desires, needs, and entitlements, while discounting the individual's societal duties and responsibilities.

People arbitrarily and irrationally define perceived needs, societal responsibility, and *value* in a manner comparable to the value assigned to holy water. This raises the issue of the nature of *free will* and the human ability and restrictions to reason and challenge *genetically based human nature* as well as cultural influences. Einstein, paraphrasing the German philosopher Arthur Schopenhauer, remarked that "a human can very well do what he wants, but he cannot will what he wants." Exercising one's free will is to consider only a selection of items that one is able to bring to mind, a perception of free-thinking but not independent of one's limited mental capacity and creativity.

In primitive societies, the economic and social systems possess a significant degree of commonality and linkages due to the simplicity of human functions and to organizational structures and relationships that are primarily based on the kinship of families, tribes, and clans. As a culture matures, socioeconomic systems become, out of necessity, more diverse and politically oriented as less kinship-oriented socialization occurs. Consequently, an inherent reduction of trust and confidence in a more complex, stressful, threatening, and less communal society initiates a logical readjustment of cultural values toward greater self-reliance, self-interest, and independent opportunism. This evolving cultural environment and attitude also increases distrust and competiveness among individuals, leading to the creation of small sociopolitical-oriented groups that attempt to construct a more orderly, regulated socioeconomic system. Clearly, this evolving social order and its cultural traits are affected by the changing nature of society's economic reality and its inevitable sociopolitical ramifications.

One such reality is the continuous need for new knowledge and the latest engineering, technologies, and tools for the production of goods and services, whether at a primitive or a more sophisticated level. Economic successes subsequently breed modifications of sociopolitical structures and functions, shaped by various ideologies as are deemed supportive of potential economic profitability. Additionally, the achievement of socioeconomic objectives usually involves the exploitation of natural resources and human labor; the latter constituting a social subsystem that significantly affects the evolving social order. The existence of diverse political, spiritual, social, and economic philosophies within a culture, even when producing generally considered successful socioeconomic outcomes, will result in only limited satisfaction within the population. The degree of success may also be influenced by uncontrollable, unexpected circumstances and/or unfavorable probabilities for certain actions initiated by nature or society.

Eisenstadt notes that building a social order relies on cultural discourse, particularly among influential economic and political "elites" who distribute resources among societal groups; possess the "ontological concepts and visions"; manage "the production, reproduction, and control of resources; and deal with ... the regulations of power in society."

> **Through those different types of control they combine the structuring of trust, the provision of meaning, and the regulation of power with the division of labor in society, and thereby institutionalize the charismatic dimension of the social order. Coalitions of these elites exercise control through their ability to manipulate access to major institutional arenas (e.g., economic, political, or cultural).[3]**

Therein lays the major significant variables and limitations of societal development at all levels of cultural maturity.

As socioeconomic progress is achieved, cultural evolution is often described as a continuous migration over time toward greater disjointed or *differentiated patterns of social organization and function* (i.e., a more stratified or diverse social order). This phenomenon and its inherent complexity establishes the seeds of future stress, competition, and social discord as it pits proponents of maintaining traditional characteristics of the historic social order against the proponents of a new, more highly integrated and stratified social order and its symbolic value system. Over time, the gulf of this *ideological and psychological continuum* widens, representing, on the one hand, a more primitive mind-set of values, security, and survival and, at the other extreme, the temptations and opportunities for socioeconomic progress and increasing prosperity. Thus, the primitive society's initial dominant mentality continuously shifts from *fear for survival* towards a mature society's *pursuit of greed* and individual materialistic goals.

Leslie White in *The Concept of Cultural Systems* adopts the approach of analyzing the structure and behavior of cultural systems as the basis for describing and defining culture. Cultures "combine structure and behavior of systems" and engage in social, political, economic, and cultural interactions that form "new permutations, combinations, and syntheses."[4] This *integrative view of culture* results in progressively higher levels of cultural maturity (i.e., greater complexity), which involves *increasing functional specialization and structural differentiation* that progressively consumes excessive quantities of wealth-energy capital and a larger percentages of a society's available wealth. White's concept is that "culture traits act and react among themselves in accordance with the principle of cause and effect. Thus, culture determines and causes culture; culture is to be explained in terms of culture."[5] The totality of a culture's properties, including its structural and behavioral characteristics, continuously undergoes change, producing a more highly integrated social order with the passage of time. Durkheim describes cultural traits as "a whole world of sentiments, ideas, and images which, once born, obey laws all their own. They attract each other, repel each other, unite, divide themselves, and multiply."[6]

The time frame or the rate of appreciable cultural change is usually longer than a human life span and is dependent on multiple factors. Thus, a given generation usually senses little cultural change during a lifetime.

> **Clearly, however, the life of human beings involves many other phenomena which cannot figure into this film of [*current*] events: the space they inhabit, the social structures that confine them and determine their existence, the ethical rules they consciously or unconsciously obey, their religious and philosophical beliefs, and the civilization to which they belong. These phenomena are much longer-lived than we are; and in our own lifetime we are unlikely to see them totally transformed.**[7]

The view of a culture as a unified system continually striving to become more secure, more economically successful, more technologically mature, and more influential externally requires a high level of cooperation, loyalty, interdependence, coordination, and political control among the various subsystems of a society. Such human motivation and objectives involve aspects of human behavior and culturally learned traits that significantly affect major societal priorities and decisions. This raises the question as to how cultural traits and behavior are learned and assimilated within a culture.

The anthropologist Morton Fried in *The Evolution of Political Society* states: "Culture may be defined as the totality of conventional behavioral responses acquired primarily by symbolic learning."[8] It is instructive to examine his thesis regarding how the members of a society acquire and transmit cultural traits, as this may provide additional understanding of a culture's functionality and the evolution of its social order. Fried emphasizes the importance of *symbolic learning,* learning which, by definition, is not based on a direct stimulus or an original situation but on complex symbols that serve as substitutes for direct stimuli capable of instilling learned behaviors. This learning mechanism is in contrast to *situational learning* based on one's direct experience and to *social learning* based on one's perceived experiences of others. Symbolic learning represents complex representational behavior that is characterized as "supremely time–binding" and produces general behavior that is determined and controlled from one generation to another.

Given Fried's thesis of the importance of this indirect learning mechanism in transmitting cultural traits to a society's membership, it is a short extrapolation to the consideration of Spengler's perspective of *cultural soul* that would embrace all social, political, and economic disciplines as well as cultural attitudes, values, and practices. Fried states, "Most political norms exist, not because they are sanctioned by force, but because they are conveyed to the young as part of the process ... of enculturation."[9] He defines the social organization as comprising "the totality of patterned relations among the members of a society, the subgroups formed in the course of these relations, and the relations among these groups and their component members."[10] Braudel expresses a similar view of Fried's "enculturation," "an unexpressed and often inexpressible compulsion arising from collective unconscious"; his statement provides the context:

> **Dictating a society's attitudes, guiding its choices, confirming its prejudices, and directing its actions, this is very much a fact of civilization. Far more than the accidents or the historical and social circumstances of a period, it derives from the distant past, from accident beliefs, fears, and anxieties which are almost unconscious—an immense contamination whose germs are lost to memory but transmitted from generation to generation. A society's reactions to the events of the day, to the pressure upon it, to the decisions it must face, are less a matter of logic or even self-interest than the response to an unexpressed and often inexpressible compulsion arising from the collective unconscious.**[11]

It is useful to consider the extent to which the emerging twenty-first-century understanding of transgenerational genetic transfer and genetic predisposition relates to Fried's "symbolic learning" and to a population's amalgamated human nature. This also applies to Braudel's perspective that "dictating a society's attitudes, guiding its choices, confirming its prejudices, and directing its actions ... derives from the distant past, from accident beliefs, fears, and anxieties which are almost unconscious—an immense contamination whose germs are lost to memory but transmitted from generation to generation."

According to Braudel's definition of time frames, such learned social structures and cultural traditions are examples of a "quasi-immobile time" period that persists for a long time relative to the human life span.[12] This notion led Fried to state that "in all societies the single most significant complex of social-control apparatuses is ... the system of education, both formal and informal means,"[13] from which he deems symbolic learning to be the primary educational mechanism. Acceptance of this learning theory of such an "enculturation" process recognizes a cultural force or cultural inertia existing within a society capable of retarding attempts to promote progressive change of individual and group attitudes, values, and objectives. This tends to support various perspectives of cultural self-destruction based on the human inability and/or unwillingness to adapt or change in order to avoid cultural stagnation. For example, Edward Gibbon views the decline of societies as due to an unwillingness to adapt to an ever-changing environment, while Schumpeter cites a society's inherent inability to adapt in the face of improving technology, communication, and transportation as the economic landscape continually changes.

The concept of a rate of change and the time scale for a civilization's existence and major transitional periods is often misused or ignored. It is noted by a number of authors, for example White,[14] that cultural systems are inherently thermodynamic systems; thus, the net changes between an initial and a final state, representing a given cultural transition, is independent of the time elapsed for such a change to occur. An alternative description is that many different mechanisms or processes representing different periods of time and/or conditions could hypothetically produce the same outcome from the same initial state. Thus, while time is not a factor (i.e., a driving force for cultural change or transformation), it is of importance as a benchmark relative to the life span and interests of human beings. Braudel offers three time frames for cultural processes. The first is a "quasi-immobile time," a reference appropriate for social structures and cultural traditions that change little during a human lifetime and is more of an appropriate environmental time frame. Second, an intermediate scale of "conjunctures" is slow but measurable and perceivable during a few human generations, an appropriate reference of time for social history. Third is a "rapid time scale" of about the duration of a human lifetime, an appropriate reference for historical events.

Irrespective of the time required for any given cultural transition, civilization studies authors make passing reference to the possibility of a society approaching, and perhaps achieving, a condition referred to as a steady state, an equilibrium condition, or a stable state. These terms must be more rigorously and precisely defined than is found in the literature and be based on accepted scientific concepts. For example, Melko uses the term "steady state" in the following passage:

> **Some civilizations never attain a high degree of integration and some remain at a well-integrated stage for a long time. Perfect integration is approached but never achieved. If it were, change would be impossible—a condition that is apparently attained in some primitive cultures, which are in what physics call a "steady state," involving a minimum of adjustment, as distinguished from the "stable state" of an artifact.[15]**

A *steady state* denotes a culture's lack of change over a given time period; such a socioeconomic system could be viewed as inherently lacking vitality and an inability to prosper. As will be more fully described in a later chapter, this would correspond to Toynbee's "cultural stagnation" stage of cultural evolution. On the other hand, a *stable economy* may be interpreted as being active and consistently productive, normally contributing to social stability and a progressive social order. However, the meaning and appropriate use of the term "cultural integration" should first be established prior to the discussion of the concept of steady state.

Civilizations are composed of an almost infinite number of functional, mostly interrelated systems, each being *integrated* to an extent that will vary from time to time and among component subsystems. Sorokin uses the term "*system*" in such a manner, while Kroeber uses the term "patterns" as in multidimensional structures or a matrix. Melko prefers to describe patterns as "strands of a woven rug" that are an integral part of a system.[16] Thus, such cultural systems have their own identity, are normally somewhat interdependent, and as a whole, generally affect the net success or failure of a whole society. For example, the overall economic system is composed of a multitude of private, nonprivate, and governmental subsystems that may or may not possess a high degree of collaboration and commonality. Areas of cooperation may exist that could lead to mutual benefits for some individuals and groups as well as for society as a whole.

However, other areas inherently exist within the economic system for which cooperation is inherently inappropriate or undesirable. Thus, the term "*degree of integration*" is used to describe how loosely or tightly interrelated these cultural subsystems are and the degree of their functional connectiveness and interdependence. The extent of such system integration may vary with time within the extremes of having very little interaction or mutual impact to having significant interrelationships affecting the system.

It is appropriate to view the integration process as a fluctuating phenomenon; that is, the degree of integration of a given subsystem fluctuates with time. Melko considers the interdependence or integration of a society's many systems and subsystems to be in a constant state of "fluctuation" or transition between "periods of formation and ... disintegration."

> **The development of all systems can be viewed as a fluctuation between crystallization and transition, between higher and lower periods of integration.... Periods of formation and periods of disintegration are both periods of transition, and both may be occurring at the same time.**[17]

Unfortunately, the reader is left without a precise understanding of just what is forming and what is disintegrating. Also, descriptive phrases appear in the literature such as the "ashes, not fire" representing complete cultural disintegration, and Toynbee's term "ossification" is described as "freezing at a crystal stage."[18]

The response to the issue of what precisely is undergoing "fluctuation" or "oscillation" is the formation and disintegration of a society's organizational structure or orderliness and the nature and rates of functional processes that represent mechanisms of numerous subsystems within, for example, an economic system. This "fluctuation" may therefore be described as a variation over time in the level of complexity, organizational connectivity, and functionality, or the degree of disorder among a society's formal systems of existence.

From a thermodynamic point of view, such models, definitions, and concepts outlined by Melko and other authors are representative of, and consistent with, the mechanistic-thermodynamic paradigm. The continual fluctuations, as described by Melko, represent a formation mechanism (crystallization) and a disintegration mechanism (reversion) between higher and lower degrees of cultural integration representing different energy content. This vision is also consistent with the concepts of molecular reactivity and thermodynamics used by scientists to describe the nature of chemical, biological, and physical processes.

Modern urbanization is an example of a highly integrated cultural macro-system consisting of an extremely large number of broadly diverse societal processes and subsystems continually undergoing modification by the addition and elimination of subsystem elements designed to meet the expanding needs and desires of a continuously changing culture. The urbanization process continuously creates a higher degree of structural and functional integration; consequently, the rate of wealth and energy consumption per capita increases. Thus, with the expenditure of sufficient wealth-energy capital, a society may extensively expand its functional specializations, as the structural differentiation of modern urban development illustrates. However, this is achieved only at the expense of the creation of increasing societal complexity and disorder, as required by the entropy concept.

Returning to the concept of the *steady state* condition as applied in the cultural studies literature, it is noted that authors using this term frequently refer to its scientific origins. To fully explore the appropriateness of this usage, the concept of dynamic equilibrium is illustrated as applied by scientists for physical, biological, and chemical processes. Assume the chemical reaction $A + B \Rightarrow C + D$ will proceed toward completion, defined as obtaining a quantity of products C and D that ultimately remains constant with time. In reality, there may be some amount of A and B present at completion, but perhaps hardly detectable.

The rate of the reaction is dependent on reaction conditions such as temperature and pressure, as well as on the concentrations of reactants. In addition, a catalyst could be added, which would alter the mechanism of the reaction, increasing or decreasing its rate of progress. At some point in time, there will appear to be no macro-change in the system, as evidenced by the lack of observable changes in the concentrations of reactants or products; thus, the reaction can be declared to be at the *equilibrium state*. This endpoint is based on the determination that, for example, *the rate of formation of product C appears to be zero* (i.e., it doesn't change). However, the equilibrium condition or the *steady state may not be* a state in which all molecular activity ceases, and such a state is appropriately referred to as a *dynamic equilibrium* condition. Thus, reactants may be reforming from the products at a rate equal to the rate that products are being formed from reactants, but no net change in concentration of any one chemical species occurs over time. Since no net chemical change or transformation can be detected in the system, it may erroneously appear that the system is *at rest,* but activity is occurring in this dynamic system. Alternatively, the system could be referred to as being in a state of thermodynamic equilibrium (i.e., the most stable thermodynamic state), and thus, for this process, the free energy change is zero; the system is in a thermodynamically balanced state, unable to perform further useful work. Thus, the concept of a net steady state for such a *closed system* as the usual laboratory chemical reaction is consistent with a low level of molecular activity that does not accomplish useful work. Conversely, it should be noted that while this chemical system may be appropriately described as being in a "stable thermodynamic state," signifying the lowest energy state, the word "stability" used in the descriptions of social and economic stability may constitute a very different meaning, signifying a degree of functional orderliness.

An existing equilibrium condition may be destabilized by the addition of energy to the system or changing reaction conditions. Thus, if a system at dynamic equilibrium is subjected to a stress (e.g., a temperature change or additional reactants or products enter the system), La Chatelier's principle specifies that the system will respond. The equilibrium will shift in a direction that will tend to counteract the applied stress. The result is that reactant or product concentration will increase, depending of the reaction's particular thermodynamic properties.

In an analogous fashion, many changing variables (e.g., the creation of economic issues) may produce counteractive societal reactions and effects that could disturb the dynamics of a society as it strives toward a steady state condition. Hypothetically, this could also initiate the rejuvenation of a faltering social order.

Thus, the dynamic equilibrium concept may be appropriately applied to the fundamental operational processes of societal systems that have been shown to constitute thermodynamic systems. That is, a culture consists of an almost infinite number of dynamic social, political, and economic subsystems continually functioning in potential support of a society's daily existence, and all are dependent on available wealth-energy resources.

One important distinction should be made between the chemical system above, which undergoes a reaction in a *closed system (e.g., within a laboratory flask),* and the *inherently open system of societal systems and processes.* In the closed chemical reaction system, no additional material is permitted by definition to enter the system during the course of the process, and usually conditions such as temperature and pressure are maintained as a constant environment. For the societal *open system,* such as a continuously functioning economic system, material and energy resources are being continuously added to the system, and conditions vary widely and unpredictably.

Functioning societies require that additional processes, conditions, materials, and energy sources continually impact operational systems, producing change and affecting short-term outcomes as well as long-term cultural viability. Thus, in contrast to the closed chemical system illustration consisting of a fixed quantity of material and maintaining constant conditions, societal systems are open systems and must be treated mechanistically and thermodynamically as such. This discussion leads back to the issue of what is precisely meant when the term *steady state* is applied to such an open societal system? It refers to the net rate of cultural growth reflecting the subsequent changes in social, political, and economic *stratification and integration* of cultural functions seeking dynamic equilibrium. In a stable economy, energy and material are constantly being added to the system, which provides a continuous disruption of the drive toward achieving the equilibrium condition.

The scientific concepts of dynamic equilibrium and steady state are useful in portraying the mechanisms and energy components of societal development. Conceptually, the almost infinite number of potentially interrelated social, political, and economic cultural subsystems are defined by processes that are analogous to open chemical systems, with each having material and energy requirements, initial and final thermodynamic states, specific but varying processes conditions and rates of progress, and explicit mechanisms. However, contrary to the illustrated chemical laboratory reaction, societal process conditions and material content may be altered at any time by the whim of society. Thus, the "ontological concept" and vision of cultural development is one of a very large network or a multidimensional matrix of interrelating processes that becomes increasingly more complex and difficult to manage with increasing cultural maturation.

This visualization provides a context for the concept of the "degree of societal integration" as a percentage of connectivity and communication among potential interrelationships of cultural processes. However, given the dynamic nature of the vast number of processes, some of which are capable of positive as well as negative consequences, the term "perfect integration" lacks precise meaning, as does Melko's conclusion that such a condition would lead to a *steady state*. "Perfect integration" may refer to a highly organized static operational structure of subsystems that may, to some extent, effectively or ineffectively interrelate among the total number of cultural subsystems, depending upon the nature and quality of communication, cooperation, and commonly accepted values and purposes.

Therefore, economic, social, and political change occurs over time by "integration" or the continual addition of new processes, new materials, and new wealth-energy resources to produce a continual flow of new activities and new outcomes, in some cases to replace the old. Thus, "perfect integration" has a functional implication and a structural connotation suggesting that increasing cultural complexity, inherent in continued socioeconomic prosperity, may also restrict economic, social, and political advancement. Consequently, to convey meaning, the term *integration* should be specifically defined in terms of structure and function and be related to social, economic, and political systems. Conceivably, structures may be highly integrated and restrictive, while the programmatic aspects of the functions emanating from those systems may or may not be effective and capable of productivity, profitability, and progressive change. However, being highly integrated may also limit the ability to change due to cultural paralysis.

Regardless of the potential vagueness of the term *cultural integration*, the view of a mature civilization's operational scheme is one of a very complex, confusing network of highly interrelated social, economic, and political activities that may concurrently be undergoing processes of both integration and deterioration. Such systems are continuously and simultaneously in states of structural and functional modification, growth, decay, and extinction. The necessary flow of available material and wealth-energy resources and the nature and quality of management are major variables that determine outcomes.

Given the inherently imperfect nature of the human species, it should be expected that the efficiency, effectiveness, and consequences of a society's attempts to achieve progress via implementation, coordination, and integration of its numerous fundamental subsystems would vary over time and among societies. Therefore, it is reasonable to assume that some of the subsystems contribute positively to societal progress, while others detract from achieving desirable outcomes. For example, the economy could be thriving as a result of greater subsystem productivity, while sociopolitical programs could be less successful and constitute a drag on social stability, economic productivity, and cultural advancement. However, the summation over positive and negative outcomes for all cultural operational subsystems could potentially be a positive, negative, or a zero sum as they impact on society over any particular time interval. This could create minor to major detectable "fluctuations" in cultural growth or decay.

Therefore, a social order's net success and prosperity is determined by a multitude of highly integrated and interdependent subsystems, some of which may be contributing positively while others may be detracting from socioeconomic progress. Melko states:

> **Civilizations are large and complex cultures, usually distinguished from simpler cultures by a greater control of environment.... Usually ... [they] incorporate a multiplicity of cultures and languages ... [and] must have a certain degree of integration. Their parts are defined by their relationship to each other and to the whole ... more interdependent economically. There will be pervading aesthetic and philosophical currents ... composed of a multitude of integrated "systems."**[19]

This mechanistic vision is interpreted as a civilization being defined by the *whole* (i.e., the summation of "interdependent" systems of social, economic, political, cultural arts, etc.). Thus, at any given time, the "multitude of integrated systems" that are interdependent, but still possessing some degree of independence, may be at different levels of success or failure. The individual societal processes represent dynamic subsystems, each being driven thermodynamically at some finite rate toward a dynamic equilibrium condition. Normally, society's continuous addition of material and wealth-energy resources as well as modifications of process conditions incessantly displaces the progression of cultural systems toward its intended, momentary thermodynamic equilibrium. The extent of this displacement from the equilibrium or steady state endpoint at any given time represents the potential for accomplishing useful work. However, to cease feeding the system will bring specific processes to a standstill (i.e., no useful work produced or contributions to society), and the system becomes nonfunctional and collapses.

To summarize the functional processes of societal existence: The rates of society's mechanistic processes depend on the availability of wealth-energy resources to support an endless progression of subsystems toward a continuously redefined dynamic equilibrium endpoint. Members of a social order perform useful work, as thermodynamically defined, by utilizing a transfer or flow of energy. This raises the issue of the validity of narrowly focused, non-energy-based criteria by which comparative historians and others have analyzed the successes and failures of past great civilizations. Arbitrary judgments and rationales of societal development and major cultural transformations have been constructed; reflecting narrowly conceived ideological, philosophical, and sociopolitical frames of reference. Given the absence of a unified theory of civilization development that includes quantitative components, representing a broad scope of recognized cultural variables, the existence of disagreements among the most respected comparative historians is not surprising.

Kroeber defines culture in terms of a society's philosophical and aesthetic characteristics, using the excellence and capacity of its arts and letters productivity as qualitative measures of societal success and prosperity.[20] This methodology considers culture as reflecting a society's intellectual, emotional, and artistic soul as expressed by the quality and quantity of its fine arts, theater, philosophy, music, and literature. When a culture possesses great vitality, the arts and letters manifest that spirit of the social order as exhibited by its members. The logic of Kroeber's methodology for qualitatively charting the success or lack of success of a civilization is thus rooted in the assumption that a period of considerable creative, talented, and original productivity in the cultural arts reflects a high point of a civilization's development.

In contrast to Kroeber's approach, Fernand Braudel, a self-proclaimed structuralist, utilizes a much different and broader academic methodology. His approach is to comprehend history by "identifying the major problems in the world today; problems of every kind—political, social, economic, cultural, technical, scientific. In a word, what is required goes beyond the double historical approach ... to distinguish the essential from the peripheral."[21] Historical events per se have little meaning for Braudel, which he refers to as "crests of foam that the tides of history carry on their strong backs." To him, the unfolding of historical events does not contain or explain the essence and rationale of civilization development; "history is the keyboard on which these individual notes [*historical 'occurrences'*] are sounded." He refers to the methods of Toynbee and Spengler as consisting of "over-simple theories" and "sweeping explanations."

Matthew Melko reviews the work of major civilization studies authors in his volume *The Nature of Civilizations*. He analyzes the various substantive as well as inconsequential conceptual differences while also offering his own viewpoints. Refreshingly, Melko projects a broader perspective of civilization studies than most comparative and cultural historians and also avoids the mysticism of the social philosophers. Specifically, Melko summarizes the major issues of contention among six leading civilization studies authors: Spengler, Toynbee, Kroeber, Bagby, Coulborn, and Quigley. The lack of consensus among these authors is noteworthy and illustrates the inadequacies of the traditional views in identifying the major driving forces and fundamental properties representing cultural development. Authors have arbitrarily developed their own criteria for a society qualifying for a particular societal classification and status. Sharp disagreements exist regarding the exact number of legitimate historic civilizations, as well as the time frames for their existence. Spengler and Sorokin even debate the elemental concept of defining a civilization.[22]

Such fundamental variances and vagaries of the scholarship associated with the etiology of the great civilizations is highlighted by the popular issue of whether a particular society was, by virtue of differing definitions, a culture or a civilization. This is illustrated by the debate as to whether China prior to 1500 B.C. qualified as a culture or a legitimate civilization. Additionally, time frame debates exist regarding the assigned period of existence for a given civilization, which often rests on the issue of whether a given social history was a succession of individual and separate civilizations or, more simply, just one continuous civilization possessing "fluctuations" or periods of more or less societal progress, as suggested by Sorokin and others. Authors are uncertain as to whether major cultures are inherently and intimately associated with each other or intrinsically separate entities. Spengler and Toynbee consider Russia as a separate civilization; Coulborn and Quigley treat it as a component of the Byzantine civilization; and Kroeber considers Russia as European.[23] Finally, various authors use descriptive terms for cultures such as "organisms" to imply characteristics of "living systems" that possess a "soul."[24] Interestingly, the one major point that authors apparently agree upon is that no civilization has escaped ultimate cultural deterioration; none can be shown to have created a continuous and unending period of socioeconomic growth and continuous social stability. All civilizations, regardless of assigned labels, descriptions, and time frames, eventually "stagnate," "ossify," or collapse.

Some of these issues discussed in Melko's review do not significantly contribute to an understanding of the major driving forces, mechanisms, and properties of the growth and decay of the great civilizations. However, they comprise secondary topics of interest; thus, it is useful to explore these perspectives and identify a number of important omissions and deficiencies.

First, the major omission, and most blatant deficiency, in the civilization studies literature is the neglect of the role of energy principles and the associated perspective of national wealth in economic policy, modeling, and practices. While such authors as Leslie A. White, Claude Levi-Strauss, George Grant MacCurdy, and Wilhelm Ostwald highlight the importance of energy to societal survival, economists and historians have generally ignored its central role during both the successful and troubled periods of societal existence. White, in particular, crafts an insightful vision that positions energy in its appropriate paramount position in the life span of a culture's existence. He links man's "life processes" to its energy requirements and to the "primary function of culture," defining culture as "an elaborate thermodynamic, mechanical system."[25] Thus, White's vision of culture "rests upon and is determined by the amount of energy harnessed and by the way in which it is put to work."[26]

Interestingly, White's wording is analogous to more scientific thermodynamic expressions cited in previous chapters. That is, in the initial state of a thermodynamic process, energy content is consumed, is transformed, and reappears in a final state and, depending on the thermodynamic efficiency of the particular process conducted, becomes divided between useful work—the "amount of energy harnessed"—and wasted heat energy. White concludes: "Energy itself is meaningless. To be significant in cultural systems, energy must be harnessed, directed, and controlled" (i.e., energy must flow). He recognizes three important factors in the societal processes of existence: the amount of energy consumed and harnessed per capita per year, the efficiency of energy use, and the extent of human needs that are met.[27] These factors are useful in defining, comparing, and contrasting simple primitive cultures, the moderately advanced societies, and the advanced but fading mature civilizations.

White notes that the more primitive cultures appear to be relatively standing still while more mature cultures have a faster pace, creating greater disorder and social imbalance. The disorder appears in the form of cultural complexity, unresolved issues, and social stress and conflict, as society consumes increasing amounts of material resources to support an increasingly more mature culture. Such outcomes are a major limiting and controlling factor for societal development. Levi-Strauss notes a "social imbalance" as modern societies produce "both much greater order—societies work like machines—and much greater disorder … in relations between people." The context is that cultures:

> **which produce little disorder—what doctors call "entropy"— and tend to remain indefinitely as they originally were … look to us like societies that lack both history and progress. Whereas our [modern] societies are powered by a difference of electrical pressure, as it were, expressed in various forms of social hierarchy.… Such societies have managed to establish within them a social imbalance which they use to produce both much greater order—we have societies that work like machines—and much greater disorder, much less entropy, in relations between people."[28]

Indirect, subjective evaluations of a culture's level of social and economic stability and productivity, relative to its "social imbalance" and disorder, have been utilized to evaluate the degree of net cultural progress. Spengler's research is focused on the quality and productivity of a society's arts and philosophy as a reliable measure of cultural vitality and advancement. He assumes an independence of these cultural properties from other internal and external influences. This cause-and-effect methodology, also embraced by Kroeber, would appear to be a reasonable, albeit extremely narrow, approach, as other factors have had significant influence on the success and prosperity of civilizations, particularly for the less advanced and nonaggressive social orders. Melko comments on such "cause-and-effect political history":

> Many historians write today with a sharpened awareness of cultural integration and characterizations. They seek relationships between politics, economics, and aesthetics, and they dismiss cause-and-effect political history as out of date, something that belongs to the nineteenth century. What they have rejected in the system builders is their dogmatic periodization. The basic concepts have stood. Civilizations do have meaningful inner relationships, they can be characterized, they can be distinguished from one another.[29]

Spengler envisions a cyclical nature of history, classified by developmental stages of a society's high culture rather than by the usual method of historical events utilized by his peers. The high cultures, as "living things," pass through stages from birth to death with the high point being the *culture stage,* which is transformed to the decay of the *civilization stage* and its inevitable loss of vitality and "soul."[30] While his concept of a cyclic birth-to-death mechanism is reasonably based on history, it does not address *identifiable variables* that directly affect the degree of social, economic, and political stability as it impacts the general population. Kroeber, recognizing the "cause-and-effect" limitation of their research methodology, appropriately refers to the high-culture developmental approach as "behavioristically factual rather than explanatory."[31]

A significant disparity exists among authors relative to the number of civilizations each has certified. Arnold Toynbee identified thirty societies based on his "intelligible fields of study," which were grouped into categories of civilizations and primitive societies. Spengler, Kroeber, and Coulborn, based on their individual criteria, each established their versions of certified civilizations, ranging from eight to fifteen.[32] This diversity of the criteria and number of identified civilizations leaves one unimpressed by the imprecision and lack of consistency of such characterizations. Additionally, their methodologies add little to an understanding of the fundamental dynamics of societal development.

As previously pointed out, the literature is devoid of a unifying concept and supporting rationale capable of representing the evolution of cultural life in terms of fundamental variables that constitute driving forces of civilization development. Many influencing factors and consequences are found in the literature, but usually in splendid isolation from other obvious variables. Pitirim Sorokin allocates great weight to cultural value systems based on his sociological criteria. Kroeber defines his patterns or "growth configurations" based on the events of history, whereas others, including Spengler, resort to a more subjective, intuitive methodology. In Kroeber's words, regarding Spengler's approach, "He is interpreting history as part of a philosophy which he feels passionately."[33]

From a broader perspective, Melko appropriately calls for "a sharpened awareness of cultural integration and characterizations" and the "relationships between politics, economics, and aesthetics" in the analysis and understanding of past civilizations. It has become obvious that a truly more interdisciplinary basis is required to represent an appropriate, multivariable model of cultural maturation, which specifically considers a society's consumption of wealth-energy capital. This model and representative variables must include the forces of societal change that society members are bound to encounter as they pursue the socioeconomic realities of the daily hurdles implicit in the human experience. Overcoming these challenges may produce short-term cultural advancement of the social order, which, in the long term, may accumulate and contribute to sociocultural transitions. Recall that Braudel recognizes three time frames of reference or viewpoints for the processes of cultural change: *Historical change* is viewed from the perspective of human events during one generation; *social history* reflects minor but perceivable changes over a few human generations; and *social structures and cultural traditions* are relatively constant over the human life span. The enduring adaptations of the primitive social order contribute to a cumulative social history, which defines the culture in terms of evolving social structures, functions, and traditions.

These aspects of cultural evolution employ the same mechanistic elements and energy requirements that pertain to biological evolution. Maturation of both types of systems requires capturing and utilizing energy in accordance with the laws of science, and both evolve into more complex, organized, and energy-dependent entities with the passage of time. In the pursuit of survival and prosperity, societies must continually acquire appropriate knowledge, technology, and economic and sociopolitical capabilities. Additionally, material and wealth-energy resources are also necessary to achieve and maintain economic productivity and profitability. Thus, cultural advancement, from ancient Greece to the twenty-first century, is quantitatively "measured by the amount of energy harnessed and put to work per capita per year, other things being equal."

> **The whole evolution of cultural systems is greatly illuminated by considering it in terms of energy harnessed and put to work.... The degree of cultural development is measured by the amount of energy harnessed and put to work per capita per year, other things being equal.... Historical records amply support the theory that when and as the amount of energy harnessed and put to work is rapidly and greatly increased, so will the culture grow and develop rapidly and greatly.**[34]

Harnessing wealth-energy resources for useful work that culminates in furthering worthwhile societal goals and outcomes requires the continual pursuit of improving economic production. This objective is the main focus of sociopolitical structures and functions influenced by various moral and philosophical ideologies deemed supportive of economic profitability. The achievement of established socioeconomic objectives has, since ancient times, necessitated the exploitation of natural resources and human labor and has defined the realities of social structures and systems capable of achieving wealth and sociopolitical ambitions. Thus, the fundamental aspects of human nature and behavior, established within a specific culture, constitute guidance and direction inherent in the social structures and cultural traditions that provide the restraining and motivating forces of cultural change. However, from a global perspective, the diversity of political, spiritual, and economic philosophies and practices among differing cultural traditions, mentalities, and social structures has provided formidable human challenges. In addition to innate suspicion and hostility toward differing ideologies, both internal and external, people must address and survive foreign threats of global economic competition and potential military aggression. Thus, an intense motivation exists, particularly in mature civilizations, to acquire the rewards of internal and external economic and political domination. The competitive human environment requires the most advanced tools, devices, and techniques whether applied to agriculture productivity or to successfully waging warfare. This necessitates a relentless search for the most advanced knowledge and related technologies.

Consequently, the persistent development and improvement of tools, machines, technological applications, production techniques, and inventions has been as historically significant to socioeconomic advancement as the discovery of new enabling sources of wealth and energy. These rudiments of production are key factors in the ease, speed, and efficiency of energy transfer and utility associated with the mechanistic steps of production processes. Thus, as more technologically advanced tools, machines, and devices are created, the probability increases for a positive rate of economic growth. The technological evolution of machines, tools, and devices, while an indispensable component of socioeconomic advancement, also represents a potential challenge for a civilization's long-term advancement. That is, it should be recognized that a normal technological transition pattern exists, migrating from the genesis and initial development phases through an induction phase of rapid adoption, to a saturation or plateau phase, and ultimately to a declining phase. The latter represents technological obsolescence or being supplanted by new tools and techniques. This effect is illustrated by automotive assembly-line production making the transition from the dependence on human labor in the 1960s to the automation of computer-assisted manufacturing of the twenty-first century. The continuous application of human creativity and inventiveness is necessary in order to maintain technological advancement, an essential ingredient for a productive economy.

Subsequent to their invention in the most simplistic versions, devices, machines, and tool concepts evolve and undergo gradual improvements up to the limit of their inherent efficiencies and capabilities. For example, the primitive mechanical hand saw evolved over time to become electrically and then gasoline driven as the consumer's needs became more advanced and diverse. In more modern times, the initially large, slow, and simple-tasked mainframe computer evolved into the mini, micro, and powerful desk-sized versions as well into special-function computers with specialized application software dedicated to complex computations, massive information storage and retrieval, and computer-aided design and manufacturing. Creativity stimulates continuous technological innovations in the form of more advanced inventions, tools, and applications to enhance economic utility and profits. This is illustrated by the computing emphasis shifting over time from a hardware emphasis on processor speed and memory capacity to specific software applications, to small-scale wire-based connectivity, and to worldwide telecommunications. At the beginning of the twenty-first century, communication emphasis is on the consolidation of multiple personal functions onto a single wireless device that serves all of an individual's personal and business audio, visual, and text information and communication needs.

Commercially, devices, tools, and machines (e.g., the saw and computer) evolve technologically through stages of utility and value until reaching a maximum functionality or societal utility and economic benefit. A stage of limited functionality may be due to size, power or work capacity, speed, limited commercial profitability, or the introduction of a radically new product or technology that is cheaper, higher quality, or faster. Alternatively, economic success may be due to more creative commercial packaging, salesmanship, financing, or advertising.

It is noteworthy that societal development and economic productivity not only face the challenge of securing a continuous supply of sufficient wealth-energy capital to fuel its socioeconomic system, but they must also respond to the challenge of continuously developing the latest technology to insure a sustaining economic profitability. Hypothetically, a society could be faced with an overabundance of wealth-energy capital but experience a no-growth economy as a result of its inadequate, noncompetitive products and declining profitability due to its limited, deficient, or outdated technologies.

To summarize, since the beginning of time, the primary variables influencing continuous culture advancement include (a) the nature and availability of material and wealth-energy resources; (b) a skilled and motivated labor force; (c) the thermodynamic efficiency and capacity of available tools, machines, and techniques related to economic productivity, and (d) human nature, values, attitudes, and behavior employed in managing short-term sociopolitical affairs and long-term cultural change. These primary variables affect the creation, development, and potential success of the organizational structures and functions of the social order, ultimately responsible for human social history.

New commodities, representing wealth-energy capital, emerge during a given era and are assessed by their value in marketplace exchange according to consumable energy content and utility as food and by society's machines and socioeconomic processes as fuels and materials. Thus, when a society has access to the most advanced and appropriate knowledge base, production techniques, and applied technology and unrestricted wealth-energy capital, it is best able to maximize its potential for socioeconomic advancement. Such probabilities are enhanced when the dominant cultural mentality is one reflecting spiritual, self-disciplinary, and communal values that benefit society as a whole. The degree to which these factors consistently or inconsistently exist over time will determine the potential limits of socioeconomic growth or decay, but subject always to chance and the realities of human strengths, weaknesses, and behavior. Generally, significant advancement in cultural development has occurred when a social order has been able to increase its per capita wealth and energy consumption per year and to focus its priorities and goals on constructive socioeconomic progress in the absence of appreciable internal and external conflict. The early development and evolution of the science and technology of agriculture and animal husbandry serve as an example of this developmental pattern.

Historically, initial primitive efforts to domesticate animals and cultivate crops resulted in significant success in generating *value* (i.e., both *wealth and energy*) on a per capita annual basis. This era of human history is characterized by the rapid acquisition and progression of new knowledge, skills, and technologies related to what became known as the agriculture industry. This agricultural revolution was highly successful in generating increasingly higher levels of available, consumable resources that were effectively utilized to generate new forms of wealth-energy capital. This evolution of agricultural advancement involved millennia of human discoveries, inventions, and technical improvements that included selective plant and animal breeding, irrigation, fertilizers, crop cultivation techniques, tools, machinery, and farm management strategies. The economic consequences of the agricultural evolution shaped momentous and lasting sociopolitical changes. As a result of significantly higher production of food per capita per year, populations increased and societies that once had difficulty surviving because of inadequate nutrition and protection from environmental threats began to acquire a primitive form of disposable income. However, economic change has social consequences and implications, both positive and negative. Thus, the social system becomes modified and conditioned by the forces of progress as more advanced inventions and discoveries alter a society's technology and production capabilities.

In the most primitive cultures, economic systems are synonymous with social structures and kin, but during the agriculture revolution, economics became more diffuse, competitive, and sociopolitically oriented, involving greater interaction with outsiders. The inherent consequences of the initial phase of primitive agriculture prosperity were to increase the number of products, to expand trading regions, and to seek competitive advantages in order to capture business and increase personal rewards. Social systems change and address the realities of the prevailing and evolving economic system. This is motivated and controlled by ambitious production and goals, attempting to meet the ever-changing, materialistic aspirations of society. Consequently, new, progressive technologies must be a continuous, planned outcome from a progressive, profitable economy, which inherently modifies and breeds subsystems and strategies and tactics to stimulate change and generate socioeconomic success.

However, during the agriculture and animal husbandry era, cultural advancement was not continuous and indefinite. The limits of existing technological capabilities and available energy resources were eventually reached, gradually dissipating economic productivity and social progress. Further advancement of human history would require more efficient and effective technologies and more advanced tools and machinery applicable to new forms of yet-to-be-identified fuels, possessing higher energy content. This new era was born in the late eighteenth century with James Watt's major technological advancement of the steam engine and the new high-energy source of coal. Subsequently, major inventions and discoveries expanded beyond the steam engine utilizing the fuels of coal, oil, natural gas, and nuclear materials, initiating and supporting major socioeconomic advancements.

White's correlation of society's success in harnessing energy from the agriculture revolution with cultural advancement reveals a significant rate of achievement from 8000 B.C. to 1000 B.C., followed by a progressive leveling-off period until the "Fuel Age" of A.D. 1800.

> **After a period of rapid growth, the upward curve of progress leveled off onto a plateau. The peaks of cultural development in Egypt, Mesopotamia, India, and China were reached prior to 1000 B.C. ... And from that time until the beginning of the Fuel Age, about A.D. 1800, no culture of the Old World surpassed, in any profound and comprehensive way, the highest levels achieved in the Bronze Age. There were innovations ... and many refinements of already existing traits. But, taking cultures as wholes ... the cultures of Europe between the disintegration of the Roman Empire and the rise of the Power Age were in general inferior to those of the ancient oriental civilizations.[35]**

Since 1800, the World has increasingly consumed record amounts of wealth-energy capital per capita per year, as the Industrial Revolution and capitalism provided a foundation for expanding economic opportunities and profitability. As a result, the world's population drastically increased and became highly concentrated in urban centers, requiring substantial readjustments of sociopolitical organizations and functions. This global transformation from a historic agricultural society to the industrial age gradually diminished and ultimately eliminated the relevance of feudalism and the aristocracy, as a new and more rewarding economic system required a more integrated and complex social order. The need for a better educated and trained labor force created a new class of urban-based industrial worker as the modern economy's need for traditional peasant and slave labor gradually diminished. Likewise, financial and business leaders, who dominated the new economic system, became the modern political force, replacing former elites of the agricultural economy.

Western technological and economic advancement during the twentieth century provided materials and wealth-energy resources, enabling military power to become an instrument of economic and political advantage. Since ancient times, such societal accumulation of wealth has produced opportunities for human conquests and the rewards of additional territories, political power, and wealth-energy resources.

Western economic success during the latter part of the twentieth century brought a significant shift in societal values and priorities and an escalation in the production of nonessential consumer goods and services. In the United States, this consumer economy absorbs an increasingly larger percentage of national wealth, while incurring a corresponding record consumer debt. By the end of the twentieth century, Western civilization's rapid increase of socioeconomic success had reached a plateau phase, as competitive Asian nations successfully assimilated elements of industrialization and capitalism into their traditional cultures.

Thus, Western civilization prosperity gradually created a broadening of its social classes from a cheap labor force to the wealthy elites, resulting in a continuously expanding gulf of societal income and wealth. Typically, as wealth becomes more concentrated in fewer individuals and groups within a society, sociopolitical forces begin to retard economic growth. Consequently, such important societal characteristics found within a successful socioeconomic system as creativity, initiative, self-discipline, self-reliance, civility, and wise use of natural resources significantly diminish. This aspect of cultural regression is also characterized by a decline in inventiveness, discovery, professionalism, and the rate of technological advancement, in addition to a general decay of cultural values. Additionally, the economic and political elite increasingly engages in personal and professional wasteful consumption of wealth-energy capital while ignoring the need for prudent, long-term societal-based investments. The mature civilization stages of ancient Greece, Rome, Egypt, Imperial Spain, and the British Empire as well as twenty-first-century America exemplify such characteristics and their ramifications.

Entering the twenty-first century, Western civilization has exhausted its former industrial and economic global advantages and opportunities, declining from its peak of cultural advancement into a stagnation phase. While experiencing economic prosperity, the West's innate limitations and decaying moral values have contributed to an inevitable socioeconomic deterioration, as suffered by all previous civilizations. However, other major factors include a social order reaching the limitations of its current science, engineering, technology, machinery, and knowledge and skill base and being unable or unwilling to evolve to the next generation of economic capabilities necessary for successful production. This translates into an inability to effectively compete in the global marketplace, fostering a gradual internal social and economic instability and cultural deterioration.

As with past great civilizations, the human and financial toll of excessive military spending for a global agenda of economic and ideological dominance has contributed greatly to the deterioration of twenty-first-century America. Additionally, Western nations, particularly the United States, will increasingly suffer severe socioeconomic repercussions from the emerging global inadequacy in the supply and distribution of energy resources to meet rapidly escalating demands in support of a worldwide industrial expansion. This issue has become more severe for Western nations, as the historically undeveloped countries become more industrialized, dramatically increasing their energy consumption while making the transition from being net energy exporting nations to net importing nations. The availability of energy resources is, and will remain, a major limiting factor of the dynamics associated with cultural evolution and the advancement of the human social order. Consequently, the World awaits the *next evolutionary fuel phase* and the associated *new generation of tools, technologies, techniques, and machines* to catalyze economic innovation and productivity for the next major global socioeconomic transformation.

Wealth, energy, and human values are the basic dynamics of a social order that people may constructive utilize or belligerently abuse, the proportions of which will define the quality and longevity of the social order.

Chapter 6
References and Notes

1. White, Leslie A. and Dillingham, B. *The Concept of Culture*. Minneapolis: Burgess. 1973. 1.
2. Ibid., 363.
3. Eisenstadt, S. N. *The Political Systems of Empires*. New Brunswick, N.J.: Transaction Publishers. 1993. xxxiv.
4. White, Leslie A. *The Concept of Cultural Systems*. New York: Columbia University Press. 1975. 3-10.
5. Ibid., 6.
6. Durkheim, Emile. *The Elementary Forms of the Religious Life*. Glencoe, Ill.: The Free Press. 1947. 424.
7. Braudel, Fernand. *A History of Civilizations*. London: Penguin Press. 1992. xxxvi.
8. Fried, Morton H. *The Evolution of Political Society*. New York: Random House. 1967. 5-7.
9. Ibid., 7-8.
10. Ibid., 8.
11. Braudel, Fernand. *A History of Civilizations*. 22.
12. Ibid., xxiv.
13. Fried, Morton H. *The Evolution of Political Society*. 5-7.
14. White, Leslie A. *The Concept of Cultural Systems*. 18.
15. Melko, Matthew. *The Nature of Civilizations*. Boston: Porter Sargent. 1969. 9.
16. Ibid., 10.
17. Ibid., 54.
18. Ibid., 53-57.
19. Ibid., 8-9.
20. Kroeber, A. L. *Configurations of Culture Growth*. Berkeley: University of California Press. 1944. Chapter XI.
21. Braudel, Fernand. *A History of Civilizations*. xxii-xxiv.
22. Melko, Matthew. *The Nature of Civilizations*. 17-19.
23. Ibid., 21-22.
24. Ibid., 12.
25. White, Leslie A. *The Science of Culture*. Toronto: Doubleday Canada. 1969. 367.
26. Ibid., 367-368.

27. Ibid., 368.
28. Braudel, Fernand. *A History of Civilizations*. 17.
29. Melko, Matthew. *The Nature of Civilizations*. 2.
30. Ibid., 12.
31. Kroeber, A. L. *Configurations of Culture Growth*. 7.
32. Melko, Matthew. *The Nature of Civilizations*. 15-24.
33. Kroeber, A. L. *Configurations of Culture Growth*. 833.
34. White, Leslie A. and Dillingham, B. *The Concept of Culture*. 58-60.
35. White, Leslie A. *The Science of Culture*. 373.

Chapter 7
The Characteristics and Dynamics of Sociocultural Transitions

"We know that viewed in their empirical aspect, all sociocultural phenomena change incessantly, without any exception whatsoever. The question arises: Why do they change but do not remain unchangeable? Why this relentless becoming instead of everlasting permanency? ... Where shall we look for the roots of change of sociocultural phenomena and how shall we interpret it? Shall we look for the 'causes' of the change of a given sociocultural phenomenon in the phenomenon itself, or in some 'forces' or 'factors' external to it?" Pitirim Sorokin[1]

In the previous chapter, the mechanisms and energetics associated with the functions and organizational structures inherent to the human social order were presented, which serve as a precursor for the examination of the characteristics and dynamics of sociocultural transitions.

Pitirim Sorokin's volume *Social and Cultural Dynamics* contains an insightful presentation of his "immanent theory of sociocultural change", which begins with the statement: "We know that viewed in their empirical aspect, all sociocultural phenomena change incessantly, without any exception whatsoever." He then asks rhetorically, "Why do they change but do not remain unchangeable?" Sorokin then raises an important issue with a second question: "Where shall we look for the roots of change of sociocultural phenomena and how shall we interpret it? Shall we look for the 'causes' of the change of a given sociocultural phenomenon in the phenomenon itself, or in some 'forces' or 'factors' external to it?" His subsequent response is that a culture "bears in itself the seeds of its change"[2] and thus is the source of its own particular evolving cultural characteristics and traits.

In contrast to some authors, Sorokin minimizes the influence of factors external to the sociocultural system, focusing instead on the nature of the culture itself as the change agent. As others have also postulated, Sorokin acknowledges that all cultures, in order to exist, must undergo continual change during the full course of their existence. "We do not know any empirical sociocultural system or phenomenon which does not change in the course of its existence or in the course of time."[3] Sorokin's cultural subsystems represent social, artistic, philosophical, economic, intellectual, and political functions that symbolize the culture's system of values. Sorokin views a culture's value system as the basis for its "most profound changes" rather than "institutional factors" or "actions of great men."[5] Additionally, he acknowledges a thermodynamic reality: "Life can never be in equilibrium. Complete equilibrium is never attained and would be fatal if it were attained, as it would mean stagnation, atrophy, and death."[4]

There is thermodynamic justification for Sorokin's view of "cultural dynamics" based on a society functioning as an open thermodynamic system and driven toward, but never achieving, dynamic equilibrium as a consequence of the forces derived from its continuous processes of living. It has been previously noted that a multitude of functional systems supporting culture's existence must continually consume wealth, energy, and other resources in order to avoid the fate of dynamic equilibrium (i.e., cultural stagnation).

The immanent theory of sociocultural change asserts that change occurs "by virtue of its own forces and properties. It cannot help changing, even if all its external conditions are constant. The change is thus immanent in any sociocultural system, inherent in it, and inalienable from it."[6] Sorokin's perspective of a cultural transition process is that a "whole integrated culture as a constellation of many cultural subsystems changes and passes from one state to another ... because each of these is a going concern and bears in itself the reason of its change."[7] These "seeds of its change" include the human elements of values, ethics, and integrity as reflected by the quality and intensity of its social, emotional, economic, intellectual, and political assets and liabilities. The totality of cultural traits guides a society as it "generates consequences" that will determine the nature, quality, probability, and extent of inevitable transitions.

> **As long as it exists and functions, any sociocultural system incessantly generates consequences which are not the results of the external factors to the system, but the consequences of the existence of the system and of its activities.**[8]

It should be emphasized that Sorokin also recognizes the real but secondary influence of external factors: "The endorsement of the immanent principle of change does not hinder recognition of the role of the external forces in the change of the sociocultural system."[9]

> **In a study of a transformation of any sociocultural system, the partisan of the immanent theory of change will look for reasons or factors of the change first of all in the internal properties (actual and potential) of the system itself, and not in merely its external conditions.**[10]

It then follows from the concept of immanent change that consequences are produced by a functioning society that predominantly originate from within its internal structure of the society and its membership. Furthermore, the *attribute of self-determination or self-articulation* of a society's destiny becomes an inherent and major focus of Sorokin's approach to cultural transitions. He defines this motivating drive of self-determination toward a defined destiny as a:

> potentially given course of existence of a sociocultural system.... The activities of its life career or destiny are determined mainly by the system itself, by its potential nature and the totality of its properties.... External conditions can crush the system or terminate an unfolding of its immanent destiny at one of the earliest phases of its development, depriving it of a realization of its complete life career; but they cannot fundamentally change the character and the quality of each phase of the development; nor can they, in many cases, reverse or fundamentally change the sequence of the phases of the immanent destiny of the system.[11]

The maturity level of a society's organizational and functional capabilities and its degree of socioeconomic integration would appear to influence the extent to which self-determination and self-articulation could flourish within a given cultural environment. The greater the extent of a culture's expression of self-determination and internal autonomy, as represented by organizational control and self-direction, the greater its potential socioeconomic success. This cultural environment will define specific goals, policies, and practices, and a political power structure will evolve. These influences will shape and guide the characteristics, properties, developmental stages, and evolutionary pattern of a social order.

The sequences of cultural transitions are systematically catalogued as recorded history and accomplishments, useful in providing instructive insights into a civilization's successes and failures. However, such empirical data do not represent fundamental parameters that reflect the driving forces of societal development. The use of observed historical events to explain fundamental driving forces of cultural change is analogous to telling a child that the reason a car moves down the street is because, "as we see, the wheels go around and around and move the car." The obvious flaw is that reporting observations of a dynamic event and its consequences does not necessarily identify the nature and source of dynamic forces responsible for the event. Historical events, while providing useful clues of the dynamics of cultural processes and transitions, do not in themselves constitute driving forces of cultural change.

Authors have used history and observed human behavior to project peaks and valleys of cultural progress that appear associated with economic growth and social advancement. Such an example is Kroeber's identified "abstract activities" used to detect high cultural developments, which permit charting societal progress by virtue of the ups and downs of "cultural patterns." However, while this impressive body of data and information constitute interesting historical insights into the quality of high culture within a society and may serve as a barometer of the health of the cultural arts, the methodology is unable to directly provide insights into the fundamental driving forces of cultural transitions. Artistic accomplishment is an outcome, a reflection of culture, not of a dynamic force driving its evolution. Kroeber does acknowledge this limitation.[12]

Historically, the quest for identifiable, descriptive cultural transition stages, phases, and periods appears, in many cases, to constitute labeling exercises of historical observations associated with periodic fluctuations in the rate of socioeconomic change. The concept of "reconstitution," as used by Melko,[13] is an example of such a label. First, it should be expected that the productivity of a society's individual functional processes as well as its general operational systems would normally fluctuate with time as cultural conditions vary. Second, it appears obvious that a civilization, required to be in a constant state of change, must either be advancing at a finite rate toward a more mature, integrated cultural form or declining and reverting to a less mature or advanced societal condition and classification (i.e., reconstitution). Alternatively, this phenomenon may be described as a normal *dynamic fluctuation* of positive and negative rates of cultural growth as represented by a general array of cultural variables. The intensity of such *fluctuations* will be dependent on particular cultural circumstances at any given time. This is analogous to the long-term positive and negative fluctuations of the stock market, as some stocks advance while others decline, typical changes in a functioning economy, and as new stock offerings emerge while others disappear. The long-term pattern of economic growth and decline represents various transitions due to a large array of internal and external social, economic, political, and psychological factors.

A consensus appears to exist among a number of authors that a civilization comes into existence and matures over time by virtue of becoming an increasingly successful multifaceted, complex, and sophisticated social order, leading to gradual sociocultural transitions. An almost infinite number of intricate and interdependent social, economic, and political functions may, by chance or design, combine in varying constructive and destructive manners to reinforce or detract from each other. An appropriate analogy would be the combination of wave action from such sources as sound and light or from pebbles dropped into a pond. Depending on conditions, such waves may constructively or destructively coalesce, resulting in observed waves being larger or smaller depending on how the individual waves (i.e., wavelengths) combine. The waves may come together in a manner that either detracts from each other (destructive interference), producing only a ripple effect, or reinforces each other (constructive interference) to produce a larger wave of greater impact.

Given the large number of different processes operating within a culture over any given time frame and under varying constructive and destructive internal and external influences, the overall maturation process of a civilization will produce a continuous fluctuation of observable outcomes. A specific category of cultural outcomes (e.g., economic) will exhibit variations in characteristics and properties over time, a function of rational and irrational variables, resulting in outcomes that may be unpredictable, uncontrollable, and irreversible. The resulting pattern of net cultural progress will be influenced by elements of both *societal progress* and *decline*. An observer, at any point in time, will evaluate the evolving social order relative to the individual's values, experiences, and aspirations.

Thus, a society's dynamic processes of complicated, mutually interconnecting systems and subsystems define the capability and potential for cultural growth as well as for major transitions. New subsystems are continually created and injected into the social order, while others become inoperable and functionally nonexistent. Accordingly, the nature of the mix of societal processes and related operational conditions, priorities, resources, and outcomes are continuously changing. As a result, the nature and rate of societal progress will fluctuate with time, with a persistent, large-scale change capable of inducing a sociocultural transition.

Authors have assigned such labels to cultural transitions as crystallization, disintegration, ossification, petrifaction, and reconstitution. These descriptive labels have been assigned to historical periods based on the extent to which the rate of cultural growth abruptly or continually rose or fell. Often, this occurred in conjunction with highly visible socioeconomic changes associated with significant historical events and unanticipated influences such as war, famine, financial insolvency, change in leadership or government, and disease. The foremost question is, What fundamental factors, including the availability of natural resources and such human characteristics as attitudes, values, and behavior, affect and control the pathway toward a cultural transition (e.g., crystallization or reconstitution)? How are the phenomena functionally related to the dynamics of such general cultural transitions?

Interestingly, these descriptive terms (e.g., crystallization) often possess and convey more precise, useful meaning if considered in the context of mechanistic and energy concepts related to their representative process. In the cultural context, *crystallization* depicts a mechanism of *creating increased orderliness within a society,* whether this order is organizational or functional or applied to social, economic, or political processes. An example is the transition from producing commercial goods by hand to the adoption of the latest technology of electromechanically based, automated assembly-line production. Such a major economic transition would be expected to have broad long-term corporate and cultural ramifications affecting profitability, employment qualifications, marketing techniques, geographical regions, safety regulations, etc. That is, more extensive institutional structures and functions are necessary to implement the next phase of production technology in order to be competitive, which generally requires incremental investments of capital.

The creation and maintenance of orderliness and stability within a culture is crucial to long-term sociocultural success but involves creating additional complexity and energy-consuming processes. The dynamic term *crystallization* implies that *things* line up and become more organized. A familiar physical example of the crystallization process is liquid water freezing to form ice crystals. The system of liquid water molecules in random thermal motion is systematically cooled, resulting in liquid water molecules becoming less mobile and more orderly arranged in a solid lattice of ice. The thermal energy content of the liquid water is gradually reduced via an external cooling process until attractive intermolecular forces are capable of immobilizing the water molecules in a continuous three-dimensional latticework. This mechanism produces a more orderly system of molecules (i.e., a crystallization mechanism), as energy is transferred from the water system to its surroundings. Thus, the thermodynamic process of crystallization has a specific mechanism and an energy transfer requirement.

Economic processes and the physical crystallization of liquid water both conform to thermodynamic principles and have entropy ramifications. First, the physical example involving the removal of heat content from liquid water to form solid water or ice requires that external energy be consumed by some cooling machine during the process of cooling the water. Heat content is extracted from liquid water, and the process is associated with waste energy and material related to the external cooling device. Consider a refrigerator's ice-making capability that utilizes electrical energy originating from a coal-burning generator. Economic processes involve segments of society creating pockets of *structural and functional orderliness* within its production processes, utilizing human labor and technology, such as computerized, mechanized assembly-line production. While creating order for some limited socioeconomic benefits for society, such economic processes are also responsible for more societal disorder and consuming greater amounts of energy than can possibly result from the outcomes of the production processes. Thus, the term *crystallization* represents an important element of cultural processes that utilize the same mechanistic and energy principles that scientists employ to explain the physical process of liquid water forming ice.

Previous chapters have emphasized the historical significance of the continuous technological advancement of tools, machines, and devices in increasing economic productivity and therefore serving as a catalyst for sociocultural transitions. However, Toynbee, using the Mayan civilization as an example, contends that a society's lack of development and use of "technique" or technology does not necessarily result in less societal growth or maturity. However, rather than using such examples to demonstrate the lack of "evidence of a positive correlation between an improvement in technique and progress in social growth,"[14] his examples merely demonstrate that any one influencing factor *may contribute* to "progress in societal growth." Of the large array of potential variables, which factors will ultimately dominate and have a major influence on the rate of socioeconomic progress, and create significant cultural changes, will depend on the particular circumstances affecting a given society. This is consistent with the representation of societal growth as the summation of an exceedingly large number of complex and sometimes independent social, economic, and political processes that ultimately will characterize and define a unique society.

Toynbee selectively uses historical examples to discount technology and other developmental factors as possible contributors to growth, when they are only one of many potential influences. Thus, technology, as with any other societal growth variable, may be a major influence for one society's growth pathway while being a minor or relatively nonexistent influence for other societies. This is illustrated by Toynbee's example of "technique" in his comparisons among the Mayan, Mexic, and Yucatec civilizations.[15] According to Toynbee, in two civilizations "technique" was "a criterion of growth," while in the third it was not, although this social order developed to a more advanced civilization. While this may indicate that technology is not a universal requisite for all societies to develop to an advanced status, historical evidence demonstrates that it has been; thus, it may not be ruled out as a *potential influencing factor* of cultural advancement.

The history of the development of technique, like the history of geographical expansion, has failed to provide us with a criterion of the growth of civilizations.[16]

The same argument applies to other civilization growth factors (e.g., geographical expansion, which Toynbee also rejects based on selective examples and the presumption that any legitimate growth factor *must affect the progress of all societies*). In a later chapter, it will be shown that Imperial Spain's excessive and overambitious geographical expansion to secure much-needed new wealth was instrumental in its initial era of significant success, but also in its ultimate deterioration. The large number of variables potentially influencing cultural growth, which may be carried to excess, may also be factors in cultural degradation.

Toynbee and others, such as G. D. H. Cole,[17] reject mechanistic arguments "to analyze and explain society as mechanism, on the analogy of biology." However, by so doing, they ignore the reality of societies as complex entities subject to operational processes that involve specific functional mechanisms. Civilizations rely upon socioeconomic processes, all of which consist of wealth-energy-driven mechanisms. In dismissing these science analogies as "harmful and misleading," Toynbee instead focuses on "the right way of describing the relation between human societies and individuals"; apparently, his dynamic forces of cultural development are defined in terms of such "relations." He states: "Society is a 'field of action' but the source of all is in the individuals composing it."[18] This narrow viewpoint appears to be inconsistent with the implied "mechanistic" nature of his argument. That is, community members providing a "field of action" would manage a large number of functional, communal processes consisting of mechanisms requiring wealth-energy capital. This requires a human element in action (i.e., "Society is a field of action"), executing thermodynamic processes that possess specific functional mechanisms, resource requirements, and conditions, subject to the laws of nature and the influences of mathematical probability. How else could a society's mechanistic, energy-dependent process be represented than as a "field of actions"?

Societal transitions are often portrayed either as a series of separate, individual cultural growth periods or, contrastingly, as merely normal fluctuations or blips of one continuous, generic process of societal progression. Kroeber notes the "frequent habit of societies to develop their cultures to their highest levels spasmodically: especially in their intellectual and aesthetic aspects, but also in more material and practical aspects. The cultures grow, prosper, and decline, in the opinion of the world."[19] He stresses: "One of the recognized characteristics of human culture is the tendency of its success or highest values to occur close together in relative brief periods within nations or limited areas."[20] His methodology is to analyze culturally "abstract activities" to detect "high cultural developments" (i.e., "flowering" or "florescence" of such disciplines as philosophy, sculpture, painting, music, and literature). He views an entire "historic culture" as possessing, to some degree, these abstract activities of high culture. "High value" is defined as possessing "qualities of human productions which normally appear in historic configurations thus limited and shaped." His objective is to seek "the recognized" but unexplained "cultural patterns."[21]

Kroeber's historic work involves the analysis of an enormous quantity of historical information, particularly the patterns of growth and decline in the production of highly valued "human culture" of the arts, sciences, and letters. This constitutes the basis for documenting his "pulses" and "lulls" in cultural growth configurations. Kroeber identifies "the strength and significance of cultural patterns ... reflecting phenomenal reality.... Products of high value appear definitely limited, even concentrated, in time and space, with a usually continuous rise and fall of value."[21] A number of authors have shown great interest in the debate as to whether the rise and decline of a great civilization actually consists of a number of mini-cultures rising and declining or simply a pattern with "pulses" and "lulls" representing periods of good times as well as not-so-good times. One of Kroeber's conclusions is the following:

> It is evident from these instances that the pauses, lulls, regressions, or barren intervals are only the minor part of a set of problems the center of which lies in the question of intrinsic relation to each other of the productivity periods of a nation or civilization: and the further question arises, how far some of these productive periods might not be reckoned as separate cultural growths rather than as pulses or phases of one growth. It is plain that we are dealing with an integrating series of phenomena, without either measure or touchstone.[22]

Kroeber states further that "cultures inherently tend to progress.... Fluctuations of cultural vigor are normal.... The reason for the fluctuation is that ... with successful development they accordingly become exhausted; and that these must be a breakdown or abandonment and reformulation of patterns before the culture can go on to new high achievement. That there is some tendency for the several patterns of one culture to form, to culminate, and to dissolve or atrophy simultaneously, is, I think, obvious."[23] It is instructive to pursue Kroeber's observation that "with successful development civilizations become exhausted, ... a breakdown or abandonment and reformulation of patterns before the culture can go on to new high achievement." Historically, such a "breakdown," as recognized by a number of authors, is associated with a high degree of cultural complexity requiring incremental resources that eventually exceed the social order's available wealth-energy capital.

Regardless of the cause, it follows that these recognized "patterns" (i.e., cultural mechanisms) become inadequate or mismanaged, or suffer from insufficient resources. The resulting social and economic instability and societal dysfunction produces the condition that Toynbee also refers to as "breakdown" as well as "abandonment." This would seem to indicate that Kroeber indirectly recognizes a limited mechanistic vision of a functioning society, as would his "organisms" analogy for differing cultures. In his words, "Cultures are therefore as distinct as personalities or even *organisms* of different genus."[24]

It is noteworthy that Kroeber defines his patterns or "growth configurations" based on historical times and defined societies, whereas others, specifically Spengler, resort to more subjective, intuitive methodologies. In Kroeber's final chapter, "Review and Conclusions," he reinforces and amplifies his previous statements regarding documented cultural patterns. First, the temporary bursts and growths in the higher valued aesthetic and intellectual activities of the civilizations analyzed are attributed to nationalistic developments, "as expressed in successful political organization and expansion."

> **It is clear that aesthetic and intellectual endeavors resulting in higher values preponderantly realize themselves in temporary bursts, or growths, in all the higher civilizations examined. The same sort of bursts or growths tend to characterize nationalistic development, as expressed in successful political organizations and expansion. Whether the phenomenon holds also for wealth and population, is a separate question.**[25]

Further, he states, "It is entirely conceivable that there may be a connection between growth of population and wealth and the achievement growths which have been analyzed" (i.e., high cultural developments). The thermodynamic-based socioeconomic model applied to social orders would assert that these "connections" as well as the patterns described by Kroeber and others are expected characteristics of a functioning society as it consumes wealth-energy capital in its normal functions of existence and particularly in times when economic prosperity is related to successful military expansion.

Kroeber's analyses of "well-unified" and "well-defined" civilizations indicate configurations of growth and decline that exhibit more than one crest with patterns approximating symmetrical or somewhat skewed normal distribution functions. Additionally, composite curves for a number of these higher valued activities (i.e., "curves for total cultures") exhibit even greater symmetry.[26] The ranges of the duration of these growth configuration or patterns vary from a few decades to thousands of years. However, as has been previously noted, thermodynamic processes are independent of time; thus, such divergent time periods, while being related to the availability of wealth-energy capital, are irrelevant to thermodynamic considerations. Thus, Kroeber describes societal patterns reflecting competency and productivity of the cultural arts as being representative of the relative progress of the entire social order. Arnold Toynbee, on the other hand, envisions a more direct and dynamic nature for sociocultural transitions.

Toynbee's approach to cultural change, "the nature and patterns of the historical experience of the human race," is found in his work *A Study of History*.[27] Referring to societies undergoing cultural maturation, he quotes Walter Bagehot: "In societies in process of civilization ... 'the cake of custom' is broken and society is in dynamic motion along a course of change and growth."[28] Toynbee notes:

> **The forces in action are not national but proceed from wider causes, which operate upon each of the parts and are not intelligible in their partial operation unless a comprehensive view is taken of their operation throughout the society.[29]... The intelligible unit of historical study is neither a nation-state nor mankind as a whole but a certain grouping of humanity which we have called a society.[30]**

It is instructive to relate the nature and ascribed characteristics of Toynbee's "forces of action" and his "certain grouping of humanity" with his rationale for the dynamic nature of cultural change. A conceptual process of change embraced by Toynbee involves "certain grouping of humanity" attempting to overcome a continuing series of societal challenges. He contends that not only does each challenge provide temporary stimulus to society, but also that continuously conquering new challenges provides a long-term momentum that propels a society along the pathway of progress. In his words, "The presentation of each problem is a challenge to undergo an ordeal, and through this series of ordeals the members of the society progressively differentiate themselves from one another."[31] *Societal differentiation* results from the diverse nature and the quality of an individual's or a group's approaches, solutions, implementation, and outcomes from the sequence of long-term challenges. As numerous societal challenges multiply over time, producing individual outcomes, the degree of accumulated differentiation or uniqueness within a social order will increase exponentially. Toynbee illustrates his concept of differentiation by using variation in artistic styles found among civilizations: "It is generally recognized that every civilization creates an artistic style of its own; and if we are attempting to ascertain the limits of any particular civilization in space or time we find that the aesthetic test is the surest as well as the subtlest."[32]

It is curious that Toynbee does not consider the discovery and development of new knowledge and its application as possible "evidence of a positive correlation between an improvement in technique and a progress in social growth." The evolution of applications of newly found knowledge, such as the taming of the horse; the hollowing out of a tree for a canoe; and the use of the bow and arrow, the long rifle, and dynamite, has generally been considered as a vital contribution to cultural transitions. Such knowledge and tools permit a society to overcome challenges, implement responses, and create opportunities, thereby exercising self-determination and an ability to prevail in a competitive environment.

Toynbee specifically identifies self-determination as a dynamic mechanism that a "grouping of humanity" with mutual interests must possess and utilize to promote societal growth and success. While Toynbee's "challenge-and-response" thesis is a reasonable methodology for providing the motivation necessary for societal progress, in reality, overcoming such challenges has positive and negative consequences. Even when the most creative responses are implemented by the most determined social orders, challenges are not always completely resolved. Whether as a result of avoidance or of well-intended but unsuccessful attempts at resolution, some challenges remain unsettled for long periods of time, sometimes indefinitely, and have the ability to breed additional issues, challenges, and antagonism. In addition, it is rare that a resolution of a major social issue or a response to an economic challenge doesn't create unhappiness with some segment of the population. Thus, the process of problem resolution and responses to the continuous flow and increasing volume of societal challenges inherently produces *an accumulation of unanswered challenges.* These often contribute to human friction and social instability within an already complex and severely challenged society. This condition exemplifies the societal entropy effect that increasingly becomes more prevalent and counterproductive in mature civilizations. In Toynbee's words:

> **When the outcome of each successive encounter is not victory but defeat, the unanswered challenge can never be disposed of, and is bound to present itself again and again until it either receives some tardy and imperfect answer or else brings about the destruction of the society which has shown itself inveterately incapable of responding to it effectively.[33]**

Fernand Braudel's view is that "Arnold Toynbee offered a tempting theory. All human achievement, he thought, involved challenge and response. Nature had to present itself as a difficulty to be overcome. If human beings took up the challenge, their response would lay the foundations of civilization ... but civilization does not always follow—at least until improved technology makes the response more adequate."[34] Thus, Toynbee views cultural change and "human achievement" as resulting from people accepting and overcoming challenge, whether it originates with nature or from other sources. Therefore, "the response" is the outcome from an opportunity afforded by the existence of a challenge to create progress. "There can be no civilization without the societies that support them and inspire their tensions and their progress."[35] Thus, the dynamics of a culture produce the *challenges*, create the *tensions and motivations*, and thereby provide an *opportunity* for members of the culture to rise to the occasion and craft meaningful change. However, nature, probability, and human self-interest and unwise behavior may work against such progress, but the alternative is cultural stagnation. The dynamic forces of the human existence, for better or worse, shape social history that is "in dynamic motion along a course of change and growth."

The "course of change and growth," while providing for human success, also unremittingly creates increasing levels of social and economic instability.

Thus, the thermodynamic foundation of the organized human existence includes the expectation of an accumulation of societal disorder or societal entropy being generated proportional to a society's socioeconomic vigor. This disorder comprises inadequate resources, environmental pollution, political dysfunction, and social and economic instability. These are among the specific contributing factors leading to Toynbee's societal "breakdown" phase, the transition to the termination of an era of positive cultural growth and a precursor to his phase of "disintegration."

> **The ultimate criterion and the fundamental cause of the breakdowns which precede disintegrations is an outbreak of internal discords through which societies forfeit their faculty of self-determination. The social schisms in which this discord partially reveals itself render the broken-down society in two different dimensions simultaneously. There are vertical schisms between geographically segregated communities and horizontal schisms between geographically intermingled but socially segregated classes.**[36]

Toynbee's "discords" and "schisms" representing the "fundamental cause of the breakdowns which precede disintegrations … through which societies forfeit their faculty of self-determination" are often self-generated, whether due to human behavior, lack of resources, or a general cultural disorder and chaos. The level of internal discord within an increasingly disillusioned society may reach an intensity whereby its members become emotionally and financially unable to cope with the escalating realities of complexity, disorder, and discouragement. More precisely, societies may not just "*forfeit* their faculty of self-determination" but rather, they may also become *humanly unable to cope* with the stifling, mounting momentum of societal dysfunction and a failing socioeconomic system. As a consequence, at some point, it no longer becomes possible to maintain the historic positive rate of cultural growth due to the declining effectiveness of society's socioeconomic processes. This results from an inability to overcome accumulating challenges and maintain a positive cultural momentum. The cumulative effect of declining self-confidence, wealth-energy capital, creativity, and concern for the common good produces a reality that the society is entering a transitional period whereby societal progress will decline. This corresponds to Toynbee's transition from the "breakdown" stage to the "disintegration" stage. Recall that a net negative rate of cultural growth exists only when the summation of the positive contributions of socioeconomic processes becomes less than the summation of negative contributions. In stock market parlance, a social order's numerous processes of existence produce more losers than winners!

An analogy to society's fluctuating rate of cultural growth may be illustrated by the progression of a *chemical reaction toward equilibrium* represented by the reaction $A + B \Leftrightarrow C + D$. Conditions may be imposed that will shift the dynamic equilibrium either toward the forward direction of D or the reverse direction of B. A positive but fluctuating rate (a pulse) results from periodically increasing and then randomly decreasing the amount of A and B added over a given time. If, at some point, no further amounts of A and B are added, the rate of the process in the forward direction will gradually decline and approach zero, which may be followed by an increase of the rate in the reverse reaction.

If C and D are added to the system, the reverse direction, representing a negative rate (regression) producing A and B, will take place. Thus, in order to continuously produce C and D, an uninterrupted flow of A and B must be added to the open system. Progress toward a dynamic equilibrium condition is thus continually thwarted or disrupted by the unremitting addition of reactants or product species.

Likewise, sociocultural transitions may be interpreted in an analogous manner to the chemical reaction illustration. Variations over time of positive and negative rates of cultural growth (i.e., fluctuations) occur due to the introduction of new processes, materials, and environmental conditions as well as to modifications of long-standing processes by human initiatives. This view is consistent with Sorokin's perspective of a "whole integrated culture as a constellation of many cultural subsystems.… Each of these is a going concern and bears in itself the reason of its change." These influences include human values, ethics, and integrity that guide a society as it "generates consequences." These continually changing variables disrupt the system's drive toward cultural equilibrium, redefining the nature of the system and perhaps the controlling conditions while producing irregular patterns of cultural growth, described by Kroeber as "lulls," "pulses," and "regressions."

To more fully comprehend *socioeconomic stagnation* and *cultural disintegration*, it is instructive to examine Kroeber's methodology and concept of "pulses" and "lulls." Kroeber questions "how far culture-growth florescence tend to follow a symmetrical normal curve in their configurations."[37] While acknowledging that some profiles exhibit a skew, he says, "for most of them the rise and fall would be substantially symmetrical." Recall that Kroeber utilizes the level of success or "florescence" of the cultural arts and sciences as a measure of cultural prosperity and advancement. A "florescent growth" period or the "flowering of intellectuo-aesthetic culture may range from a human lifetime to a thousand years. The duration of the Greek science pulse was five centuries, Italian music four, Russian music less than one, French painting less than one, English literature two, and German literature one."[38]

Kroeber notes that the during the Graeco-Roman or Mediterranean civilization expansion, "the quality of its cultural products was, on the whole, declining; that is, there was a lag of the territorial and political growth behind the intellectual and aesthetic growth." He notes a "general parallelism of growth and decline, or of expansion and contraction, in the two sets of activities, ... territorial and political growth and the intellectual and aesthetic growth." However, he comments: "Our understanding of the connecting mechanism remains vague."[39]

The "vagueness" in understanding the "lag" of this parallelism is resolved by recognizing the inherently different processes and sequencing of the two growth phases. Territorial and political growth necessitates the expenditure of scarce resources but may later result in the acquisition of new resources to support "intellectual and aesthetic growth" in a subsequent period of prosperity.

Kroeber's methodology assumes that "florescence" represents a valid measure of a society's vitality, self-determination, creativity, and stability; the wealth to support both cultural arts and societal investments; an equal concern for intellectual and social advancement; and general progress of the social order. The "connecting mechanism" of inverse parallelism between the period of intellectual/aesthetic expansion and territorial/political expansion is found in the relationships among wealth, energy, and human values and behavior outlined in previous chapters. Political and territorial priorities offer attractive opportunities for satisfying materialistic ambitions, which trump intellectual and aesthetic priorities for limited national wealth. Kroeber's profile follows the same symmetrical patterns of the rate of cultural productivity, which peaks, entering socioeconomic stagnation, followed by a cultural deterioration phase. This corresponds to the transition from the intellectual and aesthetic growth period of altruistic values during early economic prosperity to the territorial and political growth period during an era of more sensate values and increasing cultural deterioration.

As will be outlined in the next chapter, the 4,000-year history of the Graeco-Roman Western civilizations consists of three cycles of

Kroeber's approximately symmetrical profile of "growth and decline." This will demonstrate that a society undergoes continual change from its initial growth stage and eventually faces a sociocultural transition that either rejuvenates a deteriorating rate of cultural productivity or results in cultural ossification or collapse.

The mechanistic-thermodynamic principle applied to an organized culture provides a scientific basis for Sorokin's views of societal change:

> **Any system which is, during its existence, a going concern which works and acts and does not remain in a state of rest, in the literary sense of the word, cannot help changing just because it performs some activity, some work, as long as it exists. Only a system which is in an absolute vacuum at the state of rest and is not functioning can escape change under these conditions.**[40]

A society in socioeconomic decline may be able to reverse the negative rate of cultural growth, but the issue is whether the actions are timely and sufficient to counter the momentum of disintegration and avoid the next phases of stagnation and collapse. The net rate of cultural progress is a balance between the positive aspects contributing to growth and the negative aspects of cultural degradation. Obviously, the duration and severity of a net negative rate of growth (i.e., deterioration) not only affects the population's current level of prosperity, but also impacts the ability to ultimately reverse the downward trend. Toynbee describes the phases of the transition from growth through various stages of disintegration. The cession of growth is referred to as a "breakdown" that evolves into a state of "petrifaction," leading to three phases of disintegration: times of trouble, universal state, and interregnum.[41] These phases will be fully discussed in the next chapter. Toynbee traces the sequence from the growth to no-growth interface ("breakdown"), indicative of a gradual reduction in the rate of growth approaching zero, through three stages of net negative cultural growth or "disintegration," each representing a more dysfunctional society.

It should be emphasized that a positive *net rate* of cultural growth declining toward and approximating a *net rate* of zero is quite different from a society with *zero socioeconomic activity*. A net rate of growth approaching zero signifies an approximate balance of positive and negative contributions to cultural growth or productivity for a large multifaceted dynamic system. In order for cultures to exist, they inherently must undergo continuous socioeconomic activity; zero activity would imply the absence of both positive and negative elements of cultural activity. This is not possible; for as long as members of a society survive, activity will occur, producing some combination of positive and negative socioeconomic outcomes and consequences.

> **We have failed to find the immediate object of our search, a permanent and fundamental point of difference between primitive societies and civilizations, but incidentally we have obtained some light on the ultimate objective of our present inquiry: the nature of the geneses of civilizations. Starting with the mutation of primitive societies into civilizations we have found that this consists in a transition from a static condition to a dynamic activity; and we shall find that the same formula holds good for the emergence of civilizations through the secessions of internal proletariats from the dominant minorities of pre-existent civilizations which have lost their creative power. Such dominant minorities are static by definition; for to say that the creative minority of a civilization in growth has degenerated or atrophied into the dominant minority of a civilization in disintegration is only another way of saying that the society in question has lapsed from a dynamic activity into a static condition.**[42]

Toynbee's recognition of the absence of any "fundamental point of difference between primitive societies and civilizations" supports the view that both mature and immature societies function according to the same fundamental processes requiring "a transition from a more static condition to greater dynamic activity." Note that the mechanisms and energy considerations are the same for both, so that the "fundamental point of difference" is the degree of cultural sophistication and integration or stratification of organizational and functional processes. Additionally, he states, "That the creative minority of a civilization in growth has degenerated or atrophied into the dominant minority of a civilization in disintegration is only another way of saying that the society in question has lapsed from a dynamic activity into a static condition." His terminology of "static condition" is more precisely defined according to mechanistic-thermodynamic principles as the condition where the net rate of cultural growth approaches zero, as the contributions to growth are about equal to those of decline.

Toynbee's emphasis on the necessity for a society to be creative in responding to the continuous challenges presented to it and thereby creating momentum capable of generating societal gain is consistent with the view of periodic increases and decreases in the net rate of societal growth, as previously described. In any given time period, many related and unrelated factors may result in an increase or a decrease in what is generally viewed as societal gain or progress. It is noted that Kroeber describes these same irregularities in the patterns of cultural growth as "pauses, lulls, regressions, or barren intervals" and cautions that "some of these productive periods might not to be reckoned as separate cultural growths rather than as pulses or phases of one growth."

Toynbee observes, "This alternating rhythm of static and dynamic, of movement and pause and movement, has been regarded by many observers in many different ages as something fundamental in the nature of the Universe."[43] As such, it should not be surprising that "something fundamental in the nature of the Universe" would consist of mechanisms and properties that conform to thermodynamic principles. One would also expect that if "a new civilization is generated through transition of a society from a static condition to a dynamic activity," that such a society, defined as an open thermodynamic system, would be continually adding and absorbing processes and material resources, including energy.

The historic civilization studies literature appears to leave a number of major issues related to cultural transitions unresolved. How do the authors specifically and *functionally define* a society's status described as being in "a static condition" or undergoing "dynamic activity"? What are the parameters or characteristics that are measured or evaluated in arriving at such conclusions, and how are they related to societal growth processes? What are the specific processes and related driving forces that establish such circumstances as "ossification" and "dynamic activity" that the authors characterize with descriptive labels as representing various societal conditions and phases?

These questions lead to a number of comments. First, it would appear that various qualitative descriptions of developmental conditions and characteristics related to socioeconomic transitions found in the literature are conceptually consistent with and supported by mechanistic-thermodynamic principles as defined in this work. Second, it is reasonable to consider such circumstances as "a static condition" and "ossification" resulting from discontinuities over time in the rates of cultural development due to changes of dominant socioeconomic functions, conditions, and processes. Third, changes in socioeconomic subsystems such as the addition of new economic processes, new technologies, or new sources of wealth-energy capital may supply the "dynamic activity" necessary to jump-start a stalled social order. Fourth, long-term societal growth, rather than being considered a series of functional mutations, may simply be fluctuations from generic cultural processes and varying system conditions that alter the rates and outcomes of the society's operational functions. Finally, these principles explain such observed societal phenomena as the fluctuations noted by Toynbee of an "alternating rhythm of static and dynamic and of movement and pause and movement."

However, human decision-making and management functions associated with a culture's "dynamic activity" must be considered in a different context than production processes governed by the precision of the science of mechanistic-thermodynamic principles. A broad range of competencies, functioning in a very competitive and unpredictable environment, brings an appreciable degree of unpredictability to basic cultural processes. Society pursues many desirable and undesirable goals that are subject to the limitations of human intellectual capabilities and moral frailties, in addition to the restrictions and controls of Mother Nature's laws. People are society's agent of change, and the human element with its strengths and weaknesses must be recognized as a primary manager of its "dynamic activity."

The expression of a group's self-determination to accomplish societal goals encompasses many individual and group characteristics that reflect human attitudes and behavior including the quality of leadership, a common purpose and agenda, a strong work ethic, integrity, and civility, as well as individual and societal vitality. Other factors affecting the group's efforts include unforeseen opportunities and events, geography, external friends and foes, new technologies, and limited wealth-energy capital.

Toynbee emphasizes the importance of a *society's self-determination* or self-articulation and creativity to confront and overcome a continuous series of ever-present challenges, which provide a cultural momentum, referred to as a "repetitive, recurrent rhythm." The driving force or the vitality of a society required to achieve this momentum is signified as "élan vital, ... which carries the challenged party through equilibrium into an overbalance which exposes him to a fresh challenge and thereby inspires him to make a fresh response in the form of a further equilibrium ending in a further overbalance, and so on in a progression which is potentially infinite."[44] Toynbee's treatment of cultural momentum, equilibrium, and "overbalance" has been previously described and illustrated in thermodynamic terms by the *dynamic equilibrium of an open system.*

The potential union of Toynbee's elements of creativity and "élan vital" defines "dynamic activity" sufficient to disrupt an approaching "static condition" and carry the system "through equilibrium into an overbalance," enhancing the net rate of cultural growth. This also pertains to human characteristics such as motivation, inspiration, persistence, and dedication.

Toynbee's description of the ordeal faced by the city-states of ancient Greece between the years of 725 B.C. and 325 B.C. illustrates how human factors are able to guide developing societies along differing pathways.[45] The critical issue for the citizens of ancient Greece was fundamentally that of an imbalance between the size of its population and its ability to feed and sustain the population; a wealth-energy resource problem. One city, Athens, met the challenge by creating and adapting to new and innovative business practices that involved, for the first time on a large scale, the systematic manufacturing and exporting of specialty goods. This new endeavor established an economic specialization that had long-term constructive ramifications for the quality of life in Athens. Greater economic prosperity resulted from a renewal of self-determination that was outwardly expressed by the adoption of innovative ideas, sound business practices, and bold actions that addressed an existing threat and created new opportunities capable of resolving critical issues. New and more sophisticated operational concepts and techniques were implemented that brought cultural rejuvenation to Athens.

Other examples over the centuries attest to the significance and the dynamic nature of self-determination and creativity in producing a "repetitive, recurrent rhythm" and, as described by Toynbee, the ability "to grow a civilization.... Self-determination is the criterion of growth."[46] Conversely, the loss of a group's self-determination will deter further progress of a successful social order and is capable of becoming a contributing factor to cultural deterioration. This includes psychological consequences affecting innovation, creativity, and self-confidence, further illustration of *the societal entropy effect* that may develop from an initially successful socioeconomic era.

Toynbee makes reference to successful growth being the work of the "creative personalities and creative mimesis."[47] The author considers mimesis or imitation as "a generic feature of all social life" that appears in primitive societies; "imitation" is found in the living elders and in historic cultural beliefs and expectations. On the other hand, Toynbee contends that a maturing civilization seeks such imitation not so much to honor the past but to gain a more forward-looking viewpoint of what the culture might become. This places a premium on leadership and group acceptance of a vision of future strategies, priorities, and initiatives that are not only creative, but also capable of instilling hope and inspiration for developing a more prosperous future. A continuous infusion of such creativity and morale building within a society invariably generates "new dynamic forces" capable of affecting a "whole existing set of institutions, and in any actually growing society a constant readjustment of the most flagrant anachronisms is continually going on." The new forces "operate in two diametrically opposite ways simultaneously, ... perform their creative work ... at the same time they also enter, indiscriminately, into any institutions which happen to lie in their path."[48] Thus, as previously presented, such a characterization of the processes of maintaining the human existence is consistent with an open thermodynamic system that undergoes continual disruption of its dynamic equilibrium condition due to fluctuations in the nature, conditions, and rates of the intrinsic processes that constitute and support a society. Such changes of a functioning system may add to or detract from societal momentum, resulting in significant as well as insignificant adjustments of the direction, rate of advancement, and ultimate fortunes of a society.

Given our emphasis on the proper role of wealth-energy resources and the mechanisms of the processes of human cultural existence leading to sociocultural transitions, the following summary points are noteworthy. First, generic societal processes face a continuous series of challenges, each of which may, or may not, receive a response capable of contributing to societal progress and a desirable long-term sociocultural transition. Second, in addition to natural resources providing the energy for dynamic activities, inherent human capabilities and physical and mental labor must overcome challenges and create solutions to promote prosperity. Third, successful societies possessing a high level of self-determination and creativity and altruistic human behavioral characteristics must construct and implement successful socioeconomic systems. Fourth, the literature of societal development generally ignores thermodynamic principles involved in a society being driven or fueled by a continuous source of wealth-energy capital. A continuously increasing consumption of wealth-energy resources per capita eventually becomes unaffordable, necessitating inevitable cultural transformation, including cultural collapse. Fifth, a society's depletion of wealth-energy capital has a cascading, multiplying effect on a community's social, economic, and political life, undermining its structures and functions. This may be illustrated by Toynbee's historical observation: "Most often geographical expansion is a concomitant of real decline and coincides with a 'time of troubles' or a 'universal state'—both of them stages of decline and disintegration.... The history of almost every civilization furnishes examples of geographical expansion coinciding with deterioration in quality."[49] Toynbee's phrase "times of trouble" is a metaphor for major socioeconomic instability as a culture faces further deterioration; the term "universal state" indicates a society in a last desperate stage of even greater dysfunction.

Geographical expansion provides an illustration of the interrelationships among different societal variables and mechanisms associated with Toynbee's "time of troubles" that produces very different types of outcomes including the production of intense societal discord. Extreme geographical expansion is normally associated with costly militarism, which has major social, economic, and political consequences for both the invading and the invaded societies. Toynbee describes such activity as a "perversion of the human spirit into channels of mutual destruction" and notes "the most successful militarists become, as a rule, the founder of a universal state" (i.e., a transition in the latter stage of disintegration).

Significant military expenses accompanying extensive territorial expansion are often justified by supposed moral and religious ideologies and missions. However, the acquisition of newly discovered or conquered wealth is usually a major objective. Such imperialistic adventures are often motivated by a critical need for additional wealth-energy capital to survive financial insolvency and the final stages of cultural collapse. Historically, such wealth has not normally benefited the members of the expansionist society for purposes advantageous to a majority of its population. As an example, a significant majority of the wealth taken from the Americas by Imperial Spain in its waning years was to repay foreign governments that provided military services and materials in support of the empire. Ultimately, Imperial Spain, during its last stages, could not acquire new wealth fast enough and in sufficient quantities to meet the increasing expenses of the empire approaching chaos.

In the course of history, such expansionist aspirations have resulted in negative cultural ramifications, creating major societal issues including massive human suffering and a general deterioration of the social order. Militarism, described by Toynbee as "by far the commonest cause of the breakdown of civilizations during the last four or five millennia,"[50] involves the consumption of large quantities of national wealth, massive loss of life and suffering, and a fatal corrosion of once prosperous civilizations. Interestingly, Toynbee doesn't appear to associate his self-described "times of trouble" and the socioeconomic impact of insufficient wealth-energy resources with his identification of an "alternating rhythm of static and dynamic" societal activity.

The characteristics and dynamics of cultural transitions contained in this chapter provide a preparation for an in-depth review of the etiology of the great civilizations (i.e., from a successful growth phase to stagnation, deterioration, and ultimate ossification).

Chapter 7
References and Notes

1. Sorokin, Pitirim A. *Social and Cultural Dynamics*. Boston: Porter Sargent. 1970. 630.
2. Ibid., 633.
3. Ibid., 634.
4. Sorokin, Pitirim A. *Social and Cultural Dynamics.* New Brunswick, N.J.: Transaction Publishers. 1991. Introduction.
5. Sorokin, Pitirim A. *Social and Cultural Dynamics*. Boston: Porter Sargent. 1970. 635.
6. Ibid., 633.
7. Ibid., 638.
8. Ibid., 639.
9. Ibid., 638.
10. Ibid., 633.
11. Ibid., 639.
12. Kroeber, A. L. *Configurations of Culture Growth*. Berkeley: University of California Press. 1944. 12.
13. Melko, Matthew. *The Nature of Civilizations*. Boston: Porter Sargent. 1969. 33-35.
14. Toynbee, Arnold J. *A Study of History*, ab. Sommervell, D. C. New York: Oxford University. 1957. I-VI: 192.
15. Ibid., 194.
16. Ibid., 198.
17. Ibid., 211.
18. Ibid., 211-212.
19. Kroeber, A. L. *Configurations of Culture Growth*. 5.
20. Ibid., vii.
21. Ibid., 762.
22. Ibid., 769-770.
23. Ibid., 822-823.
24. Ibid., 826.
25. Ibid., 838-839.
26. Ibid., 773, 841.
27. Toynbee, Arnold J. *A Study of History*. II: Editor's note.
28. Ibid., 49.
29. Ibid., 3.

30. Ibid., 11.
31. Ibid., 3.
32. Ibid., 241.
33. Ibid., 363-364.
34. Braudel, Fernand. *A History of Civilizations*. London: Penguin Press. 1992. 11.
35. Ibid., 15.
36. Ibid., 365.
37. Kroeber, A. L. *Configurations of Culture Growth*. 773.
38. Ibid., 805.
39. Ibid., 817.
40. Sorokin, Pitirim A. *Social and Cultural Dynamics*. 1970. 634.
41. Toynbee, Arnold J. *A Study of History*. II: 360.
42. Ibid., 50.
43. Ibid., 51.
44. Ibid., 187.
45. Ibid., 4.
46. Ibid., 207.
47. Ibid., 276.
48. Ibid., 280.
49. Ibid., 190-191.
50. Ibid., 190.

Chapter 8
Periodicity of Human Advancement: Ancient Greece to the Twenty-First Century

> "Some cultures, like the Graeco-Roman and the Western, have been able to make such a shift several times [restoration]; some others could not. The first cultures continued to live and to pass through the recurrent rhythm studied; the others either perished and disappeared, or were doomed to a stagnant, half-mummified existence, with their hollow and narrowed truth, reality, and value becoming a mere 'survival' or 'object of history,' instead of being its creative subject. Such cultures and societies turn into mere material for others—more creative and alive—cultures and societies."
>
> Pitirim Sorokin[1]

Previous chapters have presented a thermodynamic perspective of the dynamics and sociocultural transitions associated with relatively short-term cultural trends. This chapter contains specific concepts, characteristics, and properties comprising and linking recurrent stages or phases of 4,000 years of human social history. The mechanistic-thermodynamic principles controlling a social order's initial stages of creative and prosperous socioeconomic growth and advancement correspond to the same dynamic forces responsible for cultural deterioration and for the cyclical periodicity of cultural progress and failure since the Minoan civilization. This very long-term history, consisting of a triple cycle of the sequential stages of genesis to cultural growth and disintegration, relies on the same mechanisms and energy transfer principles inherent in the everyday processes of human existence. Additionally, this periodicity is consistent with Joseph Tainter's thesis that society's inevitable, well-intended but fatal overinvestment in societal complexity (i.e., "hierarchical specialization") is the best explanation of collapse.[2] More specifically, the periodicity of the cycles of recurrent advancement and deterioration of Graeco-Roman Western civilizations from 2000 B.C. to A.D. 2000 will be shown to be multiple cycles of Toynbee's four stages of genesis to deterioration. The elements of wealth, energy, and human values and behavior are representative of the multitude of variables affecting this rise and fall of major civilizations since the time of ancient Greece.

One of the major figures of civilization studies literature, Oswald Spengler, describes culture as being "morphological," possessing the form and structure of an organism with predictable life cycles, super life forms, and organic characteristics.[3] Others, such as Giambattista Vico, Nikolai Danilevsky, and Friedrich Nietzsche, share this viewpoint.

Toynbee, while rejecting Spengler's "organism" model, does view a society as "completing its life course." Matthew Melko states: "Civilizations are living systems; men are living systems too.... If they do develop, it is because something inside them drives them to it, or because qualities they have are allowed to develop by the culture in which they live."[4] These authors and others, while assigning a life-like nature to the existence of cultures, do not address obvious mechanisms and energy requirements associated with all living entities. Others, for example, the German philosopher Ernst von Lasaulx, note that developing societies follow sequential patterns that require a lifetime's quantity of energy, analogous to the life span of animal and plant species.[5] Thus, although lacking in appropriate details or based on any derivation, the similarity between the life-dependent processes of the human body and that of a social order has been noted by various authors. This raises the question of a *commonality of requirements for the life cycles of the animal and plant worlds with those of cultural life cycles and specifically the periodicity of the Graeco-Roman Western civilizations*. The processes of cultural decline, during the last 4,000 years, have repeatedly led to cultural collapse and ultimately to a gradual rebirth (e.g., the collapse of Greek civilization and the emergence of Roman civilization).

Therefore, the parallel between the requirements of animal life and the *living systems of a society* will be developed by pursuing the sequence of mechanistic processes and energy requirements at the cellular and organism levels of the animal world as well as those for primitive and mature civilizations. Both systems rely on structure and function, which define operational processes consisting of mechanisms that require material resources, including energy. Thus, it is desirable to identify these mechanisms, wealth-energy resource requirements, and other major variables affecting primitive society formation and subsequent evolution through mature civilizations to their ultimate decay and potential rebirth in a new, more primitive form.

Spengler views social orders as "organisms," whereas Toynbee describes them as capable of "completing its life cycle":[6] The underlying mechanistic-thermodynamic factors interwoven with human decision-making processes create properties that characterize a society in some relative phase of prosperity or decay. However, without any apparent rational foundation, authors ascribe descriptive characteristics to societies as *organic in nature* and *living systems*. Accordingly, it is useful to identify and compare the properties, characteristics, and scientific principles attributable to animal and plant life with those of cultural growth, decay, and rebirth recognized by the civilization studies literature.

The systems of the human body (e.g., the circulatory system) represent physiological functions that maintain the processes of life. Survival is the body's most important objective and begins at the cellular level with the necessity for a healthy cellular environment, or a condition defined as homeostasis. More than a century ago, the French physiologist Claude Bernard discovered that human cells survived best in a prescribed healthy environment including temperature, pressure, and chemical content. Later, another physiologist, Walter B. Cannon, applied the term *homeostasis* to this relatively constant cellular environment required by a healthy body. The term *homeostasis* is derived from *homeo*, meaning the same, and *stasis*, meaning standing. *Standing or staying the same* is the literal meaning of homeostasis. However, homeostasis, as applied to a dynamic system of cellular activity, is not intended to imply inactivity, as cellular conditions are continually undergoing change. At the cellular level, a fluctuating range of prescribed conditions are expected during normal, active bodily functions, but within healthy ranges. Consequently, staying within a range is considered as a normal condition. Thus, *"staying the same"* translates into the individual remaining alive, healthy, and progressing through one's own unique life cycle, assuming that one is able to acquire and maintain the appropriate required resources. This concept of homeostasis requires that the body, at the cellular level, be maintained and function within an acceptable range of environmental parameters in order that the body's interrelated systems may adequately function and sustain life processes. Failure to do so may reduce the body's ability to respond to life-threatening diseases and environmental threats. The individual's attitude, values, and behavior in utilizing society's resources, opportunities, and support systems to maintain a healthy cellular environment, particularly as it affects the body's major organs and their functions, will influence the individual's quality of life and longevity.

For an analogous cultural environment, the concept of *cultural homeostasis* would imply that a society must also possess *morphological* structure appropriate to a societal "organism," as ascribed by Oswald Spengler, and also possess functional operational systems capable of providing a healthy and stable environment. The parallel with the cellular environment recognizes the need of the cultural environment to maintain a stable population of emotionally and socioeconomically "healthy" people, happily participating in the social, political, and economic processes of a society's everyday existence. Conceptually, *"standing or staying the same" for a cultural environment also translates into a dynamic environment of a social order remaining alive, healthy, and progressing through its unique life cycle, assuming it is able to acquire and maintain the appropriate resources.* The good news is that the human body, governed by Mother Nature's principles, will perform its required complex functions very efficiently and effectively. The bad news is that society's functional processes, controlled by an often-flawed human judgment, are much less effective as a result of frequently unwise, arbitrary, and self-serving behavior and decision-making.

Consideration of the fundamental mechanisms by which humans and social orders each consume energy provides another parallel for human life and cultural existence. The processing of the sun's stored energy during human consumption of food and during a society's consumption of fuels for socioeconomic processes relies on similar classifications of chemical reactions that conform to the same laws of thermodynamics. All processes follow a spontaneous, directional mechanism that may achieve intended useful outcomes as well as the unintended but inherent generation of waste materials and dissipated energy.

At the human cellular level, the sun's energy is stored as carbohydrate food material, which is consumed and converted into sugars. A chemical process converts sugars into a variety of products including a high-energy-content molecule, adenosine triphosphate (ATP). Subsequently, the ATP combines with water and is hydrolyzed to form adenosine diphosphate (ADP), also releasing energy that provides fuel for important cellular reactions. Included among these subsequent reactions is the production of additional energy-rich ATP molecules. It should be noted that *the body is capable of producing more than one energy-rich ATP molecule for each ATP molecule that is consumed by the body from its food source,* an efficiency accomplished by Mother Nature that humans do not duplicate with economic production processes. The physiological economics of this process involves the body indirectly receiving the sun's energy and very efficiently producing new energy resources in the form of ATP, which the body utilizes to insure continued performance of its functions. In economic terms, this cellular performance would be represented as a favorable "marginal return on investment" that provides the human machine with energy and continuing human development until body parts wear out or are abused, or disease intervenes.

It is noted that the *marginal return on investment* of wealth-energy capital utilized by society's processes is inferior to the effectiveness of the human cell. This is illustrated by the use of coal and iron to produce steel for its ultimate use in generating new wealth-energy capital. Successful human cellular function and societal socioeconomic prosperity both depend on the ability of their respective processes to efficiently consume and effectively use energy and other resources for desired purposes. Unfortunately, society's decision-making processes involving wealth and energy consumption do not always result in beneficial socioeconomic outcomes. But fortunately, human cell functions benefit from Mother Nature's superior wisdom, precision, and discipline.

Human cellular functions rely on self-reproduction, or biosynthesis, involving a series of energy conversions that require elaborate instructions from a self-contained, coded hereditary information system. Each duplicated cell inherits and retains a communications and information system for use by future generations in the cell replication and growth process. An analogous societal *hereditary information system* provides less precise and more divergent multigenerational cultural guidance and management, particularly in the more primitive societies. Toynbee specifically attributes "creative mimesis" to the growth and success of more primitive societies: "Growth is the work of creative personalities and creative mimesis." He considers "mimesis" or imitation as "a generic feature of all social life ... both in primitive societies and in civilizations, in every social activity,"[7] whereby societal elders or religious sects rely on historic cultural ideals, beliefs, and successful experiences to manage the affairs of a society. Recall from the previous chapter Morton Fried's emphasis on *symbolic learning* which, by definition, is not based on either direct stimuli or original situations but on complex symbols pasted on from one generation to the next and capable of instilling learned behaviors. This cultural hereditary information system intended to guide members of a society serves as a functional analog to the genetic information system of human cell functions. While both systems evolve over time based on what works best (i.e., survival and adaptation), the societal system, as a software type of information system, is quicker to adapt than the hardwired information system of the human cell. Nevertheless, the human body and the social order are both guided by information systems whose effectiveness is subject to the probabilities of mechanistic events and the variables and conditions of cultural growth and decay processes.

Diseases such as cancer and diabetes, as well as self-inflicted injuries and illnesses influenced by environmental conditions and questionable human judgment and behavior, reduce human effectiveness and often result in premature deaths. Societies also face a constant threat from *societal diseases and self-inflicted debilitations* such as infectious greed, corruption, and wars that are capable of producing societal degradation and collapse. Thus a society, as with the human body, possesses an "organic" or "morphological" nature and faces a longevity that depends on the efficiency and effectiveness of its structures, functions, and wealth and energy consumption as well as on the use and abuse of human sensibilities and opportunities. Additionally, it is recognized that historical events and occurrences, precipitated by human decisions and acts of nature, are influenced by mathematical probability and chance, often referred to as acts of God. Famine, disease, weather conditions, and improbable chance occurrences and opportunities have, in the course of history, paved the way for significant and long-lasting ramifications affecting the rise and fall of individual cultures and the course of human history.

From this review of the nature, structure, organization, and functions of the human body and cultural systems, it is clear that common scientific principles apply to fundamental life processes of both systems. In addition to this parallelism, it is noted that there exists in nature a recognized annual renewal, or life cycle, for plants and a normal human life span. Likewise, the functionally similar "organic" culture is considered by sociologists as "completing its life course," ultimately forming a new, more primitive cultural entity, the next cultural generation (e.g., the birth of Roman civilization from the collapse of ancient Greece).

Accordingly, the sequential cultural transitions leading to the completion of one life cycle of a civilization depend on a multitude of variables that gradually change with observable effects normally being within a longer time frame relative to the human life span. Civilizations exhibit minor, short-term, random positive and negative fluctuations of cultural productivity that lead to patterns, cycles, and transformations. The multivariable nature of societal growth, decay, and rebirth and the characteristics and properties of these transitions have historically been debated. While the existence or absence of such growth is normally based on arbitrary economic parameters that measure equally debatable definitions of financial success, it is recognized that related sociopolitical factors contribute to a broader concept of cultural achievement. Appropriately, the parameter of *cultural productivity* is adopted, representing an unlimited number of potential cultural variables capable of affecting the numerous cultural systems that generate outcomes that influence both progress and disintegration.

Cultural productivity is defined in terms of a society's success in achieving socioeconomic prosperity including security, social and economic stability, and a general sense of well-being for a significant majority of its members. A culture's productivity at any particular time represents the reality that both positive and negative consequences (i.e., progress and deterioration) are being simultaneously created within a social order from challenges, opportunities, and forces affecting human existence.

$$\text{Cultural Productivity} = \sum \text{Elements of Progress} - \sum \text{Elements of Deterioration}$$
$$\text{Equation 8-1}$$

The forces affecting cultural productivity and advancement may be imposed by the society itself, by nature, or from external sources.

These forces affect political and socioeconomic progress and include the following:
- Limitations of knowledge, creativity, initiative, and humanistic behavior
- Probability related to the activities of man and nature
- Availability of wealth-energy capital and other resources
- Quality, efficiency, effectiveness, and appropriateness of a culture's organizational and operational systems
- Competitiveness, objectives, attitudes, values, and actions of a society's external environment
- Spiritual doctrine, leadership, and practices
- Cultural heritage, leadership, values, practices, maturity, and morale
- Technological maturity, particularly dealing with economic productivity, transportation, communication, and military strength
- Available natural resources and the physical and geographic environment

The *net rate of cultural productivity*, a function of the social order's successful economic growth as well as its cultural deterioration, will ebb and flow over time depending on the significance of numerous ever-changing factors. Collectively, these dynamic variables represent forces producing both positive and negative effects on socioeconomic prosperity as well as contributing to long-term cultural transitions. The fluctuating nature of cultural productivity will be considered within the framework of mechanistic-thermodynamic principles, while also embracing important concepts presented by civilization studies authors, most notably Toynbee, Spengler, Kroeber, and Sorokin. Their views of civilization development and specifically the sociological aspects of the cultural maturation process are generally consistent with these energy-based concepts applied to societal fluctuations, transitions, and periodicity.

The cultural maturation sequence is represented by phases or stages in Figure 8-1, illustrating the gradual transition from a sustaining net positive rate of cultural productivity for a successful primitive social order to a net negative rate of productivity for a deteriorating mature civilization. This sequence depicts the progression from rapid, constructive advancement of Stage I, Genesis, through stagnation and ultimate cultural decay of Stage IV, Deterioration.

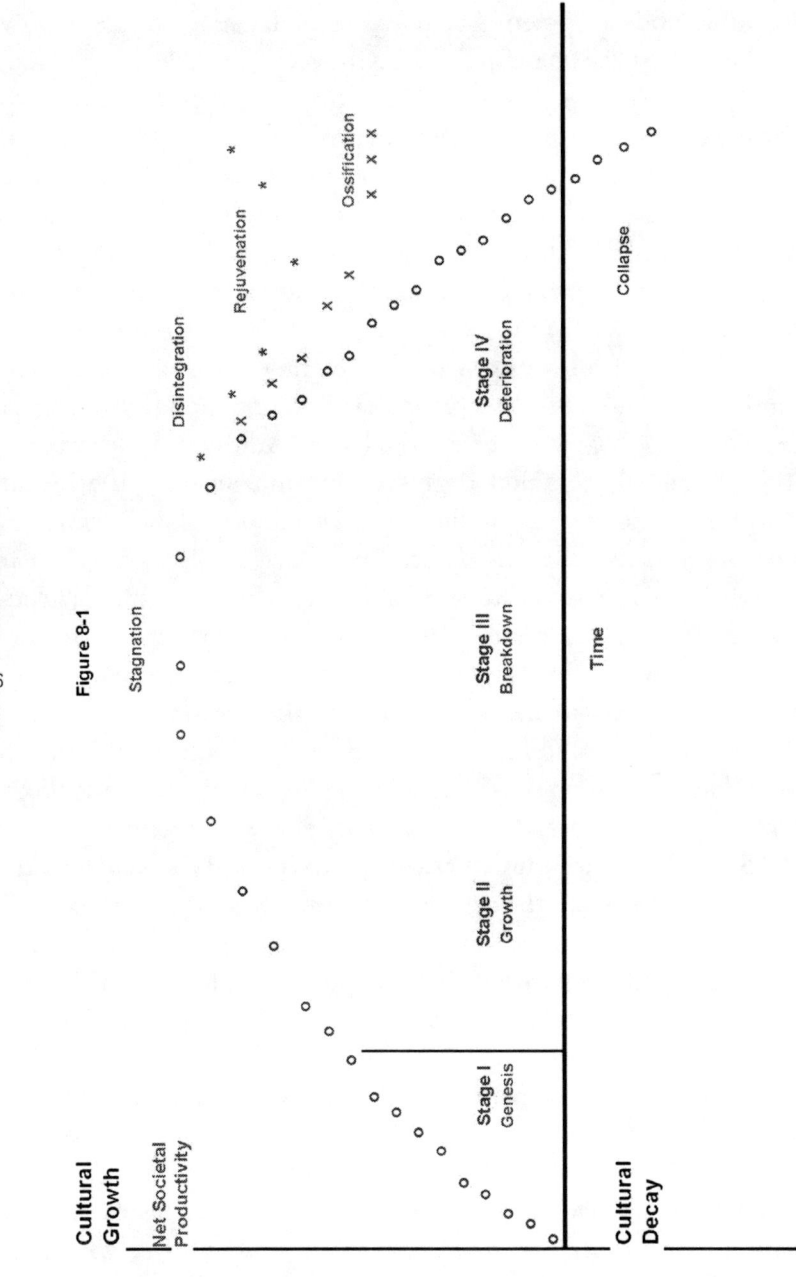

Figure 8-1. Etiology of Cultural Evolution

Stage I. Genesis and Initial Development: This stage represents the coalescence of the basic elements of a primitive society and the initiation of developmental activities at a relative rapid rate of growth. Historically, this phase relies heavily on renewable energy sources and on conservative energy consumption practices. Since a relatively small number of simplistic and efficient cultural operational systems are created and implemented during this phase, a minimal level of societal instability is created relative to a high level of socioeconomic achievement. The human values, attitudes, and behavior employed during this phase exemplify Sorokin's Ideational mentality of altruism, to be discussed later in the chapter.

Stage II. Adolescent Growth: In the early phase of this stage, significant cultural progress provides socioeconomic advancements that benefit a large segment of the population. A minimal level of societal disorder and dysfunction begins to accumulate from the inevitable cultural complexity that produces initial symptoms of increasing societal instability, detectable only by the more astute observer. The earlier portion of this phase consumes wealth-energy resources more effectively and achieves high productivity while consuming a relative small portion of national wealth for luxury items and other unnecessary goods and services. This constitutes a favorable *marginal return on investment in societal complexity* and future progress. However, during the late phase of this stage, while maintaining an increase in socioeconomic progress, *the rate of cultural growth or productivity* gradually declines.

Stage III. Maximum Cultural Productivity and Breakdown: This stage represents the zenith of cultural productivity that provides a prosperous economy and a satisfying sociopolitical environment for a considerable portion of the population. However, this stage is associated with a *minimal positive increase in the rate of productivity* (i.e., approaching a zero rate). This corresponds to the onset of cultural "breakdown" (i.e., stagnation), as defined by Toynbee, at which point the accumulative effects of societal entropy create massive socioeconomic stress and disorder for a large segment of the population. It should be emphasized that *this phase symbolizes the achievement of a culture's maximum productivity, but also represents a significant decline in the rate of productivity*. A declining net positive rate in productivity (cultural progress) passes through a minimum rate of zero in a transition to a negative rate of productivity (cultural deterioration). A zero rate of cultural productivity is the result of the dynamic equilibrium between constructive and destructive outcomes emanating simultaneously from abundant, complex processes of cultural progress and deterioration. Refer to Equation 8-1. This cultural status is often referred to as an *arrested civilization*.

Human behavior characteristic of this stage corresponds to the completion of a dominant shift in the cultural values from a more primitive, altruistic mentality to one that reflects more self-serving, individualistic, materialistic priorities, referred to as "sensate" by Sorokin.

Stage IV. Societal Deterioration: At this stage, the elements of cultural deterioration overwhelm those of progress. Thus, the social order completes the transition from a dominating environment of cultural growth through cultural stagnation to predominately cultural deterioration. Under the circumstances, the immediate prospect for the re-emergence of a net positive productivity is slim. Toynbee offers four potential escape attitudes for dealing with a disintegrating society: The *archaist* attempts to return to the past, the *futurist* mentally eliminates the present and jumps to a projected future, the *philosopher* withdraws into the soul, and the *transformer* seek a more positive future usually via spiritual pathways. Of the four, most authors agree that the spiritual mentality and organized religions provide the most effective catalyst for creating positive societal transformations and an escape from a deteriorating cultural environment.

Figure 8-1 illustrates that those societies in a state of deterioration or negative productivity may (1) undergo rejuvenation, returning to a positive rate of growth, (2) ultimately collapse from a continuing negative productivity and cease to behave as a unique civilization, or (3) maintain a relatively slight positive productivity over time, eventually entering a "petrified," "calcified," or "ossified" condition.[8] The latter state, while constituting a sick culture, represents socioeconomic fluctuations between slightly positive and slightly negative periods of productivity over a prolonged period of time (i.e., a culture treading water).

Societal rejuvenation or a reprieve from cultural disintegration may result from such sources of hope and stimulation as new spiritual directions, cultural leadership and direction, creative economic initiatives, technical inventions, discoveries, and newly acquired wealth-energy capital. Toynbee places heavy emphasis on cultural "mimeses" or guidance being transferred in times of such social difficulty from that of human leaders to religion and God.

Depending on circumstances within and external to a civilization, there is no theoretical limitation to the number of times a civilization is capable of converting a stagnating, nonproductive condition into a rejuvenated, productive society. Realistically, slight, periodic negative fluctuations of productivity (i.e., blips) are more easily rectified than significant, continuous declines, which may gain momentum in a roller-coaster fashion, depleting the hope and confidence of the population and precipitating cultural collapse. Socioeconomic fluctuations are a normal consequence of societal existence; however, it is prudent to aggressively manage the growth-to-deterioration balance, as shown by Equation 8-1, in order to prolong the Stage II Growth stage. Thus, periods of societal decline are inevitable, but not all societies are destined to disappear, but perhaps just become stagnant or petrified. Toynbee points out that a civilization may ultimately decline after a sequence of ups and downs (i.e., a "rout" followed by a "rally"). His mechanism of uncharacteristic ups and downs of ancient Egypt illustrates an evolving social order undergoing periodic fluctuations of growth and decay based on a broad array of dynamic functional variables. Ancient Egypt experienced a long period of growth, which eventually deteriorated via sequential mechanistic steps that Toynbee refers to as "the rhythm of growth and the rhythm of disintegration" (i.e., the "times of trouble," "universal state," and "interregnum"[9]). Ultimately, Egypt was rejuvenated and became more vital for about 2,000 years prior to becoming nonexistent as a unique civilization.

The continuum of cultural evolution from Stage I, Genesis, through Stage IV, Deterioration, illustrated by Figure 8-1 projects the three possible ultimate outcomes of rejuvenation, ossification, and collapse. The following key points are emphasized:

Other than an expectation for a given society to generally follow the sequential stages of Genesis, Growth, Breakdown, and Deterioration, predictions of a more precise nature regarding societal evolution are extremely uncertain. A large number of only moderately independent variables affecting cultural change are influenced by mathematical probability and chance events as well as by human impulsiveness and imperfect ethics, values, and behavior. Thus, the predictability of specific cultural change in a given time frame measured in short-term patterns of hundreds of years or in longer-term patterns of periodicity is analogous to that of rolling dice.

However, the observed characteristic profile of the four stages of cultural evolution depicted in Figure 8-1 is a phenomenon resulting from the normal functions of one "life cycle" of human social history. The sequence of these four stages generally represents increasing societal advancement and national wealth-energy consumption per capita per year, inherent in supporting the continuous creation of additional socioeconomic and political functions, bureaucracies, and social and economic opportunities and conflict.

It should be noted that from the latter part of Stage I through Stage III in Figure 8-1, there exists a *gradually declining rate of cultural productivity* (i.e., a decreasing slope of the plot of productivity vs. time). Significantly, this time frame parallels an increasing level of cultural complexity and a decreasing *marginal return on investment* in society's efforts to create and maintain a positive socioeconomic momentum.

The shift to increasing complexity, undertaken initially to relieve stress or realize an opportunity, is at first a rational, productive strategy that yields a favorable marginal return. Typically, however, continued stresses, unanticipated challenges, and the costliness of sociopolitical integration combine to lower this marginal return.[10]

Consequently, until reaching Stage IV, Deterioration, cultural productivity increases but concurrently with a systematic, gradual decline in the rate of growth, representing a declining rate of return on society's investment of national wealth in its advancement. This corresponds to a declining amount of regenerated wealth-energy capital, which restricts further investments in socioeconomic progress.

It is the inevitable consequence of society's life-supporting processes and the thermodynamic principles of energy consumption that all known civilizations eventually experience a sequence representative of these four stages of cultural development. Human behavior and cultural circumstances only influence the specific manner, time frame, and details of the cultural maturation process leading to the ultimate fate of cultural deterioration. The question is, why do social orders invariably follow this general configuration from genesis through disintegration in a growth or productivity pattern that approximates a normal distribution function? Additionally, what factors and conditions enable a collapsing society to improve its circumstances and achieve rejuvenation? To address these questions, a number of important concepts presented in previous chapters will be summarized for emphasis.

First, a civilization's economy increasingly achieves a lower effectiveness (i.e., a lower ratio of wisely and efficiently utilized wealth-energy capital relative to the total amount consumed) as its financial burdens escalate, due to increasing societal complexity and a high priority is placed on production of nonessential goods and services. The consequence is referred to as a *declining marginal productivity*; that is, too little capital is being invested in growth potential capable of generating new forms of wealth that are increasingly necessary to maintain socioeconomic growth. Joseph Tainter points out that "complex societies with large, well-developed economies have historically been able to sustain only rather inferior rates of economic growth."[11] Also, he notes: "There is in complex societies a recurrent and seemingly inexorable trend toward declining marginal productivity in hierarchical specialization.[12] Additionally, Galbraith's concept of "social balance" necessary for the realization of the "good society" reflects the same point of view.[13]

Second, a social order constitutes an open thermodynamic system, which, by definition, permits materials, including wealth-energy resources, to enter the system from the surroundings at any desired time. From society's vantage point, this allows great flexibility in altering the material content, processes, and conditions of society's operational functions, assuming that society is able to acquire the necessary materials and is capable of effectively managing the processes. Thus, conceptually, a society's essential and nonessential production is fueled and limited by its internal wealth-energy capital and from any external sources it may acquire and bring into the system. Mature civilizations that experience prosperity typically devote increasing portions of resources to nonessential, as opposed to essential, production. *The inherently declining trend of a society's marginal productivity combined with the requirement for continuously increasing wealth and energy consumption* to maintain the momentum of social advancement and to meet expanding human expectations is not sustainable indefinitely.

Third, recall that the multitude of societal functional processes are highly integrated, stratified, and dynamic and are in a continual state of change due to people's ability to alter and manipulate operational systems. As such, these societal processes are each continuously seeking a redefined dynamic equilibrium state due to incessant adjustments and unplanned systemic disruptions. Society and nature continually bring intended and unintended constructive and destructive elements into the processes of a social order that affect the rate of productivity, short-term outcomes, and long-term, often irreversible consequences.

Fourth, these production processes achieve a *thermodynamic efficiency* of less than 100 percent; consequently, only a fraction of the energy consumed actually accomplishes society's useful work. This thermodynamic efficiency combined with some level of human capability to successfully manage the energetics of useful work will establish an overall societal *economic effectiveness*. The wealth-energy capital *imported* and consumed by a mature civilization has historically been acquired through commerce, but quite often by conquest and other hostile, often desperate, sociopolitical and military tactics to avoid financial insolvency. Internal energy is *extracted* from within a society at a cost to the environment and to the quality of life of its membership. Thus, the wealth and energy acquisition and consumption processes intended to advance a social order inherently generate many forms of societal entropy and consequences detrimental to the interests of the social order.

The continuing effects of these science-based principles on cultural evolution and major transitions constitute contributing factors that explain why the positive rate of cultural productivity from genesis through disintegration invariably follows a general form approximating a normal distribution function, as illustrated in Figure 8-1. Essentially, the effectiveness of successful but limited socioeconomic processes from the genesis through the growth stages steadily declines. Despite rapid and productive societal expansion, there emerges during the late growth stage a significant depletion of financial, social, emotional, and natural and human resources and an inability to manage creeping, long-term, and escalating cultural dysfunction. Consequently, an associated socioeconomic and political chaos creates emotional and psychological stress as well as social and economic instability. The escalating cost of increasing societal complexity and the insatiable appetite of the masses for a higher social and economic achievement ultimately produces an unmanageable social order. Competitiveness, self-interest, and the lure of socioeconomic benefits breed incivility, a sense of entitlement, and less social responsibility and accountability. This describes Sorokin's Sensate Supersystem.

Ultimately, Stage III emerges, representing the "Breakdown" of the social order and its capabilities to cope with the frustrations, complexities, and wealth and energy requirements of its advanced structural and functional social, economic, and political systems. The society, whether ancient Greece or Imperial Spain, reaches a limiting societal entropy accumulation, despite the particular era's advancements of knowledge, technology, resources, and human capabilities; consequently, societal chaos, stress, and an escalating pace of cultural change becomes intolerable and an insurmountable challenge.

It is noteworthy that Toynbee links societal "disillusionment" with the observation that a "society that has become incurably divided against itself is almost certain to put back into the business of war the greater part of those additional resources, human and material, which the same business has incidentally brought into its hands." He illustrates, as a "clue" of the process of disintegration, the impact of the "money-power and man-power" accumulated from Alexander's conquest being ill spent on civil wars by his successors. The wealth gained from Roman conquests of the second century B.C. was consumed by civil wars; that is, "the spectacle of that division and discord within the bosom of a society."[14] Toynbee uses this history to illustrate "that the fundamental cause of the breakdowns which precede disintegrations is an outbreak of internal discords through which societies forfeit their faculty of self-determination."[15]

This raises these questions: First, what is the root source for this *disillusionment* (i.e., "an outbreak of internal discord"), and second, how can "internal discords" and loss of self-determination be construed as a "fundamental cause"?

Interestingly, a rationale exists for unfavorable sociological and psychological consequences resulting from mature civilizations utilizing a major portion of national wealth-energy capital for nonproductive uses (i.e., luxury goods and wars). Thus, "disillusionment" is a natural human reaction to a society's long-term inadequate and/or ineffective investment of national wealth in its socioeconomic system that negatively affects the daily lives of a large segment of the population. Disillusionment is one symptom of an infectious malaise created from the gradual loss of faith in a culture that is expected to provide unabated prosperity for all members but fails to do so. A classical example is the progressively greater wealth distribution inequities typically associated with economic prosperity of mature civilizations.

The root cause of Toynbee's disillusionment becomes more evident toward the end of Stage II Growth. Socioeconomic success inherently creates a continuous expansion of problems, issues, and challenges, all of which become increasingly greater in number, more complex, and more difficult to adequately address and financially afford. In order to maintain its positive productivity, a society must secure and maintain a continuous flow of wealth-energy capital and become more energy efficient and effective in its production capabilities. However, historically, civilizations in Stage II Growth increase energy consumption per capita per year, demonstrate less concern for energy conservation, and invest smaller portions of national wealth into the renewal of wealth-energy capital. In addition, if the population succumbs, in a substantial way, to the allure of an expanding menu of luxury goods and services associated with economic success, an even smaller proportion of wealth-energy capital becomes available to invest in the production of new wealth-energy capital. Consequently, history demonstrates that mature civilizations, assuming continual cultural expansion and prosperity, eventually become incapable of procuring sufficient wealth-energy resources within their own society and from external sources. Ultimately, socioeconomic disillusionment results from anticipated economic losses, a fear for physical survival, and a disappointment over the failure of political and spiritual leaders and their faith-based ideologies.

Stage II Growth evolves into Stage III, Toynbee's "Breakdown," as the degree of cultural dysfunction reaches the level of financial insolvency and fear for socioeconomic, if not physical, survival. Typically, there is little choice for society's leadership and its socioeconomic elite but to readjust priorities in attempts to restore social and economic stability at a basic survival level. Historically, such readjustments usually disadvantage the masses by increasing centralization, bureaucratic control, geographical expansion, and military activity. Such cultural priorities and the resulting hardships on the people usually exacerbate the decaying state of affairs, as infrastructure and public services expenditures receive lower priorities. If efforts to revive the vitality of the culture are not soon successful, harsh conditions will intensify and aggravate a stagnating condition, described by Toynbee as the "time of troubles." The culture soon fully assumes Stage IV Disintegration.

Toynbee's view of the nature of disintegration appears to be consistent with the concept that, from genesis through growth and into disintegration, cultural evolution conforms, in the mathematical sense, to a continuous multivariable function. However, Toynbee takes the position that the transitions from genesis, to growth, and to disintegration are each the result of "a new problem on its own account." His conclusion is based on, first, the "fact" that "a number of 'arrested' civilizations which had solved the problem of genesis had failed to solve the problem of growth." Second, "the fact that certain civilizations after breakdown have suffered a similar arrest but became revitalized subsequently entering on a long period of existence in a state of petrifaction or ossification." Toynbee concludes that since different transitions "suffered a similar arrest" and entered "a long period of petrifaction," disintegration is a new problem and not "a natural and inevitable sequel to breakdown."[16]

While one could agree with Toynbee that no particular "inevitable sequel" is mandated as a society passes into the breakdown stage, the same potential fundamental factors or "problems" continually face a society, albeit perhaps in a different form and intensity, at any stage of evolution. For example, insufficient, inefficient, and inappropriate use of national wealth-energy capital is historically a primary reason for the deterioration and collapse of mature civilizations. Hence, a wealth or energy issue that may face a society at any stage may not necessarily be a new problem, perhaps just a greater intensity or a new form of a typical, general issue. Toynbee offers the example of the Egyptiac Society at the breakdown phase, passing through the three phases of increasing degrees of disintegration: "This apparently moribund society then departed unexpectedly and abruptly, at a moment when it was apparently completing its life course." He considers this society as having uncharacteristically departed from "the standard pattern," somehow avoiding collapse and extending it existence for almost 1,000 years as an inert, arrested, and "petrified" state.[17]

An alternative explanation for the long "petrified" phase is that during this period, the society was successful in only barely overcoming historic socioeconomic problems but was able to periodically attain a slightly net positive cultural productivity, approximately balancing the elements of culture progress with those of failure. The culture is capable of just avoiding complete collapse via periodic, temporary improvements in the nature, efficiency, and quality of economic processes and/or resolving long-standing sociopolitical deficiencies, inaction, or abuses. In 1946, Toynbee cited the Far Eastern Society in China, the Jains in India, and others as examples of "various fossilized fragments of otherwise extinct civilizations."[18] However, by the twenty-first century, China and India had undergone remarkable socioeconomic success and cultural "rejuvenation."

Sorokin also utilizes a human behavioral basis for detecting patterns of civilization growth and decay as he seeks the nature of the forces that explain cultural change and transitions. As with Spengler and Toynbee, he professes the view that mature civilizations inherently face self-inflicted forces of cultural deterioration. Thus, it is useful to explore the extent of compatibility of Sorokin's vision of cultural change, phases, periodicity, and decline with mechanistic-thermodynamic principles.

The nature of culture itself and of shifting human values as being allied with behavior as the primary agents of societal change, conceptualized by his immanent theory of change, is the foundation of Sorokin's approach. He associates cultural change with the concepts, ideals, and roles of human values central to the beliefs, traits, and practices of a given social order.[19] Thus, a culture continually changes as a result of the dynamics and consequences of shifting human values. It will be demonstrated that Sorokin's view of the "seeds," "forces," and "consequences" of cultural change, while conceptually valuable, is too narrowly focused on variations and cycles of human behavior. The inclusion of a broader array of variables, particularly restrictive aspects of wealth and energy consumption of socioeconomic systems, would provide a more insightful, fundamental, and unifying vision of cultural development.

Sorokin, as does Kroeber, applies a methodology that relies on empirical data from the visual arts, philosophy, science, religion, literature, and ethics for his analysis of the interrelationships among cultural subsystems. Accordingly, he utilizes an enormous quantity of historical material reflecting values, ideals, and beliefs considered representative of cultural traits during the period prior to 1000 B.C. into the twentieth century. This raw material was analyzed and converted into data sets considered adequate representations of evolving human values and behavior that characterize a social order's generalized cultural traits and environment.

Sorokin defines "supersystems" as the largest cultural units possessing similar values and assumes that all segments of the supersystem undergo similar changes over time.[20] For example, the visual arts and philosophical thought could experience similar values modification during a given period. Since such a supersystem is considered representative of the culture's membership and its value system, it is therefore possible to accurately portray the nature, vitality, and changing characteristics of the culture as a whole. However, this approach, while seemingly serving as an appropriate assessment of the nature and vitality of the cultural arts, raises the obvious issue of being a narrowly focused methodology for adequately representing the fundamental "seeds" or "forces" of cultural change. Most certainly, the success or lack of success of the cultural arts and letters may coincide with societal transitions emanating from a broader scope of more fundamental and influential socioeconomic forces.

Yet, despite Sorokin's narrowly focused variables, his representation of the patterns of human behavior associated with cultural growth and decay is generally consistent with the variation in the availability and the effective use of wealth-energy capital. Both considerations have important ramifications for human attitudes, behavior, and socioeconomic success. Sorokin's techniques, which have been referred to as scientific, utilize scientific methodologies for gathering, sorting, and analyzing empirical data for the purpose of detecting cultural trends that lead to societal transitions. Nevertheless, such a laudable scientific methodology applied to the social sciences differs considerably from relying on scientific principles that may determine fundamental causes of cultural evolution. Brander, referring to Sorokin's *The Crisis of Our Age,* describes the effect of Sorokin's sensate culture:

> **Sensate culture in its overripe and rotting condition has debased mankind and now is destroying him and his environment as well. Spontaneous forces inherent in the system have stripped it of fundamental values capable of commanding allegiance.**[21]

Sorokin concludes that civilizations evolve to become decadent and self-destructive, ultimately evolving into a state of disintegration. However, he is unable to identify "spontaneous forces inherent in the system" or to provide a fundamental rationale for the sequential pattern that inevitably leads to cultural deterioration, as represented in Figure 8-1. His impressive data linking human values, behavior, and general societal characteristics with fluctuations in socioeconomic progress and historical events correspond to only one aspect of a larger and more complex cultural environment. Thus, he does not answer the fundamental question as to why cultural deterioration has historically been the eventual outcome of successful civilizations.

Sorokin's sociological approach to cultural change is to monitor shifts in his "supersystems" via fluctuations in the productivity of the cultural arts and letters that are assumed to reflect the periodic realignment of a population's cultural values and traits. This methodology identifies shifting and competing cultural mentalities between the *extremes of spiritualism and materialism* that dominate sociocultural development. The diversity of characteristics and properties associated with human values and behavior affecting cultural change includes (1) the cultural *mentalities* of religion, philosophy, and science; (2) the *truths* of faith, reason, and senses; (3) the *Ethics* of Principles and Happiness; and (4) the prevailing, supporting *ideology and discipline* of a particular mentality.[22] At any particular time, a given population will possess a broad distribution of mentalities, truths, ethics, and ideologies in the same fashion that a population has a representative distribution of heights and weights. Nevertheless, the prevailing mentality of the minority in leadership may or may not reflect the mentality of the majority; be effectively employed in the implementation of social, economic, and political priorities; or be capable of leading a society to survival and prosperity. This same diversity of mind and soul may be expected to generate a multitude of views, attitudes, and behavior regarding political and socioeconomic priorities, decisions, and desired outcomes, thereby generating winners and losers as well as additional societal tension and instability.

> **All the main compartments of culture mentality which have been analyzed in the present work—arts; systems of truth (science, philosophy, religion), moral systems, systems of law; forms of political, social, and economic organizations; and so on—are not only the phenomena of mentality, but also the phenomena of behavior, in the most overt, "behavioristic" sense.**[23]

Sorokin views sociocultural transitions as shifting between contrasting supersystems over long time periods from a period of predominately spiritualistic and altruistic values to a mentality of extreme self-interest and materialism, leading to increasing social and economic instability. He formally defines the "spiritualism" phase or mentality as the *ideational supersystem,* which represents an adherence to religious ideology, ethical principles, and spiritual faith. Thus, one seeks guidance and knowledge though the spiritual realm as opposed to relying on one's senses or reason. The ideational supersystem is characteristic of primitive cultures as well as the direction taken when disintegrating mature civilizations attempt to avoid collapse by seeking refuge in a new, more orderly, socially responsible, simpler, and disciplined way of life.

The *sensate supersystem* or mentality is a materialistic culture symbolized by the last phase of ancient Rome and is the direct opposite of the ideational mentality. The sensate mentality seeks pleasure and luxury, replacing ethical principles with greed and ruthlessness; science displaces theology as the senses replace guidance from one's faith. Sorokin also defines a third mentality as the *idealistic supersystem,* a mixture or halfway point between the two extremes of spiritualism and materialism. As a result of the success of this balance of the two extreme cultures, some historians have referred to this third category, a midpoint mentality, as marking the high point of Western civilization. Thus, the idealistic phase may be considered as the most productive phase of Sorokin's continuum of mentalities, as it represents the most productive elements of the two extreme worlds of spiritualism and materialism. However, perhaps cultural progress is only perceived to be at its highest level when socioeconomic productivity is impressively high, but the effects of cultural degradation have not reached a generally recognizable and debilitating sensate condition. This would correspond to the midpoint of Stage II, Growth. Table 8-1 illustrates the characteristics and properties of the two extreme supersystems.

Table 8-1
Sorokin's Three Cultural Mentalities

Ideational <———> **Idealistic** <———>**Sensate**

Ideational	Sensate
Altruistic, spiritual values	Selfish, ruthless, greedy, crime
Reliance on spiritual leaders	Irreligious society seeks a Caesar
Values and principles oriented	Luxury and happiness oriented
More devoted to values, truths	Ignores values for profit and self
Socioeconomic simplicity	Structural/functional complexity
Lives of minimal stress	Bureaucratic chaos and high stress
Slow pace of change, civility	Rapid pace of change, incivility
Narrow view of needs	Insatiable appetite for "wants"
Societal-oriented priorities	Self-oriented priorities
Low per capita energy use	High per capita energy use
Emphasis on renewing wealth	Wasteful use of wealth-energy
Puritan-type work ethic	Emphasis on pleasures of life
Self-reliant for survival	Society provides entitlements
Energy conservative	Energy waste and pollution
Authoritative social control	Social control collapses
Self-discipline	Standardization and centralization
Anti-intellectual	Rely on science and technology

Minimum ———> **Entropy Increase** ———> **Maximum**

Ideational <———> **Idealistic** <———> **Sensate**
Culture   **Culture**   **Culture**

The mechanistic-thermodynamic paradigm provides valuable insights into Sorokin's continuum of cultural mentalities, specifically the interpretation of the relative degrees of cultural disorder representative of the ideational mentality as compared with the sensate. The ideational mentality is a lower entropy condition, characterized by socioeconomic simplicity, self-reliance, less complex and formal controls, slower pace of change, less wealth and energy consumption per person, and greater energy conservation. Conversely, the sensate mentality is characterized by greater societal disorder exemplified by numerous bureaucratic structures and functions, extensive socioeconomic complexity, higher per capita energy consumption, and significant environmental issues: all typical contributors to cultural deterioration. It is noteworthy that societal maturation inherently proceeds from a more orderly primitive, ideational-oriented society to a more complex, advanced, sensate culture—a thermodynamically favored directionality. That is, a spontaneous migration from the more orderly state to the more random or disorderly state is representative of the cultural transition sequence of the ideational to idealistic to sensate mentalities. This process direction is in accord with the entropy principle applicable to all energy consumption processes, as mandated by Mother Nature.

The same thermodynamic principle of spontaneous directionality, applied to planting seeds, produces the same sequencing expectation that Sorokin holds for societal supersystems: "Each … cannot help changing; rising, growing, existing full-blooded for some time, and then declining."[24] The reverse sequence of nature, that is, the dying plant returning to the previous living plant form and then to its original seed, would be considered as improbable! Thermodynamics, as well as common sense, provides an unshakeable faith in the sequencing of nature's events such as agriculture production and the sequencing of the seasons of the year. Accordingly, the entropy concept of probability is consistent with the transitional sequencing of Sorokin's supersystems from the more orderly ideational cultural environment to the more disorderly sensate condition. This is the direct consequence of the nature of cultural processes providing life to a society as it consumes the required wealth-energy resources.

Thus, the entropy principle mandates that thermodynamic processes progress from the more orderly to a more disorderly state. Based on that overriding law of nature, society's mission is to manage its affairs by providing the specific timetable, resources, operational strategies and tactics, and physical conditions for a specific journey's pathway and its activities. This is the appropriate response to Sorokin's inquiry of "why the phases follow each other in the sequence: Sensate-Ideational-Idealistic."

Sorokin offers "logic" for the observed directionality, sequencing, and recurrence of cultural transition patterns that extend from the pre-2000 B.C. Minoan civilization through the twentieth-century Western civilization. First, he reasons that "many processes cannot move forever in the same direction; having reached their limit, they turn to a new direction" and must therefore find a new direction "from a limited number of possibilities."

> We have seen that many processes cannot move forever in the same direction; having reached their limit, they turn to a new direction; along this new direction they also cannot move forever, and sooner or later have to turn again, and so on.... The number of fundamental possibilities as to ever new fundamental turns in direction, essentially new forms, patterns, and appearances the system can assume is limited and bounded.[25]

However, Sorokin's logic elicits more fundamental questions than it answers. Why is it that many processes "cannot move forever" in the same direction? What precisely is being "limited and bounded," and what forces produce "this new direction" and why? Additionally, he reasons, "Supersystems cannot help changing" because "each of these forms does not stay forever at its domination.... It has to give place to the other forms of the triad." Such statements raise additional questions of what forces are acting and the why and how of cultural change processes.

> What is the reason for such a rhythm? The answer is given by the principles of immanent change and of the limited possibilities of the main integrated forms of a cultural supersystem. By virtue of the principle of immanent change, each of the three integrated forms or phases of the Ideational, Idealistic, and Sensate Supersystems cannot help changing; rising, growing, existing full-blooded for some time, and then declining. The principle explains why each of these forms does not stay forever at its domination, and why it has to give place to the other forms of the triad. It does not explain, however, why this triple rhythm with its three phases is recurrent, and why the phases follow each other in the sequence: Sensate-Ideational-Idealistic.[24]

Periodicity of Human Advancement: Ancient Greece to the Twenty-First Century

**Recurring Cultural Mentality Patterns
Graeco-Roman Western Cultures
Figure 8-2**

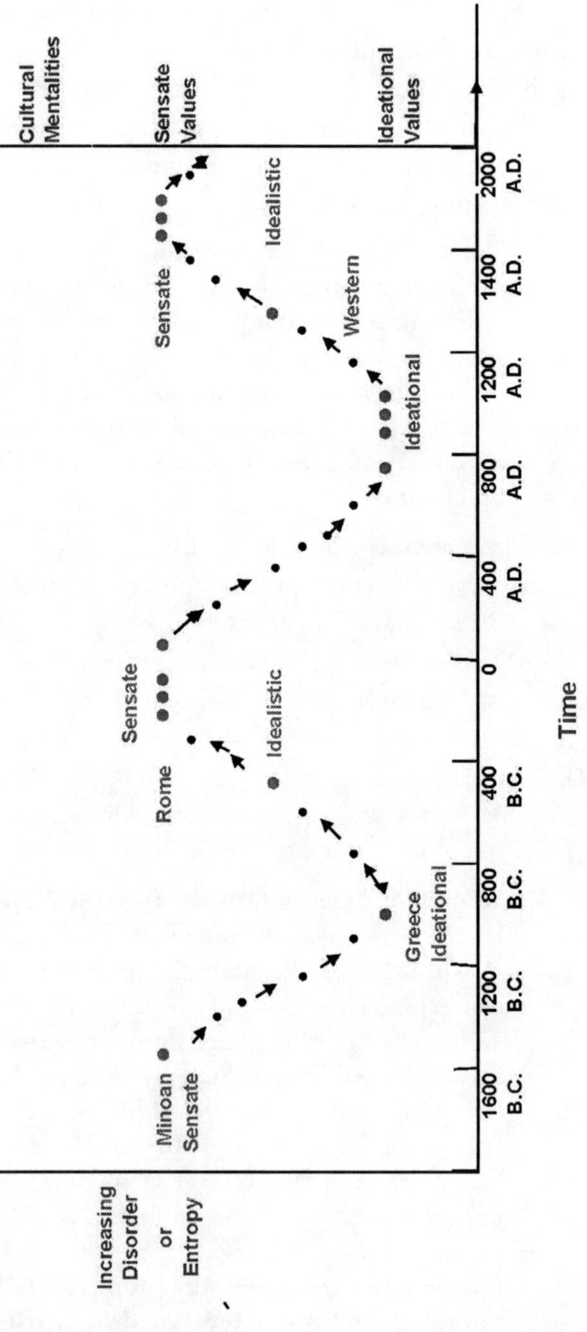

Sorokin offers only a partial explanation for the recurrent "triple rhythm" illustrated in Figure 8-2: "The recurrence is sufficiently accounted for by the principle of the limited possibilities of the main integrated forms of culture."[26] His partial response applies only to "integrated forms," which correspond to only the peaks and valleys in Figure 8-2. However, the phenomenon has a more scientific, fundamental thermodynamic basis; that is, the spontaneous directionality of Sorokin's supersystems sequencing from order to disorder (i.e., from Ideational to Idealistic to Sensate). Recall that Figure 8-1 illustrates one generic *cycle of the "triple rhythm"* (i.e., the sequential periodicity from the genesis to the deterioration stage). Cultural "rejuvenation," shown as representing one potential outcome of Stage IV, Deterioration, essentially constitutes a potential second Genesis stage that may materialize to provide a sustained new phase of Stage II cultural Growth. Successful rejuvenation could initiate a *second cycle* of socioeconomic growth and prosperity, but only under a new cultural environment of radically different human values and behavior and appropriate resource priorities and investments. Fittingly, the sensate era of the Roman culture ended as a new cultural environment of more primitive, altruistic values was born, ultimately leading to the Western culture's ideational era.

Sorokin identifies two cycles of the "triple rhythm" representing the Minoan period prior to 2000 B.C. through the twentieth century. The two-cycle illustration in Figure 8-2 is simply the repeated profile of the triple rhythm illustrated in Figure 8-1. The figure begins with the Minoan *sensate phase* (1700 B.C.) followed by the *Greek ideational phase* (800 B.C.), an *integral phase* (500 B.C.), and a *sensate phase* (100 B.C.). This was followed by the Roman Empire's quick rise and fall (A.D. 135 to 500) and an *ideational* phase; that is, the birth and growth of Western civilization (A.D. 500 to 1200), the *integral* phase (A.D. 1250 to 1400), and the current *sensate* phase that began in the fifteenth century.[27] Sorokin also reports rhythms and recurring patterns in the Hindu, Chinese, and Arabian cultures similar to those representing the Graeco-Roman Western culture's periodicity.

Thus, the fundamental explanation for "why this triple rhythm of its three phases is recurrent" is identified by considering the nature of the successful cultural "rejuvenation" or rebirth phase of the failed state that links the two triads in Figure 8-2. The emerging genesis stage or birth of

Western civilization from the ashes of the Roman civilization essentially redefined the fundamental parameters of a new culture (i.e., created a new thermodynamic system). Successful attempts to rejuvenate a collapsed socioeconomic system may be depicted as re-creating the cultural environment, properties, and characteristics of a more primitive behaving society capable of greater social and economic stability. Thus, refinements of the culture must embrace the values, ethics, resource allocation priorities, and spirituality of a more primitive society. The consequences of this more primitive society and its altruistic mentality includes greater socioeconomic simplicity, less materialism, lower wealth and energy consumption per capita, and a higher cultural priority for regenerating wealth in lieu of the acquisition of nonessential goods and services. These characteristics ultimately enable the achievement of a net positive rate of cultural growth. Essentially, this spontaneous pathway to cultural stability is successful because the process respects the laws of science that restrict and control the activities of the social order and utilizes altruistic human values and behavior, Mother Nature's wisdom, and conservative management of wealth and energy.

Sorokin's broad view of culture change may be summarized as follows: A society "changes by virtue of its own forces and properties" by passing through a "cycle of completed life, moving from childhood to senility and death," and relying on a "self-determination" that "incessantly generates consequences ... of the existence of the system and its activities."[28] Additionally, he embraces the principle of limited possibilities based on the logic that "the number of fundamental possibilities as to ever new fundamental turns in direction, an enormous number of them, if not all, seem to have a clearly limited range of possibilities of variation."[29] Sorokin establishes a maximum number of such potential possibilities as "six main economic types that have existed in history," five "forms of political organizations," a "few texts" of constitutional law, a "small number" of religions, a "few themes" of the history of philosophical thought, a "few" codes of law and ethics, and a "limited" number of known sociocultural systems and processes.[30]

Sorokin's concept of "cultural convergence" provides insight into mechanisms whereby similar cultural traits from various cultures are shared or independently appear from different origins. He concludes: "From logical and factual evidence, it is reasonably certain that an enormous number of sociocultural systems and processes have a limited range of possibilities in their variation, in the creation of new fundamental forms."[31] This pragmatic concept that a limited number of possible cultural alternatives and directions exist for future systemic cultural change is based on empirical data that provide a reasonable array of the most probable societal directions. However, it hardly constitutes an acceptable explanation for the driving forces of recurrent cultural patterns observed over the millennia. Sorokin's statement that "the recurrence is sufficiently accounted for by the principle of the limited possibilities of the main integrated forms of culture" is simply a probability argument based on an arbitrary set of limited historically observed conditions and an intuitive projection of a number of possible future directions. That is, only identified potential consequences of human behavior are considered relevant to a society's future. This approach eliminates the consideration of a period of unique cultural inventiveness, the creation of new alternatives, and the occurrence of low-probability events.

To summarize, Sorokin's principles of change, self-determination, consequences, and limited possibilities appear relevant to the cultural patterns represented in Figure 8-2. However, these principles describe, are limited to, judgments regarding past events projected as potential future causes and effects of sociocultural transitions. Conversely, the mechanistic-thermodynamic perspective provides scientific explanations and insights into the observed developmental stages of cultural genesis, growth, deterioration, collapse, and revitalization. In addition, Sorokin's description of, and rationales for, the transition from one cultural mentality or phase to another appears to be limited to influences of human behavior on cultural development. However, explanations "governing social development" require fundamental principles.

> **The value of history lies not in the multitude of facts collected, but in their relation to each other.... If the sequence of events seems to indicate the existence of a law governing social development, such a law may be suggested, but to approve or disapprove of it would be as futile as to discuss the moral bearing of gravitation.**[32]

Sorokin's identified "sequence of events seems to indicate the existence of a law governing social development" but its "value" must be determined based on a broader collection of academic perspectives and "their relation to each other." For example, Sorokin observes that during "the transitional periods, when one integrated form disintegrates while the other is not yet crystallized, ... these transitional periods exhibited a variety of combinations of the elements of the main forms, and dissimilarity from each other in a number of important aspects."[33] This may be interpreted as some processes of socioeconomic subsystems undergoing cultural disintegration as others simultaneously are engaged in the growth process of crystallization. From a mechanistic perspective this represents the simultaneous creation of order (crystallization and integration) and disorder (disintegration). That is, as a number of society's functional processes contribute positively to socioeconomic productivity others are deteriorating with negative cultural effects. The variables of such socioeconomic transitions are obviously complex and multidisciplinary, and produce simultaneously constructive and destructive outcomes.

For an entire social order (e.g., a mature civilization), the entropy concept mandates that the outcome of cultural disorder (disintegration) will ultimately dominate and neutralize the creation of order (growth). Thus, as long as a society exists, it will be required to continuously compensate for this net continuous production of societal entropy by consuming increasing amounts of wealth-energy resources and/or adopting more energy-conservative ways of living. This represents the fundamental societal dynamics that explains the continuous decline in the positive slope of cultural productivity (Figure 8-1) and its transition to a negative slope during Stage IV, Deterioration. This rationale also provides a basis for Sorokin's empirical deduction of the principle of immanent change: A sociocultural system:

> **changes by virtue of its own forces and properties.... It cannot help changing, even if all its external conditions are constant.... [It] is sufficient to answer the problem of Dynamics: Why a whole integrated culture as a constellation of many cultural subsystems changes and passes from one state to another. The answer is: It and its subsystems—be they... social, political, and economic organizations—change because each of these is a going concern, and bears in itself the reason of its change.**[34.]

In thermodynamic terms, "a constellation of many cultural subsystems changes and passes from one state to another," consuming wealth-energy capital and providing the forces of change that enable the transition (e.g., increasing cultural crystallization and integration). This process also inherently results in the inevitable societal entropy production that is the basis for cultural disintegration. The same entropy principle, responsible for process directionality and the generation of societal disorder, is relevant to Sorokin's description and characteristics of the extreme sensate cultural environment. See Table 8-1. This representation defines the state of maximum societal disorder, a precursor to cultural collapse.

Toynbee's breakdown stage represents the attainment of a maximum intolerable degree of societal complexity, dysfunction, and chaos and widespread cultural ineffectiveness, inefficiency, injustice, and financial insolvency. This environment of extreme sociopolitical and economic instability, a product of society's adoption of extreme sensate values, is a major cause of failure of the fundamental processes of the social order. The available human and material resources are insufficient to overcome a critical situation and set in motion a population seeking a sociocultural transition in an atmosphere of fear for survival. The prevailing dominant cultural motivator of economic greed is displaced by fear for survival.

> **A society that cannot counter this trend, such as through acquisition of an energy subsidy, becomes vulnerable to stress surges that it is too weak or impoverished to meet, and to waning support in its population. With continuation of this trend, collapse becomes a matter of mathematical probability, as over time an insurmountable stress surge becomes increasingly likely.[10]**

Thus, "by virtue of its own forces and properties," a culture is unable to rejuvenate or jump-start the economy in a new sociopolitical environment that requires the reversal of a culture's sensate mentality toward greater altruistic values. Social and economic instability ignites "a constellation of many cultural subsystem changes and passes from one state to another."

Figure 8-2 illustrates three peaks of *maximum sensate* cultural mentality encompassing societal disorder and instability; each is followed by a transitional period leading to an *extreme ideational* mentality. The Mesopotamian civilization, which began 6000 B.C., peaked 1700 B.C. The Greeks destroyed its Persian Empire in the fourth century, and a parallel civilization of the Nile Valley peaked 2300 B.C. and ended with the Egyptian Empire 400 B.C. Also from the Mesopotamian civilization, the Cretan civilization that began 4000 B.C. peaked at the Minoan period 1500 B.C. and ended with the Mycenaean Empire in the twelfth century B.C. The Mediterranean civilization grew out of the Cretan civilization from 1000 B.C., peaking in 400 B.C. and ending with the Roman Empire collapse in the fifth century A.D. After which, the Western civilization emerged and appears to have peaked as the American Empire declined during the latter part of the twentieth century. Other civilizations experienced such periodicity. The Indic civilization began 3500 B.C., peaked about 2200 B.C., and culminated as the Harappa Empire destroyed by Aryan invaders after 1700 B.C. From this ending, the Hindu civilization and the Mogul Empire emerged.[35]

General *sensate values and characteristics* are associated with reaching the peaks in the pattern of Figure 8-2, corresponding to cultural stagnation and a maximum cultural complexity and disorder that dominate civilizations experiencing massive cultural degradation. The momentum of overwhelming cultural disintegration will continue unless and until prevailing socioeconomic conditions are altered to stabilize the social order and thereby permit an increase in cultural productivity. The culture must find a way to create a more orderly, altruistic society and to more wisely utilize available human and wealth-energy resources in order to re-create a more viable social order. This is the objective of cultural rejuvenation or rebirth, to seek a new beginning, characterized by a more primitive, altruistic mentality that embraces social and economic stability. The first step to such a goal is for the culture to reduce its current rate of cultural deterioration and to systematically increase economic growth. Successful rejuvenation of a social order necessitates "changes by virtue of its own forces and properties." Sorokin's "immanent change" provides the mechanism: "The answer is: [*The culture*] and its subsystems—be they ... social, political, and economic organizations—change because each of these is a going concern, and bears in itself the reason of its change." To exist requires maintaining the processes of living which entails seeking the material resources and behavioral characteristics needed to implement successful change and secure a socially and economically stable social order. This applies not only to the potential rejuvenation of a mature civilization in Stage IV Deterioration but also to a primitive society in Stage I Genesis. Thus, the transition from a peak to a valley is a consequence of decreasing sensate values as an ideational mentality increasingly influences a new social order in its early growth stage (e.g., the transition of the mature, collapsing Roman civilization to the primitive Western culture). The valleys in Figure 8-2 signify the end of a period of increasing ideational values as a sensate mentality increasingly affects the social order.

Survival and hope are the human motivators to seek a new pathway for achieving cultural rejuvenation. In the face of societal collapse, people in large numbers begin to search for radical personal and societal approaches to physical and financial survival, safety, and economic and emotional stability. This goal becomes the highest priority, replacing previous materialistic objectives and clearly requiring a significant alternation of human attitudes, behavior, and socioeconomic strategies. As a result, predominant sensate cultural values are displaced by a gradual long-term shift toward more ideational characteristics as the population recognizes that their way of life is in jeopardy. Fear predominates, thus reason, faith, and spirituality begin to increasingly capture people's minds and souls, replacing greed, ruthlessness, and resource wastefulness. As a consequence, the culture actively shifts its values toward an ideational mentality, attempting to deal with the chaos, insecurity, and loss of hope.

The designation of the maxima in Figure 8-2 as a maximum sensate cultural condition is consistent with authors who have observed that for some reason, cultural breakdown is the beginning of a new cultural phase of creating societal orderliness out of chaos. Out of the necessity to survive, the population begins a reversal of fundamental attitudes, values, and behavior toward the more orderly, less complicated, more spiritual, and conservative ways of life reflective of a more primitive life style. Thus, the society slowly begins to search for, and embraces, the characteristics of the Ideational Supersystem of ethical principles, conservative economic investments, self-discipline, socioeconomic simplicity, and communal-oriented priorities. This transition of values and behavior tends to decrease the rate of wealth and energy consumption and societal entropy production. A new direction for the revised social order is initiated out of necessity; self-initiated change in a spirit of self-determination will make revival and survival possible.

> **Hence the dilemma for the respective society and culture: either to continue such a dangerous drift, and dry up and perish, or to make a great effort and restore a fuller and more genuine truth and system of values.... Some cultures, like the Graeco-Roman and the Western, have been able to make such a shift several times; some others could not. The first cultures continued to live and to pass through the recurrent rhythm studied; and others either perished and disappeared, or were doomed to a stagnant, half-mummified existence, ... becoming a mere "survival" or "object of history" instead of being its creative subject.**[36]

The fear of not surviving cultural collapse and the terror of, or faith in, God and potential damnation appear to provide intense psychological motivation and emotional support for actively seeking and embracing cultural change and a more ideational mentality. The desperate search for a new direction after the collapse of Roman civilization coincided with the new success of the Christian spiritual ideology, providing a secure, emotional pathway to fortify the struggle for human existence and to replace the Mediterranean civilization of Rome with the Christian-based Western civilization. This sociocultural transition is illustrated by the peak of one maximum sensate culture era, 200 B.C. (maximum social and economic instability of Rome), evolving to a minimum disorder (A.D. 1000) at the early growth stage of Western civilization. This transitional direction from the Roman sensate culture to the new ideational culture of Western civilization required a reversal of societal attitude and behavior from the era of materialism and its seduction.

The triple rhythm in Figure 8-2 illustrates the course of human history, repeatedly experiencing an ideational period motivated by a fear for basic physical and spiritual survival, followed by a sensate period motivated by ruthlessness and greed. In his late-nineteenth-century work, *The Law of Civilization and Decay*, Brooks Adams outlines an energy-based economic philosophy of civilization disintegration. He states:

> **Thought is one of the manifestations of human energy, and among the earlier and simpler phases of thought, two stand conspicuous: fear and greed. Fear, which by stimulating the imagination, creates a belief in an invisible world, and ultimately develops a priesthood; and greed, which dissipates energy in war and trade.... In the earlier stages of concentration, fear appears to be the channel through which energy finds the readiest outlet; accordingly, in primitive and scattered communities, the imagination is vivid, and the mental types produced are religious, military, artistic. As consolidation advances, fear yields to greed, and the economic organism tends to supersede the emotional and martial.**[37]

Thus, the repeated reversals between ideational and sensate mentalities prior to 2000 B.C. through A.D. 2000, as reported by Sorokin and others, may be viewed as a shifting equilibrium between a culture of fear and a culture of greed. At any given time, depending on the perceived quality of life and prospects for the future, a society's prevailing cultural mentality will reflect a population exhibiting varying degrees of greed and fear. This reflects cultural values, attitudes, and behavior that determine the properties and characteristics of a culture. The prevailing human psychology and the ensuing culture's behavioral characteristics will determine how self-determination and so-called "free will" will successfully utilize available resources and the extent of success of civilization's operational systems.

Given the factors responsible for the periodicity of human advancement, the obvious question is, do strategies exist for counteracting the inevitable trend of humanity toward a dominant stage of cultural disintegration? Will the time arrive when it becomes impossible for the cultures of the world to recover from the maximum sensate condition? First, recall that the thermodynamic principle of entropy production mandates a spontaneous energy flow from an orderly condition to a more disorderly state and from a usable energy form to an unusable form. Assuming the availability of wealth-energy resources, a civilization that is willing and able to minimize its societal entropy production, while maintaining a positive rate of cultural socioeconomic progress, will be able to maximize its longevity and the quality of life for its population. There is a higher probability of achieving this objective of maximizing cultural longevity if a civilization conducts highly efficient and effective socioeconomic processes that benefit a large majority of the population, avoids the social and financial burdens of debt, expanded bureaucracies, and war, and practices ideational values, particularly a spiritual form of the Golden Rule. Accordingly, accomplishing this objective will maximize the success of a civilization's Stage II Growth and prolong, but not eliminate, its arrival at the Stage III Breakdown phase of cultural stagnation. Likewise, these optimal cultural characteristics also maximize the probability of a successful cultural rejuvenation process of a stagnating or disintegrating social order.

The challenge of continuous change is a given reality for a social order as the new inevitably attempts to drive out the old ... sometimes for the better and sometime for the worse.

Chapter 8
References and Notes

1. Sorokin, Pitirim A. *Social and Cultural Dynamics*. Boston: Porter Sargent. 1970. 692.
2. Tainter, Joseph A. *The Collapse of Complex Societies*. New York: Cambridge University Press. 2004. 106.
3. Spengler, Oswald. *The Decline of the West*, trans. with notes by Charles Francis Atkinson. New York: Oxford University Press. 1991. 71-73.
4. Melko, Matthew. *The Nature of Civilizations*. Boston: Porter Sargent. 1969. 25-26.
5. Brander, B. G. *Staring into Chaos: Explorations in the Decline of Western Civilization*. Dallas: Spence. 1998. 367. 41-43.
6. Ibid., 98-100, 188.
7. Toynbee, Arnold J. *A Study of History*, ab. Sommervell, D. C. New York: Oxford University. 1957. II: 276, 49.
8. Melko, Matthew. *The Nature of Civilizations*. 33.
9. Ibid., 548.
10. Tainter, Joseph A. *The Collapse of Complex Societies*. 127.
11. Ibid., 108.
12. Ibid., 106.
13. Galbraith, John Kenneth. *The Good Society: The Humane Agenda*. New York: Houghton Mifflin. 1998. 186-199.
14. Toynbee, Arnold J. *A Study of History*. 364.
15. Ibid., 365.
16. Ibid., 360.
17. Ibid., 360-361.
18. Ibid., 361-362.
19. Sorokin, Pitirim A. *Social and Cultural Dynamics*. 630-646.
20. Ibid., 20-39, 678-679, and 644-646.
21. Brander, B. G. *Staring into Chaos*. 353.
22. Sorokin, Pitirim A. *Social and Cultural Dynamics*. 20-39.
23. Ibid., 609.
24. Ibid., 676.
25. Ibid., 654.
26. Ibid., 676.
27. Ibid., 676-677.

28. Ibid., 633-639.
29. Ibid., 654-656.
30. Ibid., 656-658.
31. Ibid., 660.
32. Adams, Brooks. *The Law of Civilization and Decay*. New York: Gordon Press. 1975. v.
33. Sorokin, Pitirim A. *Social and Cultural Dynamics*. 677.
34. Ibid., 633, 638.
35. Quigley, Carroll. *The Evolution of Civilizations*. Indianapolis: Liberty Fund. 1979. 81-82.
36. Sorokin, Pitirim A. *Social and Cultural Dynamics*. 692.
37. Adams, Brooks. *The Law of Civilization and Decay*. iv.

Chapter 9
Properties and Characteristics of Sociocultural Stagnation

> "Cultures can evidently deteriorate, in the sense that their total content as well as their highest values may shrink; but there is nothing to show that the process can go on to the point of extinction.... Specific cultures, that is, particular, geographically limited forms of culture, can and do die. They die not only by complete extinction of the population which carry them ... but by absorption of societies into larger societies which carry different cultures; and even by displacement of one culture by another, without annihilation, absorption, or fundamental damage to the underlying population. Primitive cultures are dying in this way every year before the 'impact of civilization.'"
>
> A. L. Kroeber[1]

The history of Graeco-Roman Western civilizations and other less successful social orders illustrates that cultural growth leading to inevitable societal deterioration passes through a stagnation phase that ultimately evolves into cultural extinction, ossification, or rebirth as a radically different, more primitive social order. Logically, the properties and characteristics of this *stagnation phase*, the culmination of unsuccessful human efforts over the last 4,000 years to achieve *sustaining socioeconomic prosperity and fulfillment*, should provide useful information regarding the fundamentals of the cultural degradation process. Additionally, those deteriorating cultures that successfully reconstitute themselves into more effective, albeit more primitive, social orders provide important insights into constructive attributes that minimize elements of cultural deterioration.

Significant agreement exists among respected authors regarding gradual sociocultural transitions that sequentially progress from simple, orderly societies to more complex, disorderly, and stagnant socioeconomic and political systems that eventually deteriorate. This applies to all major cultures since the Mesopotamian era. Using various methodologies to detect socioeconomic fluctuations and sociocultural transitions, societal progress has been shown to approximate a symmetrical profile of cultural growth, stagnation, and deterioration. Given the empirical nature and scarcity of such historical information, the narrow range of variables considered, and the inherent imprecision in the methodologies and data utilized, the level of general agreement among authors regarding such profiles is impressive.

The objective of this chapter is to examine the gradual cultural transition process just prior to, during, and following cultural stagnation. Reference is made to Figure 8-1 of the previous chapter, in which the slope of the cultural growth curve systematically declines over time, making the transition from a positive to a negative rate of growth and productivity. Toynbee defines this transitional stagnation stage as "breakdown, ... the termination of growth.... A society does not ever die 'from natural causes,' but always does from suicide or murder—and nearly always from the former."[2] Thus, an understanding of this particular transitional inflection point just prior to and following "breakdown" would appear to be instructive, as would a full appreciation of his use of the word "suicide." Appropriately, Toynbee asks, "What then causes the breakdown of civilizations?"[3] Kroeber states: "The real question is why particular cultures die." Sorokin inquires: "Where shall we look for the roots of change of sociocultural phenomena, and how shall we interpret it? Shall we look for the 'causes' of the change of a given sociocultural phenomenon in the phenomenon itself, or in some 'forces' or 'factors' external to it?"[4]

Kroeber links cultural "deterioration" to the shrinking of the "highest values," while Sorokin professes that values shift toward a more materialistic mentality in times of socioeconomic prosperity. Additionally, Kroeber's reliance on "absorption," "displacement," and "adaptation" of primitive cultures "before the impact of civilization" presupposes mechanisms with wealth and energy requirements in support of the fundamental processes of human existence. These viewpoints identify *typical variables or factors* capable of promoting cultural advancement as well as leading to deterioration.

Kroeber defines cultural "dying" as "a replacement of most of the material and patterns with new material and patterns developed within the culture, until after a sufficient length of time, the transformation is so great that it is descriptively more useful to speak of the end product as a new culture, or one different from the original one."[5] To illustrate this point, he uses the dead culture of ancient Egypt, whose culture was gradually displaced sequentially by Asiatic, Greek, Roman, Christian, and Islamic societies. Further, he raises the question of whether long-term displacement, absorption, and adaptation were "due to some intrinsic superiority" of these cultures or "had the Egyptian culture 'lost its vitality' and become 'senile'?"[6] He suggests that the failure of the reform of Ikhnaton shortly following Egypt's peak advancement of 1400 B.C. resulted in Egypt reaffirming old patterns and shortly becoming a less competitive civilization. It would be expected that restricting cultural change and becoming less commercially and militarily competitive would result in a vulnerability to both internal and external negative consequences. However, Kroeber asks, "Did other cultures outgrow Egyptian culture until they smothered it, or did it of itself age and become feeble until it was extinguished?"

He offers only "a personal opinion," that is: "It would be easy to argue at length for either view, or that both processes were operative; but very difficult to adduce conclusive evidence."[7] The proper response is that both processes were most probably involved. Accordingly, one of his alternatives is consistent with, and supportive of, the mechanistic-thermodynamic paradigm. He reasons that cultures "undergo variations in vigor, originality, and values produced.... But sooner or later, merely on the probabilities of change, it will lie exposed to competition in one of its weak phases, and one of its neighbors will happen to be in a strong, expansive phase, and then replacement will begin to occur."[8] That is, such *competition* is defined in terms of dominant wealth, energy, and human values. Kroeber embraces the concept that:

Properties and Characteristics of Sociocultural Stagnation

> "...fluctuations of cultural vigor are normal, ... any notable cultural achievement presupposes adherence to a certain set of patterns; that these patterns, to be effective, must exclude other possibilities, and are therefore limited; that with successful development they accordingly become exhausted; and that there must be a breakdown or abandonment and the reformation of patterns before the culture can go on to new high achievements".[9]

This reasoning is essentially that of Sorokin's principles of "immanent change" and of "the limited possibilities of the main integrated forms of a cultural supersystem." As presented in the previous chapter, Sorokin utilized these principles to explain major transitions and, specifically, the "triple rhythm." It is noted that the "roots of change of sociocultural phenomena" that Sorokin sought are primarily forces derived from human values, attitudes, traits, and behavior and from society's priorities for, and utilization of, available human and natural resources.

The human behavioral elements related to cultural maturation expressed by Sorokin, Kroeber, and Toynbee plus Tainter's economic concept of the continuously increasing cultural complexity being responsible for a declining marginal productivity in hierarchical specialization address Kroeber's question of "why particular cultures die." Cultures die because they "accordingly become exhausted." The people become emotionally exhausted and discouraged as a result of societal complexity and disorder; the leadership becomes morally, politically, and financially bankrupt; societal operational systems become burdensome and ineffective; and wealth-energy resources become exhausted. This is the "suicidal" environment of Toynbee's "time of troubles" that characterizes cultural stagnation and degradation as well as provides a foundation and motivation for cultural rebirth. Kroeber's atmosphere is described in a similar fashion: "There must be a breakdown or abandonment and the reformation of patterns before the culture can go on to new high achievements."[10] A culture that "lost its vitality" and becomes "senile" is often "smothered" and "extinguished" by a more aggressive and advanced culture. Thus, the people of an "extinguished" culture have no alternative but to reinvent themselves, create a new and different social order appropriate for a new era, or perish.

This sequence of events corresponds to civilizations reaching the peak of Sorokin's triple rhythm, illustrated in Figure 8-2. This era represents a low-level rate of cultural productivity transitioning into a "time of troubles," through stagnation, and into a negative rate of growth (i.e., cultural disintegration).

The *time of troubles* produces a frantic period characterized by the elite and political figures taking desperate measures to resuscitate a dying culture. These often include warfare, territorial expansion, abusing the population, forcible political unification, and frenzied searches for additional wealth in order to address chaos and financial insolvency.

Consequently, there comes a time when a significant portion of the population becomes painfully aware that severe socioeconomic conditions exist that most likely will be irreversible. The realization of the potential loss of one's livelihood, food, shelter, and security, if not life itself, generates fear and alters human behavior. People who have previously demonstrated a high tolerance for complexity, disorder, poverty, and general mistreatment become motivated and actively respond to chaotic conditions that promise little future hope for improvement. At this point, confidence in leadership evaporates as the former dominant minority (the once-creative ruling class) fails to provide stability and solutions to long-standing issues. This results in a rapid disappearance of common cause, harmony, and support within the general population and more specifically within the "internal proletariat" (i.e., the workers and internal followers). Severe and unjust taxation, inequitable distribution of wealth, and the human and financial cost of wars are often major and crippling issues that precipitate a population's emotional defection from its political leaders and their agendas. A subsequent declining level of creativity and initiative, as well as a lack of financial and moral support, incapacitates the economic system, which quickly fragments into various social and political splinter groups.

Widespread dissatisfaction and disunity lead to class conflict, open rebellion, and calls for the creation of a new and different society. Such circumstances induce a society's membership to modify values and behavior and adopt new processes, roles, and leadership criteria in order to create a more attractive societal vision in a desperate attempt to stimulate change in the face of misery and failure. While some people quietly wait for change, others actively work to promote philosophical, spiritual, socioeconomic, and political reforms. Typically, increased taxation, new military actions, increased spending, and desperate searches for additional needed wealth often highlight these "times of trouble." However, these strategies and tactics only produce additional socioeconomic strain on a society that is beyond its breaking point. Conflict among social classes, between religious and ethnic groups, and between leaders and followers produce greater discord and often open warfare, leading to Toynbee's "universal state."

One of the most conspicuous marks of disintegration … is a phenomenon in the last stage when a disintegrating civilization purchases a reprieve by submitting to forcible political unification in a universal state.[11]

The *universal state* emerges from a declining civilization due to the actions of its "dominant minority" intending to revive a dying civilization through more extreme and desperate economic, military, and political actions and by promoting additional laws, complexity, standardization, centralization, and control (i.e., "unification by force"). Brander refers to the universal state as "a temporary shelter in the wilderness of disintegration, … a holding action."[12] Meanwhile, many persons have already mentally moved forward with a new vision of the future, seeking emotional, psychological, and spiritual solace to replace a failed culture's sense of security. Sensate values become less applicable to the realities of new survival issues and are pragmatically displaced over time; most importantly, spiritual values and emotional security are desperately sought in the search for cultural change. Toynbee reports "soul-fractures" (i.e., an inward schism or turmoil within the soul) in every decaying society he studied, with the low point for a society corresponding to a high point for religion, particularly those offering new spiritual thinking. Quite often, a new religious ideology is anxiously adopted by members of a dying culture, giving rise to the phenomenon of the "universal church," as, for example, Christianity emerging during the time of the Roman universal state.

The "interregnum" is Toynbee's final phase in the disintegration sequence when the stagnant or arrested condition prevails for a prolonged period of time.[13] That is, the *universal state* is followed by an *interregnum*, which signals the beginning of constructive efforts to create a new social order from the ashes of the old organizational structures and functions.

This search for a new social order has historically been characterized by a gradual transition from a predominately selfish, materialistic, ruthless value system to a culture with a more altruistic, communal orientation mentality.

> The Roman Empire was immediately preceded by a "time of troubles" going back to Hannibalic War, in which the Hellenic Society was no longer creative and was patently in decline, a decline which the establishment of the Roman Empire arrested for a time but which proved in the end to be the symptom of an incurable disease destroying the Hellenic Society and the Roman Empire with it.... The Roman Empire's fall was followed by a kind of interregnum between the disappearance of the Hellenic and the emergence of the Western Society.[14]

According to Toynbee, the existence of an inert or petrified civilization in a universal state and the development of a *universal church* provides the opportunity for the "recurrence of birth" as well as "barbarian war bands" to invade an already dead society, which "clears the way for the creation of new civilized societies." Brander provides a summary of Toynbee's viewpoint:

> In many ways the entire process of decline works out to what Toynbee called "palingenesia"—recurrence of birth. Social schisms of the disintegration phase lead to new creations: the universal state, the universal church, and barbarian war bands. Schisms of the soul find their palingenesia in the personal renewal of transfiguration. Finally, the savage and wanton destruction of barbaric ages clears the way for the creation of new civilized societies.... Toynbee detected an elemental rhythm: the alternating beat of Yin and Yang, whose variations are challenge-and-response, withdrawal-and-return, rout-and-rally, schism-and-palingenesia. It is the song of creation, in which destruction enters not as discord but as completed harmony. "Creation would not be creative if it did not swallow up all things in itself, including its own opposite." Some philosophies describe the cycle of life embodied in this elemental rhythm as an endless and futile wheel of existence.[15]

Toynbee categorizes the potential outcomes of attempts to create and nurture primitive cultures as *developed civilizations*, *abortive civilizations*, or *arrested civilizations*.[16] Arrested civilizations "have been immobilized in consequence of having achieved a tour de force. They are responses to challenges of an order of severity on the very borderline between the degree that affords stimulus to further development and the degree that entails defeat." This condition is also described by the analogy of "climbers who have been brought up short and can go neither backward nor forward.... Posture is one of perilous immobility at high tension."[17] His depiction of an arrested civilization corresponds to the previous thermodynamic characterization of the dynamic equilibrium condition, whereby a culture's rate of growth or progress is approximately balanced by its rate of deterioration. Depending on a large number of variables, cultural "arrest" may appear quickly to prematurely "abort" a budding culture or materialize at some later phase of growth. An arrested or stagnant condition is depicted in Stage III Breakdown of the generic profile of civilization development.

Toynbee documents twenty-six geneses of civilizations, which have resulted in five arrested, six dead, and ten surviving cultures, with the remaining having been aborted. Of the ten surviving, all but one are "under threat of either annihilation or assimilation" by the remaining one.[18] His identification of the sole survivor in the mid-twentieth century was Western civilization, which now appears in the first decade of the twenty-first century to be approaching the arrested condition.

Toynbee defines "breakdown" as the "termination of the period of growth," which is consistent with his rout-and-rally concept.[19] While the mechanistic-thermodynamic perspective is, in principle, consistent with Toynbee's descriptive, qualitative concept, it is a much more specific and quantitative model. Breakdown occurs when dynamic forces produce a rate of decay that approximates or exceeds the rate of growth, as detailed in Chapter 8. Toynbee's concept is consistent with this perspective of a continuously shifting dynamic equilibrium of an open thermodynamic system representing a civilization continually responding to new forces while purging some existing forces:

> Ideally, no doubt, the introduction of new dynamic forces ought to be accompanied by a reconstruction of the whole existing set of institutions, and in any actually growing society a constant readjustment of the most flagrant anachronisms is continually going on. But *vis inertiae* tends at all times to keep most parts of the social structure as they are, in spite of their increasingly incongruity with new social forces constantly coming into action. In this situation the new forces are apt to operate in two diametrically opposite ways simultaneously. On the one hand they perform their creative work either through new institutions that they have established for themselves or through old institutions that they have adapted to their purpose; and in pouring themselves into these harmonious channels they promote the welfare of society. At the same time they also enter, indiscriminately, into any institutions which happen to lie in their path—as some powerful head of steam which had forced its way into an engine-house might rush into the works of any old engine that happened to be installed there. In such an event, one or the other of two alternative disasters is apt to occur. Either the pressure of the new head of steam blows the old engine to pieces, or else the old engine somehow manages to hold together and proceeds to operate in a new manner that is likely to prove both alarming and destructive.[20]

Toynbee's portrayal of "dynamic forces" within a "growing society" that result in "constant readjustment" and "reconstruction" is an equivalent expression for the mechanistic-thermodynamic model of cultural change found in Chapter 8. That is, a functioning civilization is depicted as an open thermodynamic system subjected to continuous disruption of its journey toward dynamic equilibrium as new materials, including energy resources, and variations of cultural conditions affect society. Meanwhile, the social order is imperfectly managed by society as it endures the consequences of both cultural growth and decay in accordance with the laws of thermodynamics. Of these "dynamic forces," human behavior is the more unpredictable factor and the primary source of the necessary "constant readjustment" and "reconstruction" that produces fluctuating rates of cultural growth and decay. The sociology literature provides an abundance of specific human characteristics and societal conditions associated with the resulting progressive accumulation of cultural decay leading to a stagnation phase.

Quigley demonstrates a similar viewpoint to that of Toynbee, which is also consistent with the mechanistic-thermodynamic paradigm. He considers stagnation as resulting from an increase in cultural decay and emphasizes Spengler's pattern of transition as a progression from "vigorous creativity" defined as "culture" to a weakening moral fiber and selfishness that he defined as "civilization."[21] Quigley describes decay or stagnation as "a period of acute economic depression, declining standards of living, civil wars between the various interests, and growing illiteracy. The society grows weaker and weaker.... The religious, intellectual, social, and political levels of the society begin to lose the allegiance of the masses of the people on a large scale. New religious movements begin to sweep over the society. There is a growing reluctance to fight for the society or even to support it by paying taxes."[22] This depiction also links the rate of cultural growth to social and economic instability, ultimately leading to cultural stagnation.

> **Crisis ... arises from the clash between the decreasing rate of expansion and the fact that people's minds and the organization of the society are arranged for expansion.... An expanding society can become so organized for expansion that it enters upon an acute crisis if the expansion rate decreases, ... the chief result of the institutionalization of the instrument of expansion, ... something that occurs in every civilization.**[23]

It is of interest to explore such historic socioeconomic expansions of mature civilizations primarily intended to solve critical problems "so organized for expansion that it enters upon an acute crisis if the expansion rate decreases." The pursuit of additional wealth-energy capital to solve pressing socioeconomic problems often involves acquiring greater political power and territorial conquests to secure cheap labor and valuable resources to fuel imperialistic objectives and ultimately to survive. While many of these factors have been acknowledged in previous chapters, a few with particular relevance to the transitions from cultural success to stagnation and to degradation will be more fully explored.

A chapter dealing with the stagnation and degradation of mature civilizations would be incomplete without consideration of the historic role of warfare and territorial expansion. Some authors regard warfare as a *symptom* of societal decay, while others consider it a fundamental *cause* of long-term deterioration. According to Sorokin, "internal and external disturbances—revolution and war—are but logical and factual consequences of the state of disintegration of the crystallized system of relationships." He relates "inner tensions and disturbances" with normal societal functions. "In the life history of nations, the magnitude of war, absolute and relative, tends to grow in the periods of expansion ... of the nation at least as frequently as in the periods of decline."[24] Spengler considers war as inevitable and part of the transition to a civilization's breakdown designation. Toynbee states: "Times of trouble produce militarism, which is a perversion of the human spirit into channels of mutual destruction, and the most successful militarists become, as a rule, the founder of a universal state.... Militarism has been by far the commonest cause of the breakdown of civilizations during the last four or five millennia."[25]

Whether militarism and wars are "logical and factual consequences of the state of disintegration," "societal transitions" associated with an insatiable human thirst for wealth, or "the commonest cause of the breakdown of civilizations," the human and financial costs have drained major resources from civilizations, which could have been utilized to better serve humanity or at least prolonged the arrival of socioeconomic stagnation. The human and financial cost of militarism over the millennia has clearly been a major contributor as a cause and/or effect of cultural degradation.

Such discussions appear to center on whether a society at war has reached an author's defined status of cultural breakdown or has long since entered the author's self-designated condition of disintegration. Regardless, it appears that the enormous wealth and energy requirements of waging war have been a major factor in the final deflation of civilizations. Significant resource depletion to support large military expenditures has historically led to financial insolvency, sociopolitical instability, and the collapse of financial and governance functions. The case will be presented that unsustainable levels of national wealth invested in military power designed to acquire territory, power, and additional wealth and to promote political and religious ideologies are a primary cause of declining rates of productivity leading to cultural stagnation. Civilizations simply run out of fuel and spirit as unrealistic, self-serving human ambitions exceed available wealth-energy resources and the population's physical and emotional tolerance for chaos.

Paul Kennedy's volume, *The Rise and Fall of the Great Powers: Economic Change and Military Conflict from 1500 to 2000*, serves as a useful historical perspective documenting the role that wealth-energy capital and scientific principles play in "the rise and fall of great powers."[26] Kennedy provides valuable insights into the interrelationships among the rate of cultural growth, national wealth and finance, evolving global economics, military power and conflict, and new technologies. In addition, his analyses demonstrate the utility of defining and employing science-based economic foundations in the analysis of global economic change. Thus, his work will be heavily cited to demonstrate the historic influence of mechanistic-thermodynamic principles on national economics. The excesses of warfare and expansionism will be shown to be a major cause of inadequate wealth and energy investment in the creation and maintenance of stable socioeconomic and political systems. This is a primary factor in the inevitable decline in the rates of cultural growth and productivity and for the ultimate stagnation of mature civilizations. Thus, Kennedy's perspective on the rise and fall of nation-states since 1500 is of particular interest to the thesis of this work.

While Kennedy focuses on "economic change and military conflict" affecting international power, order, and economic balance, he also considers a nation's consumption of wealth and its financial solvency as major influences in avoiding socioeconomic decline. He summarizes nicely his appreciation for this fundamental concept and for the importance of the thermodynamic efficiency of a civilization's energy consumption:

> **Wealth is usually needed to underpin military power, and military power is usually needed to acquire and protect wealth. If, however, too large a proportion of the state's resources is diverted from wealth creation and allocated instead to military purposes, then that is likely to lead to a weakening of national power over the longer term. In the same way, if a state overextends itself strategically—by, say, the conquest of extensive territories or the waging of costly wars—it runs the risk that the potential benefits from external expansion may be outweighed by the great expense of it all—a dilemma which becomes acute if the nation concerned has entered a period of relative economic decline.**[27]

More generally, this statement expresses the consequences of a society's excessive and nonproductive use of wealth-energy capital that results in a declining marginal return on investment of national wealth, capable of debilitating a nation-state's socioeconomic system ("too large a proportion of the state's resources is diverted from wealth creation"). Recorded history provides evidence that those who hold positions of leadership and power within a civilization often exhibit values and behavior that lead to unrealistic visions of the acquisition of wealth, power, and domination. Consequently, "a state overextends itself strategically" as it desperately attempts to expand or maintain its wealth, glory, ideology, and territory by consuming, at the expense and well-being of its population, greater quantities of national resources than it can afford. As of the latter part of 2007, the U. S. Congressional Budget Office (CBO) estimates that the cost of the Iraq and Afghanistan wars will exceed $2 trillion, or nearly $8,000 per citizen. Prior to the 2003 invasion, the Bush administration estimated the Iraq war would cost no more than $50 billion, which the CBO now estimates to be $1.9 trillion.[28] Historians note that Phillip II of Spain and England's Elizabeth ultimately utilized as much as three-quarters of government spending for warfare and related debt.

> **Despite the great resources possessed by the Habsburg monarchs, they steadily overextended themselves in the course of repeated conflicts and became militarily top-heavy for their weakening economic base. If other European Great Powers also suffered immensely in these prolonged wars, they managed—though narrowly—to maintain the balance between their material resources and their military power better than their Habsburg enemies.[29]**

In another era, the Spanish captains correctly stated that military victory goes to him who has the last *escudo*! The leadership of the great civilizations has strived to extract the last *escudo* from conquered populations as well as from their own constituents, and in the process, they stimulated cultural deterioration.

Previous chapters provide the context of the rate of cultural growth and decay, which Kennedy employs in his view of economic growth:

"the process of rise and fall among the Great Powers—of differentials in growth rates and technological change, leading to shifts in global economic balances, which in turn gradually impinge upon the political and military balances."[30] Kennedy contends that since the sixteenth century, Western Europe has witnessed positive correlations among the elements of production, revenue-raising capacity, and military strength, "a constantly upward spiral of economic growth and enhanced military effectiveness."[31] He notes that the relative strength of the leading nations in world affairs fluctuates because of an "uneven rate of growth among different societies," which is related to innovation in technology and an ability to organize, particularly in the area of the creative financing of warfare.

> **The relative strengths of the leading nations in world affairs never remain constant, principally because of the uneven rate of growth among different societies and of the technological and organizational breakthroughs which bring greater advantage to one society than to another.**[32]

In the past, as society became more progressive, opportunistic, and complex, it became obvious that future military success would depend on larger and more technologically advanced military forces that would require modern industrial capabilities and significant amounts of additional resources. Thus, materialistic values and priorities become more significant in critical decision-making processes, resulting in higher priorities and excessive investments of national wealth assigned to the military. Invariably, the sensate mentality wins the battle of competing investment priorities for the military versus the social agenda. The appeal of the military objectives of glory, wealth acquisition, flag, and potential religious and political conversions is irresistible. History from the late seventeenth through early nineteenth century witnessed a significant increase in the cost of waging war. Opportunism stimulated financial creativity for innovative methods of financing military budgets, including banking credit mechanisms, devaluation of coinage, and selling political offices and bonds. Those nation-states that acquired more sophisticated credit systems were successful in significantly escalating the nature, scope, and technology of warfare.

> **No state in this period ... could pay immediately for the costs of a prolonged conflict; no matter what fresh taxes were raised, there was always a gap between governmental income and expenditure which could only be closed by loans.... Again and again, however, the spiraling costs of war forced monarchs to default upon debt repayments, to debase the coinage, or to attempt some measure of despair, which brought short-term relief but long-term disadvantage.[33]**

The focus of Kennedy's discussion is that the economic ramifications of excessive military expenditures by Western civilization nation-states since 1500 contributed to a scarcity of long-term, sensible investment in balanced socioeconomic systems. As a result, imperialist expansion and conquest contributed significantly to social and economic instability and ineffective sociopolitical priorities, policies, and practices. Human abuses, societal disorder, and financial bankruptcy are the products of a culture's sensate values and excessive materialistic ambitions and the consequences of aggressive, nondefensive warfare. The histories of Imperial Spain and twenty-first-century America are examples.

Imperial Spain provides an appropriate case history upon which to apply the mechanistic-thermodynamic paradigm to the documented transitions from cultural growth to stagnation to disintegration. The approach to reviewing Imperial Spain's growth to deterioration sequence will emphasize the *history of economic and social developments,* as employed by J. H. Elliott and Fernand Braudel, rather than interpretations of historical events traditionally found in the social science literature. The specific focus of attention is the socioeconomic evolution from the fourteenth century to Imperial Spain's seventeenth-century collapse. This era of Spain's rise and fall in global prominence is reviewed in terms of changing cultural characteristics, properties, and objectives that have been emphasized in previous chapters. The elements of cultural evolution represent the dynamics of the cultural maturation sequence depicted in Figure 8-1 and relevant to Spain's cultural evolution from its early impressive socioeconomic advancements to its ultimate seventeenth-century collapse as an empire. Prior to their application, five lessons from such cultural evolution and transitions are noted.

First, humankind's innate motivation is primarily to satisfy basic human impulses, particularly those related to economic objectives. The *successes* and *failures* of a civilization are influenced by human abilities, values, priorities, and behavior as fundamental societal institutions are created, revised, expanded, managed, and eliminated, utilizing available tools and resources. History illustrates that prolonged periods of *success* occur when economic prosperity is achieved that is broadly and equitably shared within a population that embraces a common purpose and enjoys personal security, the comforts of a good life, as well as a sense of participation in the fundamentals of societal affairs. In essence, this represents a collective sense of commitment to and participation in a dynamic and mutually satisfying community. Conversely, *failure* often has as its root cause a stifling socioeconomic system resulting, to some degree, from the absence of this sense of community and associated altruistic values. The latter stages of the mature Imperial Spain did not meet this human behavioral test of success and, as a result, subsequent to *its initial success, made the transition to failure.*

Second, for a society to maintain a positive rate of cultural growth, sufficient wealth-energy capital must be available and skillfully utilized, hopefully embracing human values and behavior designed for broad-based societal success. This is a necessity for creating and maintaining an effective economic system capable of providing a civilization with opportunities for long-term social stability and financial viability. Thus, the design, implementation, and management of a successful economic system should be a primary priority for a society at any stage of its cultural evolution. The particular aspects of the complementary social and political processes become, out of necessity, a secondary priority to economic profitability, designed to achieve national objectives. Yet, the overall degree of stability and satisfaction within a civilization will depend greatly on the manner and effectiveness of human behavior in achieving and equitably distributing benefits derived from adopted economic priorities and objectives. However, of all the potential factors that may affect the degree of economic success and significantly impact the political and social systems of a civilization, the availability of wealth-energy capital is dominant. Ultimately, Imperial Spain's excessive appetite for consumption of resources to support an unwise and unbalanced set of priorities greatly exceeded its own resources plus those expropriated from others, thereby producing uncontrolled social and economic instability. Spain underwent a transition initially from consumption of wealth that achieved a high proportion of regenerated wealth and cultural prosperity to excessive expenditures and debt and an unstable economy.

Third, history has demonstrated by virtue of economic success, wars, human carnage, and territorial conquests that securing wealth-energy resources, whether for further military objectives, the luxuries of a sensate lifestyle, or basic survival, is humankind's greatest motivator. However, whether a civilization has a shortage or an abundance of wealth-energy capital at any given time, human behavior and the manner in which socioeconomic processes are conducted will influence the degree of economic success and social stability. The effect on Imperial Spain's attainment of, and subsequent retreat from, global prominence is a prime example of the generic influence of Sorokin's thesis that altruistic human values are replaced by dominate sensate values as a society prospers and matures, thereby influencing human behavior and activating negative cultural transitions. Human behavior, unpredictable circumstances, and cultural challenges and opportunities influence the priorities for and the effectiveness of wealth-energy capital utilization. These variables, related to human attitudes and behavior, ultimately determine the time frame for a civilization to initially experience major symptoms of socioeconomic degradation, rather than whether cultural stagnation will eventually occur.

Fourth, civilizations, as exemplified by Imperial Spain, have demonstrated a limitless appetite for the consumption of wealth-energy resources that eventually exhaust internal and external supplies and, in the process, create significant socioeconomic and political stress and disorder within a population. Without exception, such initially successful civilizations have conducted themselves in a manner that inherently accumulates disabling levels of societal entropy production and have suffered the consequences of socioeconomic deterioration. Thus, the aspirations of the "Caesars" to conquer, dominate, and secure wealth and fame have been significantly altered, if not destroyed, by the consequences of their frantic, greedy, and relentless searches for additional wealth, initially to subsidize imperial expansion, but ultimately to desperately avoid bankruptcy and the loss of empire. Imperial Spain is an illustration of this vicious cycle of intense sensate behavior employed to desperately secure resources in order to continue functioning in an aggressive, imperialistic manner that, if successful, would then require even more resources to maintain the increasing momentum of an expanding materialistic agenda. This expanding, self-perpetuating cycle of societal entropy production endlessly escalates society's complexity, disorder, and wealth and energy consumption. Spain completed this transition from a relatively orderly primitive social order to one of greater maturity, increasing societal entropy accumulation on its pathway toward cultural stagnation.

Fifth, recall from an earlier chapter that money and credit do not qualify as legitimate wealth in the mechanistic-thermodynamic paradigm of science-based economics. While emperors create currency and bankers extend credit, Mother Nature does not issue wealth and energy credits, even to Caesar! Extensive reliance on credit and artificially manipulated currencies to achieve expansionist and materialistic objectives has contributed significantly to the demise of empires; Imperial Spain and America are no exceptions. Both made the transition from a primitive, conservative economy to a faulty, debt-ridden, and wasteful economy.

To summarize these five historical lessons, it is a wise strategy to minimize a culture's chances of progressing toward economic stagnation by prioritizing national wealth consumption so as to achieve a high marginal return on economic investments and to realize worthwhile social objectives. This requires that wealth-energy capital be invested to successfully generate new forms of wealth and to maintain a positive rate of growth. This desired societal outcome hopefully results in human survival and prosperity within the framework of a satisfying and rewarding society. As with prior successful cultures, this strategy and the inherent values and behavior successfully guided the genesis and early growth stages of Spain's cultural development and impressive achievements. However, as with previous successful social orders, abuses of wealth, energy, and humanity led to Spain's socioeconomic deterioration and eventual collapse during the late sixteenth and seventeenth century.

The sequential transitions of a primitive society result in a mature civilization that eventually reaches the stagnation and disintegration

stages. The characteristics and properties of increasing cultural stagnation include the following:
1. Cultural mentalities shift from altruistic to sensate values as increasing materialistic success and opportunity stimulate greed and self-interest.
2. Evolving economic priorities become more highly focused on production of luxury goods and services and other nonessential expenditures incapable of contributing to new forms of wealth-energy capital.
3. Increasing cultural complexity generates more costly and time-consuming political and socioeconomic bureaucracies, policies, laws, regulations, codes, daily schedules, and societal expectations.
4. Increasing expenditures of national wealth to support a higher degree of cultural complexity and society's high priority on nonrenewable types of wealth consumption ultimately produce an unsustainable budgetary burden and financial insolvency.
5. Declining economic effectiveness, resulting from inappropriate and inefficient consumption of national wealth, necessitates securing new financial resources by any means possible including excessive debt and military interventions. Such tactics may transform socioeconomic stagnation into a temporary "pulse" or "rejuvenation" before the next, and possibly last, "lull."
6. Declining morale intensifies during the sequence of "times of trouble," socioeconomic stagnation, and cultural disintegration, reflecting general despair, a loss of creativity and a sense of self-determination, and a lack of confidence in the current political leadership and prevailing spiritual ideology.
7. Distribution of wealth shifts toward the socioeconomic elite at the expense of the average citizen, accompanied by increasing hardships on the general population.
8. Declining economic and social stability with a corresponding loss of citizen confidence escalates into general cultural chaos and despair, initiating the search for potential alternatives to

the current social, spiritual, and political philosophies, practices, leadership, and vision of the future.
9. Declining cultural productivity results from overinvestment in economic and sociopolitical complexity and bureaucracy intended to advance the civilization. Consequently, the marginal return on wealth-energy investments in cultural advancement becomes socially and economically unsustainable. Thus, mature civilizations inevitably deteriorate and face collapse unless new leadership utilizing more ideational, altruistic values is able to create a less complex, less resource-dependent, and more orderly and primitive social order.

Imperial Spain will serve as a case study to illustrate the emergence of cultural characteristics and properties that gradual evolve, representative of significant cultural deterioration and an "arrested civilization."

Spain's initial economic success would appear to begin with the aggressive initiatives of the state of Catalonia's overseas expansion and conquests in the Mediterranean during the late thirteenth and fourteenth centuries. However, the fifteenth century witnessed an economic recession, plague, famine, and significant social instability including armed civil conflict. One of the economic consequences was that the wealthy withdrew as investors from critical areas necessary for economic productivity, "turning themselves into a society of rentiers."[34] This would appear as a "blip" or fluctuation in Spain's growth curve, a temporary setback of its fifteenth-century economic system.

Fifteenth-century Spain consisted of the three Christian Crowns of Castile, Portugal, and Aragon. The states of Catalonia, Aragon, and Valencia constituted the Crown of Aragon and the dynasty was Catalan. Upon their marriage in 1469, Ferdinand and Isabella became joint sovereigns of Aragon and Castile through the union of two of the five royal houses of late Medieval Spain: Castile, Aragon, Portugal, Navarre, and Granada. As a result, cultural rejuvenation slowly reappeared from the *Union of the Crowns* uniting the Crown of Castile and the Crown of Aragon. The intense vigor, larger population, and more significant wealth of Castile provided the initial elements under the *Catholic Kings* for Spain's advancement, which eventually led to it becoming a great empire.

During the sixteenth century, Spain became a great military power, acquired great wealth by colonizing vast new territories, and in the process developed creative technological improvements in communication, cartography, navigation, mining, and metallurgy. Additionally, advanced methodologies were devised to manage unique issues associated with the social, economic, and political aspects of a massive imperial expansion into undeveloped regions of the world.

The rationale for Imperial Spain's seventeenth-century collapse is related to negative consequences of cultural transitions associated with the generic maturation of social orders as reflected by the characteristics and properties of increasing cultural stagnation summarized above. The seventeenth-century emergence and influence of advancements in science, philosophy, and technology are characteristic of Sorokin's sequential shifting of cultural mentalities from faith and religion to reason and philosophy and to science and the senses. This constitutes the systematic evolutionary nature of his sociocultural transitions, which is applicable to Imperial Spain.

Characteristically of mature civilizations, Spain was ultimately unable to adapt and provide the necessary discipline, initiative, creativity, and resources to maintain its economic and military primacy among competitive forces of seventeenth-century Europe. Additionally, an over commitment to the strategy of territorial expansion and warfare greatly exceeded Spain's human and financial resources, creating a decline in morale and the abandonment by the masses of the empire's mission. "Castile had become the victim of its own history desperately attempting to re-enact the imperial glories of an earlier age."[35]

J. H. Elliott's *Imperial Spain* provides an analysis of "the ills of an ailing society" by one of Spain's "*arbitristas*" (projectors), Gonzalez de Cellorigo, who experienced the Spanish atmosphere of Castilian "national disillusionment" during the reign of Phillip III in the early 1600s. Elliott reviews his analysis:

> The fundamental problem lay not so much in heavy spending by Crown and upper classes—since this spending itself created a valuable demand for goods and services—as in the disproportion between expenditure and investment. "Money is not true wealth," he wrote, and his concern was to increase the national wealth by increasing the nation's productive capacity rather than its stock of precious metals. This could only be achieved by investing more money in agricultural and industrial development.... The Castile of Gonzalez de Cellorigo was thus a society in which both money and labour were misapplied; an unbalanced, top-heavy society, in which, according to Gonzalez, there were thirty parasites for every one man who did an honest day's work; a society with a false sense of values, which mistook the shadow for substance, and substance for the shadow.[36]

This description of Spain's "fundamental problem" essentially depicts a failing socioeconomic system and demonstrates the necessity for a civilization's economic system to conform to a science-based economic model. Interestingly, Elliott's analysis of Gonzalez's 1600s Spanish cultural deterioration that the way to resolve "the disproportion between expenditures and investments ... concern was to increase the national wealth by increasing the nation's productive capacity rather than its stock of precious metals," reflects the twentieth-century scientific approach to public finance and and the national economics of Frederick Soddy, as found in previous chapters.

The Crown was unable to meet all of its financial obligations, which exceeded available resources. Consequently, only the highest pressing priorities could be achieved, which did not include productive investments. Resources were insufficient despite military conquests and financial assistance from supportive and unwilling wealthy citizens. Credit was no longer available, as a potential financial return on investments in the Crown's possible success was highly improbable. This produced disillusionment and severe hardships among the people, but given the severity of the problem, neglecting the needs of the internal population was the Crown's pathway of least resistance. Second, Gonzalez recognized the evils of debt, unwise use of capital, and monetary manipulations, which violate the integrity of a science-based national economic system (i.e., "Money is not true wealth"). Recall that such artificial financial arrangements do not constitute value in the form of *true wealth* either as a transactional medium or as consumable energy. Wealth is a flow of physical wealth-energy resources into production systems.

Third, Gonzalez conceptually reflects the wisdom of the second law of thermodynamics, viewing Imperial Spain's wasteful consumption of wealth as highly frivolous and inefficient, generating insufficient new wealth to support a viable socioeconomic system. That is, Spanish economic profitability was inadequate and ineffective in generating socioeconomic stability. The Crown's economic policies and practices, combined with insufficient investment in new, economically sound production, failed to "yield profits and attract riches from outside to augment the riches within."

> **Surplus wealth was being unproductively invested— dissipated on thin air—on papers, contracts, censos, and letters of exchange, on cash, and silver, and gold---instead of being expended on things that yield profits and attract riches from outside to augment the riches within.**[37]

Numerous, continuous wars, expanding bureaucracies and administrative costs, extensive public services, and the hoarding of wealth systematically reduced the wealth-energy capital available for investments in a potentially sound economy and stable society. Thus, the lack of productive investments in essential goods and services in favor of excessive nonproductive expenditures, particularly for territorial expansion and military activity, created disproportionate socioeconomic and political issues, including financial insolvency. Consequently, during the latter stages of Imperial Spain's existence, aspects of cultural stagnation proliferated in the forms of inequitable distribution of wealth, extreme poverty (including starvation), and extreme abuses of privilege and position.

Emphasizing mechanistic-thermodynamic principles, it is instructive to examine Spain's history from its thirteenth-century genesis phase through its seventeenth-century deterioration and collapse. In so doing, it is possible to demonstrate that the availability of wealth-energy capital and the ramifications of human behavior maintained a commanding influence on the ultimately irreversible deterioration of the Spanish Empire. Thus, particular attention is placed on socioeconomic and political conditions whereby sensate behavior and associated imperialistic expansion resulted in unwise utilization of resources. The consequences were a fluctuating but continually increasing rate of cultural decay that culminated in cultural stagnation and collapse.

The initial reference point for the discussion of the socioeconomic transition from Spain's growth stage to its ultimate stagnation and collapse is the impressive era of advancement during the fifteenth- and sixteenth-century reign of King Ferdinand and Queen Isabella. An appreciation for these years of socioeconomic progress under the guidance of the Catholic Kings provides insight into the subsequent years of cultural deterioration under the Habsburg reign of Charles V from 1517 to 1556 and subsequently under his son Phillip II.

However, it is useful to first review aspects of the thirteenth and fourteenth century from the perspective of establishing a context for the cultural values, traits, and motivations that Ferdinand and Isabella would inherit. It is noteworthy that this early socioeconomic environment created an atypical sociopolitical culture that would prevail and become a major factor during the critical years when Spain achieved enormous economic success as well as when the empire faced collapse. While the Crown of Aragon consisted of the three states of Catalonia, Aragon, and Valencia, Catalonia was responsible for the Crown's impressive foreign expansion and prosperity during the thirteenth and fourteenth centuries. The successful nobility of the Crown of Aragon politically leveraged their commercial strengths into an unusual governance arrangement with the Crown. Elliott describes this formalized understanding:

> **At the heart of this constitutional system was the idea of contract. Between ruler and ruled there should exist a mutual trust and confidence, based on a recognition by each of the counteracting parties of the extent of its obligations and the limits to its powers. In this way alone could government effectively function, while at the same time the liberties of the subject were duly preserved.[38]**

Thus, mutual understandings were established on the strength of the nobility's political influence resulting from economic control and past financial success and the Crown's economic and political weakness due to a lack of wealth and military power. This cultural concept and its inherent strengths and weaknesses, initially formulated within Catalonia, were carried forward into the political culture of the Middle Ages and beyond. This philosophy was of vital significance in Spain's continuous struggle to secure much-needed wealth-energy resources at critical periods in its history and a primary factor in the periodic creation of civil unrest as well as the population's ultimate abandonment of the empire's problems, debts, mission, and leadership.

However, during the Middle Ages this *contract* arrangement between the "ruler and the ruled" provided the political framework for successful commercial exports that more broadly benefited the Crown of Aragon. While this governance agreement was mutually beneficial for this early stage of Spain's advancement and scope of activity, ultimately it proved inappropriate as a foundation for the unprecedented expansive imperialistic initiatives and governance responsibilities. This historic cultural understanding of the *contract* was inherently a simplistic, small-scale cooperative economic arrangement that was ill-suited for pursuing and administering a global militaristic expansion. Philosophically, the sociopolitical environment associated with the contract was unaccustomed to, and antagonistic toward, attempts to centralize authority, particularly involving financial matters and manpower solicitation for the Crown's imperialistic wars. Rather than being subjected to the rule of imposed laws and policies by a dominating authoritative king, the political culture was described as:

> **a loose federation of territories, each with its own laws and institutions, and each voting independently the subsidies requested by its king. In this confederation of semi-autonomous providences, monarchical authority was represented by a figure who was to play a vital part in the life of the future Spanish Empire: … the viceroy**[39] **[first appointed in the fourteenth century in Catalan].**

The "Cortes," a powerful institution that was a component of the political confederation, was often able to significantly thwart the initiatives of the king. Each territory had a Cortes that would meet separately or jointly at their discretion or that of the king's. In addition to voting subsidies to the king, the Cortes had legislative power, with any repeal of laws requiring the mutual consent of the Cortes and the king. This system could be viewed as highly decentralized, quite democratic, and a valuable asset during the early stages of Spain's maturation. However, in a latter period, loyalty, and moral and financial support, as well as manpower needs for the king's expansionist activities and foreign wars, would always be dependent on the existence of an attractive quid pro quo, which in the latter stages of the empire could not be provided. The long-standing culture of the loose federation was inherently less cohesive than a traditional national unity government and placed the king in a weak power position if financial incentives did not exist. Consequently, commitments to a common cause were inherently limited to those with religious objectives, local issues, and financial matters possessing mutual benefits that could be accrued from cooperation, generally gifts of land or dispensation from taxes. During the cultural evolution of the thirteenth and fourteenth century, a more altruistic, ideational mentality prevailed that slowly evolved toward a sensate, materialistic mentality representative of the more mature Spanish Empire.

The catalyst that initially catapulted Castile into becoming a very successful commercial international market was its wool industry. About 1300, a new breed of sheep was introduced into Castile, and during the fourteenth century, with the Crown of Castile providing leadership, this industry became a powerful economic force. Castile's wool industry became increasingly profitable in the European market, but warfare reduced the extent of its expansion and economic profitability. Despite these issues, enormous profits for the aristocracy provided the wealth and influence to establish the great dynasties of Castile, and by the middle of the fifteenth century, a political power base was established. Since the Crown occupied a weak political position, the Cortes was able to focus on successfully managing economic expansion while ignoring governance responsibilities, thus creating inadequate civil administration and social disorder throughout the territory.

The early fifteenth century witnessed a severe economic decline of the Crown of Aragon, attributed to socioeconomic disruptions resulting from the accession of an alien dynasty in 1412. Elliott describes the dramatic change:

> **The glittering imperialism of Alfonso V, dynamic in inspiration and militaristic in character, differed sharply from the commercial imperialism of an earlier age, and, by encouraging lawlessness in the western Mediterranean, directly conflicted with the mercantile interests of the Barcelona oligarchy. The strategies and tactics of dynasty and merchants no longer coincided, and this represented a tragic deviation from the traditions of the past.[40]**

Consequently, the earlier period's coalescence of economic and political philosophies of the Crown and wealthy commercial leaders based on mutual understandings, cooperation, and benefits disappeared, thus destroying the successful Catalan socioeconomic system. The consequences included economic depression, civil unrest, and class warfare. In addition, the aristocracy, rather than addressing the potential of a futuristic national economy, chose to remove its wealth from Spain's flow of investment capital in favor of less risky financial returns from annuities and properties. Consequently, unfavorable social and political circumstances influenced the removal of wealth-energy capital from the economy, and the system failed. This illustrates how negative economic consequences may result from changing sociopolitical circumstances in a similar cause-and-effect relationship to an increasingly prosperous economy creating positive sociocultural ramifications. In the latter portion of the fifteenth century, the Crown of Aragon's economy failed, and socioeconomic stagnation and civil war resulted. At that point, Spain's future became the province of Castile and new leadership during an era referred to as the age of the Catholic Kings.

Fifteenth-century Spain was divided among three Christian Crowns: Castile, Portugal, and Aragon. Ferdinand, king of Sicily and heir to the throne of Aragon, and Isabella, the heiress of Castile, were married in 1469. However, not until 1479 did the subsequent turmoil end and Ferdinand and Isabella became joint sovereigns of Aragon and Castile, constituting the union of two royal houses of the five segments of late medieval Spain: Castile, Aragon, Portugal, Navarre, and Granada.

The year 1492 is considered the time at which Ferdinand and Isabella embarked upon their expansion within Spain and into foreign regions, resulting in the recapture of Granada and the discovery and conquest of the New World. It is instructive to review the differing motivations among the Spanish and, more specifically, the controlling Castilian principals involved in the exploration of the New World. In 1492 when Columbus departed on his first voyage to the West, the Crown was poor, and thus one motivation for their approval and support of the voyage was the potential for the acquisition of much-needed wealth. Isabella was also motivated by the potential expansion of the Christian mission and its contribution to the ongoing anti-Islamist crusade. Additionally, within Castile there existed a long tradition of military and maritime

Properties and Characteristics of Sociocultural Stagnation

expertise appropriate for exploring foreign territories. This faction of Castilians was motivated by the potential for the acquisition of wealth from contractual rewards from the Crown and from conquered land, booty, slave labor, and other informal spoils of aggression under the guise of religious motives. The historian Bernal Diaz del Castillo, described as a companion of Hernan Cortes, is quoted as stating, "We came here to serve God and the king, and also to get rich." Elliott adds, "The conquistadores came to the New World in pursuit of riches, honour, and glory. It was greed, cupidity, the thirst for power and fame that drove forward a Pizarro or a Cortes."[41] Finally, another Castilian faction with maritime expertise had been successful in expanding commerce, as for example the development of the wool trade with Europe and the creation of productive Spanish port cities. Columbus represented the latter faction, having the objective of establishing outposts of commerce to develop mutually beneficial trade involving some unknown land possessing some unknown valuable riches. The commerce faction lost the battle of strategies to the tactics of military conquest, plunder, and enslavement, after which replicas of Castilian communities would be established. Elliott describes the motivation of the conquistador; he

> **knew that he faced sudden death. But he also knew that, if he survived, he would go back rich to a world in which riches conferred rank and power. On the other hand, if he should die, he had the consolation of dying in the Faith, with hope of salvation. The religion of the conquistadores gave them an unshakeable faith in the rightness of their cause and in the certainty of its triumph.**[42]

During the period from 1519 to 1540, Spain established its American empire while extracting great wealth but, in the process, destroying the Aztec and Inca civilizations, all with papal approval and establishing America as a legal Castilian possession.

Given that the Crown was in financial difficulty at the beginning of the sixteenth century, it is instructive to determine the methods and sources of the wealth that financed this massive Spanish imperialist expansion as well as who benefited, and to what extent, from the resulting revenues. The Crown employed a combination of private and public funding sources, with both the State and the Crown participating in financing specific projects. Some entirely private expeditions also took place, but only under the legal authority of the Crown and with the expectation that certain specified rights of the Crown would be respected. The private investor could legally acquire property, natives for forced labor, land, and titles. The Crown's lack of wealth necessitated rather generous incentives for private investors, and generally the funding of Spain's conquest and colonization was provided by private businesses, but always under the control of the Crown. This control also included provision for the achievement of religious objectives and the requirement that the State would have sole legal authority over conquered territory.

The economic importance of the Indian slaves to profitable ventures in the New World quickly became a major issue and collided with the religious objectives as stated by the Crown, more specifically those of the queen. Thus, she formally prohibited the use of native slaves for forced labor in 1500, but creative methods were quickly devised by investors to rationalize enslavement as an economic necessity, particularly for the work of the mines. One solution was based on the premise that the provision of "pastoral care," which included security, education, and religious instruction, was payment for labor, which amounted to blatant slavery in the name of profit. This practice was eventually prohibited and future native slavery was abolished in 1542; however, it was replaced by imported Negro slavery. Ultimately, all slave labor was abolished.

Generally, for the benefit of the imperialistic economy, the Crown controlled the territories in an absolute fashion that, interestingly, did not exist in fifteenth-century Castile.[43] The approach to governance adopted by Ferdinand and Isabella within Spain was in stark contrast to the authoritarian philosophies and tactics employed by other monarchs of the time, such as Henry VII and Louis XI. The Catholic Kings respected and observed the historic thirteenth- and fourteenth-century Spanish political culture and its governing structures as well as the rights and privileges of the Cortes. They scrupulously avoided imposing any meaningful modifications on the two historically different governance systems of Aragon and Castile and specifically shunned any sense of governmental amalgamation of the cultural foundations of the two Crowns. The *unity* of the Crowns in the era of the Catholic Kings applied to creating opportunities and cooperative, constructive goals for Spain, but not the manner by which each Crown was administered and governed. Elliott expresses the governing values of Ferdinand and Isabella: They "believed in royal justice, in good kingship, which would protect the weak and humble the proud. If they had a high sense of their own rights, they also had a high sense of their own obligations, and these included the obligation to respect the rights of others."[44]

Ferdinand was able to restore law and order in Catalonia utilizing historically established rights and laws rather than using civil disorder as an opportunity to impose a new system more in line with Castile's long-established authoritative approach. Elliott expresses this reality as: "Where the kings of sixteenth-century Spain would be able to behave in many respects like absolute monarchs in Castile, they would continue to be constitutional monarchs in the states of the Crown of Aragon."[45] Consequently, their ambitions for Spain would continuously be complicated by the realities of dealing with the strong-minded Cortes and the inability to change laws and raise and maintain a military force, all having resource implications that affected imperialistic strategies, tactics, and outcomes. Thus, Spain in this era was composed of pluralistic states and a group of patrimonies, each with its unique governance system under a dual monarchy that exhibited little authoritarian control over internal affairs.

The Catholic Kings' approach to governance of respecting historic cultural values and prerogatives demonstrated a rare wisdom. The Catholic Kings' "patrimonial concept of the State" reflected Sorokin's ideational altruistic values constituting an atypical, constructive environment for the ruler and the ruled. Undoubtedly, this cultural climate was a major contributing factor to Spain's impressive socioeconomic progress during the reign of the Catholic Kings. Furthermore, such an ideational governing mentality is a valuable leadership methodology that places a high priority on broadly based societal progress. Additionally, this more altruistic approach would tend to minimize societal instability and disorder within a population as well as to maximize the distribution of wealth among the population. Such a constructive environment increases the probability of greater mutual satisfaction for both ruler and general population. On a broader scale, such a cultural atmosphere is noteworthy, not just based on a population's favorable response to leadership maintaining constant historic cultural values, but also when the inevitable change of leadership occurs. This became crucial with the eventual transition of leadership philosophy to a more authoritative and forceful style when the Catholic Kings were followed by the Habsburg reign of Charles V in the second quarter of the sixteenth century. The resulting clash of the historic governance culture and its value system with the new leadership's philosophy produced negative socioeconomic consequences.

However, prior to Charles V, Ferdinand and Isabella were able to establish order in Castile and to neutralize, to some degree, the aristocracy's political power. This objective necessitated creating a positive working relationship between the Crown and the municipalities based on a common interest of reducing the influence of the aristocracy and restoring the rule of law. It would appear that the Catholic Kings embraced and applied Sorokin's altruistic ideals to create a cultural environment conducive to economic advancement and social stability. It is appropriate to review this era of Spanish socioeconomic development to analyze the strategies and tactics utilized in the pursuit of socioeconomic progress.

The Crown's attempt to effectively govern and stimulate socioeconomic advancement necessitated dealing with the populations of the two distinct cultures of Castile and Aragon, both with a politically powerful nobility, and a wealthy and influential clergy that needed to be reformed. Fortunately, an intense religious tradition provided a common unifier that the two differing political and economic cultures did not inherently offer. History demonstrates that human values, behavior, and religious zeal based on a common spiritual ideology can be effective instruments of cultural motivation for a ruler, particular if pagans or heretics are readily identifiable and greatly feared or hated. Thus, religion may assist in unifying a culture's socioeconomic and political systems and creating common causes and foes.

The power of religion as a tool that inspires, controls, and manipulates a population in order to justify irrational and brutal imperialistic behavior was liberally applied. Spain's recapture of Granada in 1492 serves as an illustration. The country "whose religious sensibilities had been heightened almost to fever pitch by the miraculous achievements of recent years … saw the kingdom of Granada crumble before them." They "realized … that they should think of themselves as entrusted with a holy mission to save and redeem the world" from Islam. However, "to be worthy of their mission they must first cleanse the temple of the Lord of its many impurities; and of all sources of pollution, the most noxious were universally agreed to be the Jews."[46] A series of repressive policies and the final expulsion of the Jews were implemented, creating an economic disaster, as a large Jewish community had made very significant contributions to the economic success of both Castile and Aragon. The Crown's strategy was to closely align the political system with a religious ideology in order to achieve a sense of unity and inspiration for successful economic initiatives that required the mutual support of the clergy and the people. However, the economic outcome of the Jewish strategy was decidedly negative. The conquest of Granada and the defeat of the Moors were followed by the Catholic Kings' edict expelling all Jews from the kingdom, a final step begun by the Inquisition. Foreign immigrants, who exploited rather than contributed to Spain's mission, were unable to replace the "dynamic Jewish community whose capital and skill had helped enrich Castile." This "victory of faith" created unity and established the political advantage of a common national purpose but at a significant cost to "the economic foundations of the Spanish Monarchy at the very outset of its imperial career."[47]

However, such an emotional and spiritual context can effectively assist a ruler in securing the necessary manpower for waging war, increasing taxes, justifying public debt, validating civilian and military casualties of war, and unilaterally altering the laws of the land. Human values are politicized and utilized, whether or not in actual conformity with personally professed spiritual beliefs in order to influence the form, priorities, and characteristics of a civilization's socioeconomic system. This in turn provides the leadership with an effective rationale for the specific design and implementation of political strategies and practices.

Ferdinand and Isabella successfully manipulated the aristocracy by insuring that town officers and the judicial system representatives were loyal to the Crown. In addition to effectively managing secular institutions, the Crown was also able to control the Spanish Church and its immense power and wealth. Typically, the clergy's financial benefits included about one third of the Church's large annual income, tax exemptions, and large amounts of property. Effectively, it was almost impossible to segregate politics from religion, whether the issue was secular, religious, or military, but it was imperative to maintain separation of the governing politics of the two Crowns.

Given this cultural context, it is of interest to examine the tactics and objectives of Ferdinand and Isabella, as well as the Spanish economic strengths, weaknesses, and potential for achieving long-term socioeconomic advancement. The Catholic Kings' attempts to advance the fundamental social and political environment of Spanish society coincided fortunately with the unexpected, enormous wealth discovered and expropriated from America. Given Spain's historic deficiency of natural resources and the Crown's lack of wealth, this sudden appearance of external wealth provided new opportunities for exploration and global dominance. It has been noted that during Isabella's twenty-nine-year reign in Castile, 128 ordinances were established embracing many aspects of Castilian life with the intent of improving the organization, functions, and effectiveness of the national economy. Such reforms have been categorized as random reactions to address fiscal issues and to solidify royal power and wealth rather than a coherent and logically developed economic program. Moreover, the Crown's economic success continued during a time of peace, thus avoiding significant military costs and allowing the Castilian aristocracy to benefit substantially from an expansion of internal and external commerce.[48]

Thus, the primary variables affecting the rate of cultural growth and productivity, and therefore economic and social stability, were quite favorable under the reign of the Catholic Kings. The society reflected a relatively high degree of altruistic values, available resources, and capable leadership striving for broad-based socioeconomic success while believing "in royal justice, in good kingship which would protect the weak."[49] However, the long-term prospects for the necessary continuous supply of human and financial resources were not highly favorable. Castile, which carried the primary economic burden for the Crown, suffered from an inadequate financial base, a lack of skilled labor, and minimally endowed natural resources and thus did not possess an impressive economic foundation. Additionally, agriculture was not a particularly successful national industry, and inferior production processes and transportation systems further contributed to an unfavorable internationally competitive economic position. Consequently, while the reign of Ferdinand and Isabella provided a major constructive boost to Spain's socioeconomic circumstances, a major lack of resources and other economic limitations complicated and restricted Spain's future imperialistic ambitions. Thus, the history of Spain is a prime illustration of the devastating and ultimately fatal socioeconomic consequences of chronically inadequate resources affecting an ambitious, maturing civilization. From a broader and more long-term perspective that includes the future agendas and challenges to be assumed by their successors, the national assets available to support Spain as it embarked on its extended, major global expansion and conquests were grossly inadequate for the mission. Most likely, this explains the relative short life of Imperial Spain as a successful civilization and world power.

With the passing of the reigns of Ferdinand and Isabella, Spain's imperial course came under the leadership of the Habsburgs. Emperor Charles V, who had inherited Austria, Burgundy, Aragon, and Castile, functioned as ruler of Spain as King Charles I from 1517 until 1556. One significant aspect of the Spanish transition to Habsburg rule was the elimination of the historic degrees of freedom and self-determination permitted under the rule of Ferdinand and Isabella. However, Castile envisioned positive self-serving economic consequences from the new leadership of such a powerful emperor, who had ambitions to aggressively pursue the wealth of the New World and to defend Christianity worldwide against heresy and infidels.[50]

Charles's approach to governing was to permit each independent entity to be considered as an individual territory within an aggregation of territories and to be governed by its own traditional laws and traditions despite being ruled by a single authoritarian sovereign. This philosophy effectively stifled economic growth and cooperation among the territories and was not conducive to development of common strategies and policies or to a sense of shared purpose. This political environment ultimately led to ambivalence among the territories regarding the emperor's overseas military actions, financial problems, and other foreign issues affecting the empire.

Expensive wars and imperial bureaucracies were the distinguishing Habsburg themes under Charles V. New administrative complexity, procedures, and supporting bureaucracies were created to manage the necessarily large numbers of duties spread over a broad expanse of the globe. This involved numerous personnel and formal councils and a vast communication network to support time-consuming, intricate administrative processes. Essentially, sixteenth-century Spain attempted unsuccessfully to adapt medieval political, social, and economic organizations and processes to the requirements of an expanded agenda of a worldwide empire. The financial demands and procedural complexities of the expanding bureaucracies had a negative impact on socioeconomic stability and productivity. Castilian agriculture was unable to feed its population while supplying the new American market, thus bringing about higher food prices and making the necessities of life unaffordable for the Castilians. Consequently, foreign goods became cheaper than the domestic variety. Gradually, it became evident that the *American opportunity* was beyond the human and resource capabilities of Castile and the general Spanish population. However, Charles was able to obtain much-needed financing from the Netherlands, Italy, and the pope based on the imperial mission of fighting heresy. Additional revenues were acquired by increasing the burden of taxation, particularly on those least able to afford it. As a result, the population became disenchanted with Charles's imperialistic mission. It would appear that Spain, at this time, entered a pre-Stage IV cultural Deterioration phase of socioeconomic stagnation.

Human nature adapts to the extreme frustration that accompanies a society's sense of an unalterable momentum plunging toward disaster by simply ignoring the situation and mentally disengaging from the realities of societal existence. Elliott expresses a typical Spanish reaction: The "second and third generations of merchant dynasties preferred the pleasures of aristocratic existence to the tedium of the counting-house."[51] Particularly troublesome to citizen morale was a lack of evidence that wealth from American silver and gold was being used to stimulate the Castilian economy or even to improve the inadequate quality of everyday life. It became obvious that Spain lacked an aggressive and profitable, broad-based economic program to create new opportunities in the New World and that the Spanish infrastructure was rapidly and irreversibly deteriorating.

As the reign of Charles V was coming to an end and Phillip II began his rule, the magnitude of the empire's yearly operational deficits and its accumulated debt, as well as its related sociopolitical problems, became insurmountable.[52] German and Genoese bankers held mortgages on future revenues, thus foreign bankers controlled the country's future sources of wealth. Castile had been saddled with the country's main financial burden, particularly the least wealthy, who were unable to buy exemptions from various forms of taxation that were established in the hope of quickly increasing revenues. Those still possessing wealth became inactive in their commercial enterprises to minimize their losses and to salvage as much of their wealth as possible. Consequently, wealth-energy capital was being dramatically removed from the financial system, which was strangling the economy.

Resisting external threats and conflicts with France in the 1520s and the Turks from the 1530s through the 1550s and combating the German heresy was very costly. Wars were no longer affordable, credit was becoming unavailable, and the looting of other territories for new revenue was no longer an option. During the 1540s, the suffering of the common people brought increasing and widespread misery. As a result, raising money from any source became impossible. Meanwhile, the nobles continued raiding the national treasury, and the emperor continued to spend. Thus, in 1557 Phillip II inherited the Spanish-American Empire and imperial bankruptcy. However, a temporary financial rescue of the empire unexpectedly emerged, a *pulse* before the final *lull*.

Precious metals were becoming scarce in Europe, and meeting this need became a primary objective of the empire's American agenda. A new mercury amalgam process utilized in Peruvian silver mines provided a significantly improved extraction technique and tripled Europe's silver resources by the mid-seventeenth century. It has been estimated that the Crown's share of the bullion was about 40 percent, but most of it was due merchants for materials sent to the New World, which was almost totally dependent on imported supplies. The king's portion of bullion imports provided sufficient wealth to satisfy the bankruptcies of 1557 and 1575, thereby permitting Phillip to pursue his imperialistic aims during the 1580s and 1590s, including the annexation of Portugal.[53] However, the financial rescue, a blip in a downward spiral, was short-lived, and by the early 1590s Aragon was in revolt and ungovernable. The Castilian economy was effectively destroyed as the new silver revenues were once more used to resolve the Emperor's debt rather than to directly assist the people. Bankruptcy in 1596 brought an end to Phillip's imperialistic activities, as he was without financial resources as well as the moral and financial support of countrymen. However, as the rate of cultural decay escalated, the internal pain and conflict experienced by the masses continued.

During this time, Spain was unable to produce goods needed in America, and as a result, an economic opportunity to stimulate new commerce and much-needed revenue was lost. Europe provided Spain with such necessities as grain, timber, and naval supplies, which were paid for by America's precious metals. Also, Spain's continued lack of investment in municipal programs and infrastructure contributed to a lack of confidence in leadership to improve conditions within the country. As the American silver revenues again declined and the colonies became more self-sufficient in producing required goods, the emperor's revenues declined further.

Elliott captures the extent of Spain's seventeenth-century socioeconomic systemic failures and the population's desperation:

> **Castilians from all walks of life had come to look, as a matter of course, to the church, court, and bureaucracy to guarantee them the living which they disdained to earn from more menial occupations, at once despised and unrewarding.**[54]

The economic system was expected to support a welfare system funded by the church and the government. By the latter part of the seventeenth century, Castile was in the last stages of complete internal collapse of its fundamental governmental and economic functions. Elliott provides the post-mortem:

> **The fatal over-commitment of Spain to foreign wars at a time when Castile lacked the economic and demographic resources to fight them with success cannot be simply attributed to the blunders of one man. It reflects, rather, the failure of a generation and of an entire governing class. Seventeenth-century Castile had become the victim of its own history desperately attempting to re-enact the imperial glories of an earlier age.**[55]

While Elliott's analysis is not challenged, there exists deeper meaning in the wording of "fatal over-commitment," "economic and demographic resources," and becoming "the victim of its own history." A broader and more generalized understanding of Elliott's analysis is achieved if one applies mechanistic-thermodynamic principles to Elliott's rationale for the collapse of Imperial Spain. The "fatal over-commitment" was based on Spain's misjudgment that its available resources were sufficient to meet the requirements of an unrealistic scope of internal and external activities. Thus, the specific nature of the lacking "resources" originates with misguided priorities, unwise behavior, and materialistic values, which results in inadequate wealth-energy capital.

Consequently, Spain, as with previous mature civilizations, was "the victim of its own history," inevitably facing collapse as the result of its inability to regenerate wealth. Rather than investments in profitable enterprises, significant expenditures were devoted to territorial expansions, the military, increasing bureaucracies, excessive public services, debt, and consumer luxuries. Finally, one of Elliott's most familiar statements requires consideration:

> **The men of the 17th century belonged to a society which had lost the strength that comes from dissent and they lacked the breadth of vision and the strength of character to break with a past that could no longer serve as a reliable guide to the future. Heirs to a society which had over-invested in empire, and surrounded by the increasingly shabby remnants of a dwindling inheritance, they could not bring themselves at the moment of crisis to surrender their memories and alter the antique pattern of their lives. At a time when the face of Europe was altering more rapidly than ever before, the country that had once been its leading power proved to be lacking the essential ingredient for survival—the willingness to change.[56]**

It is instructive to consider this quotation based upon material contained in previous chapters. Civilizations, in an advanced stage of maturation, embrace behavior reflective of Sorokin's sensate mentality, a ruthless and selfish pursuit of materialistic economic rewards. Imperial Spain's history includes the creation of cultural complexity, bureaucracies, financial insolvency, and social disorder. However, one interpretation of Elliott's famous words cited above provides a slightly different view of Imperial Spain's stagnation and collapse. That is, while it "lacked the breadth of vision and the strength of character to break with a past" and these "men of the seventeenth century" were "heirs to a society which had over-invested in empire," the culture also lost the human capability to manage the expansive, unparalleled complexity and cost of the empire. They simply did not possess the discipline, resources, tools, and skills necessary to halt the inevitable escalating momentum of a historically massive, deteriorating civilization headed toward collapse. Rather than just "lacking the essential ingredient for survival—the willingness to change," the society also lacked sufficiently advanced human capabilities to manage a massive, runaway society that had rapidly become highly sophisticated. They lost the ability, will, methods, and means to survive impossible circumstances. The critical cultural traits necessary to affect change and particularly socioeconomic rejuvenation were inadequate to the task; discipline, creativity, ambition, initiative, integrity, and hope had been dissipated by events and despair of recent history.

Finally, this pattern and the characteristics of Imperial Spain's transitions from rapid and productive cultural growth to stagnation and collapse is a recurrent pattern followed by the ancient Greek and Roman civilizations as well as by the British and American empires. The etiologies of all follow similar pathways of cultural degradation and exhibit the characteristics and properties of cultural stagnation outlined and illustrated in this chapter: the inevitable consequence of abusing Mother Nature's wealth-energy resources and of human shortcomings.

Chapter 9
References and Notes

1. Kroeber, A. L. *Configurations of Culture Growth*. Berkeley: University of California Press. 1944. 818-9.
2. Toynbee, Arnold J. *A Study of History*, ab. Sommervell, D. C. New York: Oxford University. 1957. I-VI: 273.
3. Ibid., 247.
4. Sorokin, Pitirim A. *Social and Cultural Dynamics*. New Brunswick: Transaction Publishers. 1991. 630.
5. Kroeber, A. L. *Configurations of Culture Growth*. 820.
6. Ibid., 821.
7. Ibid.
8. Ibid., 822.
9. Ibid.
10. Ibid.
11. Toynbee, Arnold J. *A Study of History*. I-VI: 244.
12. Brander, B. G. *Staring into Chaos: Explorations in the Decline of Western Civilization*. Dallas: Spence. 1998. 196.
13. Toynbee, Arnold J. *A Study of History*. I-VI: 244-246.
14. Ibid., 12.
15. Brander, B. G. *Staring into Chaos*. 213.
16. Toynbee, Arnold J. *A Study of History*. I-VI: 164.
17. Ibid., 165.
18. Ibid., 244.
19. Ibid., 272.
20. Ibid., 282.
21. Quigley, Carroll. *The Evolution of Civilizations*. Indianapolis: Liberty Fund. 1979. 130.
22. Ibid., 159.
23. Ibid., 142-144.
24. Sorokin, Pitirim A. *Social and Cultural Dynamics*. 535, 565.
25. Toynbee, Arnold J. *A Study of History*. I-VI: 190.
26. Kennedy, Paul. *The Rise and Fall of the Great Powers*. New York: Vintage Books. 1987.
27. Ibid., xvi.
28. Dilanian, Ken. "War Costs May Total $24 Trillion." *USA Today*. 23 October 2007.

29. Kennedy, Paul. *The Rise and Fall of the Great Powers*. xvii.
30. Ibid., xx.
31. Ibid., xvii.
32. Ibid., xv.
33. Ibid., 72.
34. Elliott, J. H. *Imperial Spain*. London: Penguin Books. 1990. 39.
35. Ibid., 380.
36. Ibid., 317-318.
37. Ibid., 317.
38. Ibid., 28.
39. Ibid., 31.
40. Ibid., 36.
41. Ibid., 65.
42. Ibid., 66.
43. Ibid., 72-76.
44. Ibid., 77.
45. Ibid., 82.
46. Ibid., 105-106.
47. Ibid., 110.
48. Ibid., 110-125.
49. Ibid., 77.
50. Ibid., 164-168.
51. Ibid., 198.
52. Ibid., 199-207.
53. Ibid., 268-277.
54. Ibid., 312.
55. Ibid., 380.
56. Ibid. 382.

Chapter 10
Ecological Ramifications of Modern Socioeconomic Progress: Chesapeake Bay and the World's Oceans

> "No one likes to believe that his or her personal well-being is in conflict with the greater public need. To invent a plausible or, if necessary, a moderately implausible ideology in defense of self-interest is thus a natural course."   Author Unknown

The previous chapter highlights the properties and characteristics of Imperial Spain's sociocultural transitions from successful growth to decline, stagnation, and collapse. Historically, this inevitable sequence of a maturing social order has inherently created significant, escalating higher orders of cultural complexity, disorder, and degradation as well as human physical, emotional, and psychological stress and debilitation. The purpose of this chapter is to document a parallel sequence of ecological degradation that is also inherently and inevitably associated with the achievement of a society's social progress.

In the pursuit of a rewarding standard of living, based on an expectation of continuously expanding economic productivity, society unabashedly alters, despoils, consumes, and often exhausts Mother Nature's land, water, air, and wildlife. Such inherent *environmental societal entropy production* is the inevitable by-product of economic prosperity with both reaching historic levels during the modern era of industrialization, capitalism, and urbanization. The price for "human progress" is dramatically illustrated by the ecological degradation of North America largest estuary, the Chesapeake Bay, a microcosm of the current vitality of the world's oceans. The ecological demise of the Chesapeake Bay during the last one hundred years, "in conflict with the greater public need", has been the "natural course" of an "implausible ideology in defense of self-interest" seeking socioeconomic "personal well-being."

A 2008 article in the journal *Science* reports that oxygen-depleted marine "dead zones" in coastal oceans have increased exponentially since the 1960s to about 400, representing a total area of more than 245,000 square kilometers.[1] These zones occur in historically prime fishing areas and include the Gulf of Mexico, Chesapeake Bay, the Pacific Northwest, offshore China and Norway, and the Baltic Sea. The dead zone in the Gulf of Mexico has doubled in the last twenty years and currently covers an area the size of Massachusetts. These zones are a direct consequence of people's choices of where to live, what to eat, and how to travel; the excessive use and waste of wealth-energy resources; and the proliferation of more numerous and serious environmental problems of modernity. Chemical fertilizers for agriculture and lawns, human and animal waste, and air pollution, including that from cars and trucks, generate the nitrogen products in the water that are responsible for oxygen depletion and reduced fisheries harvests.

More specifically and with a more focused geographical analysis, the extent and ramifications of society's environmental entropy production resulting from rapid socioeconomic advancement is illustrated by the systematic devastation of the Chesapeake Bay's water quality and wildlife during the last century, a direct consequence of extremely profitable regional economic opportunities. Economists refer to such undesirable by-products of socioeconomic profitability as "externalities." A parallel may be drawn between the physical, biological, and chemical degradation of the Bay as an ecosystem and the processes of cultural existence responsible for the wide-ranging stagnation and breakdown of Western civilization. Nature's ecosystems and human cultural systems both rely on the same generic functional mechanisms and their inherent wealth and energy consumption that accomplish useful, desirable outcomes while producing wasteful "externalities" in conformity with thermodynamic principles.

The Chesapeake Bay is North America's largest estuary, covering 4,500 square miles with 12,000 miles of shoreline, containing almost 20 trillion gallons of water supplied by more than 150 tributaries. The population of the Bay's six-state 64,000-square-mile watershed has almost doubled, from 8 million to 15 million people, since 1950 and is projected to reach 18 million by 2020.[2] During the last century, the social, economic, and political realities within the Bay's watershed have produced significant and continuous irreversible environmental deterioration of water quality, marine life, and the surrounding landscape. This is the result of environmental abuses and willful disregard for the long-term economic and esthetic importance of this unique and rich natural resource. The degradation of the Bay's world-renowned oyster and crab populations from excessive harvesting, water contamination, and disease has been far-reaching and harsh. Unfortunately, short-sighted economic and political priorities for quick, self-serving socioeconomic rewards will continue to insure that the health of the Chesapeake Bay continues its current rapid deterioration toward total environmental collapse. This degradation of the environment over the last century is the direct consequence of opportunistic tactics, political decisions and indecision, harvesting practices, and business initiatives designed to extract maximum profits and a satisfying standard of living from one of the world's most plentiful ecosystems. The success of this strategy is evident by the region's successful economy, which exceeds that of many states and some European countries. The commercial harvest alone from the Bay is estimated at $1 billion per year.

Industry, business, governmental agencies, and the public have, often in ignorance but always with aspirations for greater economic gain, successfully manipulated sociopolitical systems to place short-sighted economic self-interest above long-term sustainability of the Bay's ecosystem. People residing in the six-state region, representing various political and economic interests, have made ethical, political, and economic decisions resulting in tributaries of the Bay containing high levels of contaminants. These include kepone, tributyltin, parasites such as MSX (a nonindigenous species introduced by transplanted Pacific oysters), chromium, iron, lead, zinc, arsenic, mercury, nickel, and toxins that have adversely affected water quality and wildlife. As a result, the Bay is currently incapable of achieving and sustaining prior levels of water quality and fishery resources, a price paid for enormous financial gains achieved during the last century.

As Mother Nature mandates via her entropy concept, the spontaneous pathway of society's management of the Chesapeake Bay's historical rich natural resources has been, and will remain, in the direction of declining water quality, fisheries resources, and economic productivity. While marginal, temporary reductions of some limited aspects of the Bay's ecological disintegration are possible, the long-term prognosis is for further irreversible environmental disintegration. The only question is the rate of degradation that will be imposed by human behavior. This century-long process of ecological deterioration is analogous to Toynbee's cultural evolutionary phases of growth to stagnation to disintegration being affected by problematic human behavior, excessive and wasteful resource consumption, and the realities of mechanistic-thermodynamic principles. For such are the consequences for both cultural and environmental systems from socioeconomic transitions that involve everyday economic, political, and social priorities, policies, and practices. Additionally, both systems depend upon human values, motivation, creativity, and intelligence as well as available technology, tools, and resources. Interestingly, the current high rate of environmental degradation emerging from China's rapid twenty-first-century economic development is strikingly similar to that of the Bay during the twentieth century and, on a larger scale, from Western civilization's industrial development during the last century. Socioeconomic advancement of both East and West involves similar human objectives, motivations, behavior, and ecological abuses in the name of human advancement.

Given the century-long transformation of the Chesapeake Bay region, it is appropriate to examine the mechanisms and human behavior related to its economic expansion and environmental evolution from a pristine ecological condition prior to its twenty-first-century status. A principal influence in determining the environmental health of the Chesapeake Bay is the oyster, which is the keystone species of this ecosystem. This designation is not based on its harvest value but on its ability to create and maintain clear water for the ecosystem. Nature relies on the oyster's ability to strain and clean the Bay water by virtue of its capability to filter up to fifty gallons of water a day and thus remove water-borne nutrients. It has been estimated that the oyster population prior to 1900 was able to filter the estuary's entire volume of water in a few days while, because of the significant reduction of the oyster population, the same goal today would require about a year.[3]

Given clear water, sunlight is able to penetrate the shallow Bay's depths and provide energy for aquatic plant life, which is an essential element of a functioning ecosystem. The effects of the depleted oyster population extend beyond its filtering function to maintain clear water and to absorb nutrients. In the past, oysters paved the bottom of the Bay, serving as an anchor for aquatic plants and forming oyster reefs to shelter small fish and crustaceans. Such reefs provide surface areas fifty times greater than a flat bottom and an ideal habitat for marine creatures.

The fate of the oyster population has been compromised by decades of overharvesting, pollution, and imported diseases. Thus, it should not be surprising that the depletion of the Bay's oyster stock to less than 1 percent of the pre-1900 level should have a significant detrimental effect on water quality and thus on the fisheries population and the general health of the entire ecosystem. Concern for the oyster harvest and particularly for excessive harvesting was recorded as early as 1875, when the annual harvest dropped from 14 million bushels to 10 million bushels, thus prompting the Maryland General Assembly to call for an assessment of the state's oyster reserves. Lieutenant Francis Winslow of the U.S. Coast and Geodetic Survey carried out an intensive study of oyster beds in Pocomoke Sound and Tangier Island. The study determined that oyster stocks had been depleted over the previous thirty years and recommended that the state take action. Professor. William K. Brooks of Johns Hopkins University, a prominent biologist and experienced waterman, joined the attempt to protect the oyster population. In 1891, he published *The Oyster,* in which he strongly supported oyster aquaculture and was highly critical of "public fishery." "It is a well-known fact that our public beds have been brought to the verge of ruin by the men who fish them…. Soon there will be none left to replenish the beds."[4] His assessment and prophecy were accurate.

In recent years, the Bay's total annual crop has declined below 100,000 bushels, or less than 1 percent of pre-1900 harvest levels. The 2003-2004 harvest was only 26,000 bushels.[5] In past decades, the tonging, dredging, and scraping of the bottom for what appeared to be an inexhaustible supply of oysters was very profitable. Unfortunately, the natural habitat, particularly underwater grasses that support many species of aquatic life, was destroyed. More advanced harvesting techniques and tools, while increasing harvest profitability, have proven detrimental for wildlife, particularly the oyster and blue crab populations.

The blue crab has been the Bay's trademark aquatic species and seafood staple for centuries and is currently an annual $150 million industry, but its numbers have significantly declined. As with the oyster, the abundant crab population has been viewed as a *resource to be plundered rather than a long-term industry to be wisely managed,* another example of society's inept stewardship of Mother Nature's resources. As an example, the adult female crab population has plummeted by as much as 80 percent during the past ten years. Currently, restaurants and retail stores surrounding the Bay must import oysters, blue crabs, and fish native to the Bay from other countries in order to meet the local commercial sales potential of Virginia and Maryland consumers. Much of the U.S. supply of crabmeat comes from Asia and South America, is cheaper than the limited local supply, and is available all year.

Even with such a diminishing blue crab population, Virginia has permitted the harvesting of egg-bearing crabs and winter dredging, both practices prohibited by Maryland. The Virginia Marine Resources Commission, whose membership has overwhelmingly represented the politically influential and the financially prominent commercial fishing industry, guides such Virginia policy. There are approximately 5,000 licensed commercial crabbers on the Bay utilizing over 600,000 pieces of "gear," which is five times the number utilized fifty years ago. During this period, commercial harvests have steadily declined, with estimates that as many as 90 percent of the crabs capable of spawning are harvested each year.[6]

The experts agree that the blue crab population is fully exploited and near the point of collapse.[7]

In 2007, a committee of eight scientists from Virginia, Maryland, North Carolina, and South Carolina was appointed to address the issue of the continuing decline of crab harvests. Specifically, Maryland's 2006 catch of 28.1 million pounds was among that state's lowest harvests ever recorded, even worse comparatively than Virginia's low harvest of 22.5 million pounds caught in 2006. One member of the committee stated that just about everything had been tried: "We're kind of at a loss to do something that we haven't already done." The Virginia director of fisheries stated, "We have this huge regulatory package, and we're not seeing results. We've obviously missed something, but we're not sure yet."[8] However, others, for example the Chesapeake Bay Foundation, have stressed that the depletion of fisheries resources is due in large measure to poor water quality, overharvesting, and the massive loss of aquatic vegetation. It is estimated that only about 10 percent of the potential 600,000 acres of aquatic vegetation essential for breeding and nursery functions of wildlife presently exists.

The Chesapeake Bay receives fresh water from approximately 150 tributaries, and as an outlet to the Atlantic Ocean, the rising tides from the ocean bring salt water many miles into the Bay and its tributaries. Over the course of the last century, many governmental and business policies, practices, and sociopolitical strategies have contributed to the Bay's declining water quality and to the depletion of fishery resources. Periodically, large numbers of fish suddenly die within a river basin leading into the Bay, described as a *fish kill*. In 2005, one tributary of the Potomac River experienced an 80 percent fish kill of its smallmouth bass population, one of many such kills within four years in the region's rivers.[9] Investigations often lead to a specific source of pollution such as an industrial spill, commercial activity, or a sewage treatment plant problem, which may be corrected. However, unexplained declining fish populations have been recorded for decades, and dead fish appear with lesions that cannot be traced to a spot source, a particular virus, or even a particular cause of death. Often such diseases and fish kills are attributed to such vagaries as water quality and "environmental stress," the science of which is known and documented.

Nutrients from farm animals and crops and from urban populations are transported by the Bay's tributaries and feed algae that bloom, particularly during the summer heat, consuming oxygen and creating aquatic dead zones for shellfish that extend over many square miles. August 2007 witnessed such a *mahogany tide* that covered 40 percent of the lower Bay from Norfolk to Cape Henry, Virginia, and into the Atlantic Ocean.[10]

Politics, coupled with potential socioeconomics prosperity from the Bay's natural resources and scenic beauty, has played a historic role of insuring commercial success while ignoring the discipline and options of the scientific realities that could minimize environmental damage. Since 1875, when the Maryland General Assembly expressed concern for the Bay's oyster harvest declining from 14 million bushels per year to 10 million bushels and commissioned an "assessment,"[11] to the modern-

Wealth, Energy, and Human Values

day oyster harvests of fewer than 100,000 bushels, the number and costs of such assessments continue without justifiable results. Governors, state and federal agencies, universities, environmental nonprofit groups, corporations, and businesses have studied the Bay and its tributaries but have been unable to stem the tide of the Bay's declining health. Of particular note is the River and Harbor Act of 1965, directing the U.S. Army Corps of Engineers to perform a $15 million study *to assess* the Bay's environmental condition, *project* water resource needs to the year 2020, and *recommend solutions* for the identified problems. The outcomes were a seven-volume *assessment* of current conditions in 1973 and a twelve-volume report in 1977 of the *projected state* of the Bay in 2020. The third segment, *solutions to problems*, was not completed.[12]

From 1984 through 2002, $282 million in federal funds were dedicated to restoring the Bay. The Chesapeake Bay Foundation estimates a minimum of $8.5 billion is needed to meet water quality and land use goals contained in the Army Corps of Engineer's recommendations for a 2010 deadline.[13] The state of Maryland plans to spend about $630 million per year for Bay-related cleanup programming and anticipates an overall cleanup cost of $20 billion. It has been estimated that just since the early 1990s, total government spending on oyster restorations efforts alone has been about $300 million a year, during which time the oyster population has declined further.[14] Regrettably, actions of the executive and legislative branches of government during the last century, presumably to "restore the Bay," appear to be political actions intended to create the impression of doing something, albeit too little too late, while protecting the economic interests of regional business and industry. These have constituted political holding actions designed to transfer irresolvable problems to the next generation of politicians. Typically, economic interests have guided and motivated politics and public policy.

The lack of clear water and sufficient oxygen, the presence of inorganic and organic pollutants, and the continuous influx of high levels of nutrients fundamentally prohibit any appreciable return of the oyster and crab populations and the necessary aquatic vegetation for marine life. Industrial contaminants such as mercury, tributyltin, PCBs, kepone, and other similar foreign materials have entered the waterways in alarming quantities and have restricted swimming, fishing, and human consumption of fish. The Elizabeth River at the city of Norfolk is too polluted with bacteria for safe swimming, and the tributyltin concentrations are the highest levels found within the Bay system. The river's eastern and southern branches lack sufficient oxygen for the survival of aquatic life. The southern branch has the highest concentration of toxic polycyclic aromatic hydrocarbons (PAHs) in the world, and 60 percent of the riverfront's native wetlands have been destroyed in the last half century to accommodate urban development.[15]

Nutrient levels of nitrogen and phosphorus runoff into the Bay from lawns, septic system drainage, and farm animal waste enrich algae growth proliferation and reduce dissolved oxygen levels. A fish's lack of available oxygen is referred to as "environmental stress," which kills or physically weakens its resistance to disease. As a direct result of human activity, it is projected that 287 million pounds of nitrogen and 20 million pounds of phosphorus enter the Bay annually. The origin is about 25 percent from point sources, including about 300 major sewage treatments plants, and about 66 percent from ground water that includes agriculture runoff.[16] Poultry and pig farming have been major sources of pollution. A poultry farm of about 30,000 breeding chickens, as typically found in this region, produces over 350 tons of droppings each year, usually sold for farm fertilizer. Interestingly, the Commonwealth of Virginia did not establish poultry waste regulations until 1999, and state and federal monitoring of water flowing into the Bay on a routine basis is essentially nonexistent.[17]

With new scientific advancements, new chemical species arrive, bringing additional threats of contamination into the Bay's water, contributing to society's mounting "externalities" (i.e., *environmental entropy production*). For example, prescription drugs that pass from human and animal intake into the runoff chain of ground water affect wildlife, specifically the health and reproduction of some species of fish in the Bay. Endocrine disrupter chemicals (EDCs) are known to interfere with natural hormonal functions and appear to be associated with fish kills. A Potomac River investigation of such a fish kill by the U.S. Geological Survey revealed that almost 100 percent of the male smallmouth bass possessed female characteristics (i.e., eggs in the testes).[18] *Intersex imposition* (simultaneous presence of both testicular and ovarian characteristics) is considered a potential consequence of estrogen exposure. Specifically, estradiol, one of the most potent EDCs, reaches aquatic systems mainly through sewage and animal waste disposal. A University of Maryland study of ponds, rivers, sewage treatment plants, and coastal bays reports that estradiol "concentrations in the various surface waters of the Chesapeake Bay watershed on the Eastern Shore of Maryland are ... sufficient to induce estrogenic effects in aquatic organisms."[19] The Wicomico River, one of the most polluted rivers in Maryland, had the highest EDC concentration at the Salisbury site, with over half being attributed to a large poultry production facility.

With the continuous increase in contaminants and the decline of water quality, it is not surprising that aquatic diseases have become increasingly more prevalent in the Chesapeake Bay. The rockfish provides a fitting illustration.[20] The population of Chesapeake Bay rockfish, also called striped bass, was decimated by excessive fishing during the 1980s. Consequently, the Atlantic coastal states imposed a rockfish moratorium, resulting in the successful replenishment of the fish population despite the Bay's polluted condition. In 1995, the moratorium was rescinded. However, within two years a bacterium fatal to rockfish and responsible for a human skin infection appeared. The disease is a *wasting disease*; the infected fish initially appear healthy, as the disease spreads through internal organs, and the fish gradually loses weight. Recent research indicates that by age one, about 10 percent of the population is infected, which increases to almost 20 percent by age two; by the third year, some of this population have begun to migrate. The nature of the disease is currently unknown.

A decade after the moratorium was lifted, the disease had infected about three quarters of the rockfish in the Bay. In 2004, rockfish was a $300 million industry for the states of Maryland and Virginia, but the impact of the disease potentially extends beyond the Bay. Scientists determined that the Chesapeake Bay was the spawning ground for rockfish as well as for the source of the mycobacteriosis. Normally, rockfish can live for more than twenty years, and most begin life in the rivers that flow into the Bay. At three to six years of age, they migrate into the Atlantic Ocean, traveling as far north as the Canadian border, but return to the Bay in order to spawn.

Clearly, the realities of science and the sociopolitical nature of the twenty-first-century urban culture of the Bay's watershed region combine to create an irreversible pathway toward continuing ecological stagnation and collapse. A century of declining cultural ethics, integrity, and values within the population has supported governmental priorities and policies that have enabled economic interests to abuse the Bay's natural resources in the name of progress. The history of the last century demonstrates that while preaching "restoration of the Bay," practices have continuously produced insufficient oxygen, excess harvesting, and inadequate aquatic vegetation for wildlife. Thus, the dominant cultural traits and behavior of the population within the six-state watershed can be expected to continue the historic environmental devastation, with only minimal progress at the margins of these major ecological problems and issues.

The intent of this chapter is not to exhaustively review the physical and biological influences of societal advancement on the deteriorating fisheries industry and the health of the Chesapeake Bay bionetwork during the last century. The aim is to illustrate and highlight the political and socioeconomic priorities and behavior affecting the transformation of the rich natural resources of the Bay into a deteriorating ecosystem, increasingly unable to support aquatic life or a profitable marine fisheries industry. Unfortunately, this unique environmental disaster is a microcosm of worldwide sociopolitical behavior, motivated by aspirations of wealth that has contributed to a rapid global production and accumulation of *environmental societal entropy*. Unfortunately, this is an example of wealth-energy resources being unwisely utilized for the intended purpose of providing long-term economic, social, and environmental stability.

The twenty-first-century reality is that the self-interest of commercial benefactors within the six-state Bay region, with the duplicity of political entities, has resulted in the depletion of the fisheries stock and the degradation of the physical health of the Bay and its tributaries. The science applicable to the current inferior water quality, excessive contaminants, and loss of aquatic plant and animal life is well known. However, the political aspects that contribute to this irreversibly deteriorating ecosystem reflect the historic shortcomings of leadership and human behavior that also typify the deterioration of mature civilizations since the time of ancient Greece. Chesapeake Bay politics provides an appropriate illustration. The political follow-up to a 2006 recommendation by the Atlantic States Marine Fisheries Commission regarding the Bay's menhaden fish population demonstrates how business interests and political ambition have colluded and contributed to the Bay's demise.

The menhaden is a relatively unknown, little silvery fish considered unfit for human consumption, but an inexpensive and easily obtainable source of fishmeal for animal feed. These fish live in immense schools several miles long, numbering in the millions, and are easily netted in the Bay by a fleet of commercial vessels. As a source of fish oil, menhaden harvests are used to manufacture many types of commercial products. A large Texas-based company, Omega, has historically removed enormous quantities of menhaden from the Bay (exceeding 100,000 metric tons annually). Such industrial large-scale "mining" of menhaden is permitted by only two of the fifteen Atlantic states: North Carolina and Virginia. Such large catches resulted in fishermen and environmentalists publicizing the negative ramifications of these large annual harvests. As a result, the Commission recommended an annual limit on these commercial catches. In order to be implemented, this recommendation required the approval of the Virginia General Assembly and was brought before the appropriate House of Delegates subcommittee in 2006, where the recommendation died from lack of action.[21]

Interestingly, the Virginia Public Access Project reveals that the chairwoman and four other committee members received campaign contributions from Omega. The committee's decision was supported by an opinion from the commonwealth's attorney general that the Commission's recommendation was not binding in Virginia; he also received a campaign contribution. This matter was subsequently taken to the governor, who had also received a campaign contribution from Omega to assist him in defeating a rival who had received campaign contributions from Omega for a number of years. The governor took the position that for technical reasons his hands were tied and he could not act. Consequently, Omega was allowed to proceed unimpeded with its menhaden harvest. A *Virginian-Pilot* newspaper editorial entitled "If Only Menhaden Wrote Campaign Checks" noted "the ridiculous lengths that Richmond will go to protect a single campaign contributor."[77]

This history is a modern version of the corrosive influence of economic and political collusion that for decades has contributed to the Bay's demise, a microcosm of the larger national corrosive experience over the last century. Additionally, this illustrates the realities that have historically faced the implementation of appropriate non-politically based public policy recommendations from experts that challenge real estate, tourist, and industrial expansion. Consider the nature and purpose of the Commission: The Atlantic States Marine Fisheries Commission was formed in 1942 by the fifteen Atlantic Coast states, recognizing that fish do not adhere to political boundaries. The Commission serves as a deliberative body, coordinating the conservation and management of the states' shared fishery resources. It serves as a forum for the states to collectively address fisheries issues based on the premise that using a cooperative multistate approach can achieve more than individual states acting alone. The Commission's mission is to develop joint programs to promote and protect fisheries resources. It has five main policy arenas: interstate fisheries management, research and statistics, fisheries science, habitat conservation, and law enforcement. However, the Commission's membership consists of politically connected persons (e.g., the director of the state's marine fisheries management agency, a state legislator, and an individual appointed by the governor).

The general deterioration of the Chesapeake Bay as an important ecological system is unfortunately being duplicated in waterways through out the world. Based on fifty years of Japanese fishing for 15 species of large ocean fish and on American and Australian data for more than 140 species, commercial fishing has significantly reduced the diversity of fish in the world's open oceans by about half. Boris Worm and colleagues examined fish catch data for the period 1950 to 2003 within sixty-four ecosystems worldwide that produced 83 percent of the global harvest. Twenty-nine percent of species were considered "collapsed" in 2003. The extrapolated trend from this data projects "the total collapse of all taxa currently fished by the mid-twenty-first century."[22] This consequence is based on an industrial-scale commercial fishing harvest of approximately 150 million metric tons annually. The report also highlights the necessity for "ecological management" in contrast to "species management." In addition, such a reduction of specie diversity negatively impacts the general health of ocean ecosystems. Other studies have found a parallel decline in the global distribution of tiny ocean creatures referred to as zooplankton, which are distant from the large predator fish in the food chain. These findings suggest common global environmental properties affecting quite dissimilar marine creatures and bodies of water.

The United Nations reports that by 2030 the rivers of the world will contribute 14 percent more nitrogen into the oceans and seas than existed in 2006.[23] The annual "Testing the Waters" report of beaches by the Natural Resources Defense Council for 2006 indicates a fifteen-year record of 25,643 beach closings and health advisory days across the United States, an increase of 28 percent from the previous year. Eighty-five percent were based on dangerously high bacteria levels, indicating the presence of human or animal waste.[24] In a related study by the U.S. Geological Survey, it was reported in 2003 that beach sand could carry more bacteria than the water; the sand on Chicago's lakefront averaged more than ten times the bacteria level of the water.[25]

The continuous development of more advanced technology and the creativity applied to economic opportunity has escalated air, water, and soil pollution as well as the reduction and extinction of wildlife. The Mediterranean Sea has suffered similar degradation as that of the Chesapeake Bay and for the same reasons: excessive aspirations of wealth. The World Wildlife Fund (WWF) issued a 2006 report detailing the disappearance of the giant bluefin tuna in the channels that separate the Adriatic Islands.[26] While in the 1980s the bluefin tuna were abundant in these waters, by 2006 they were rarely caught by professional fishermen. The WWF reports that even with the advent of high-tech trawlers, catches have been reduced by 80 percent and concludes, "We are seeing a complete collapse of the tuna population. It could disappear and never come back." Despite the Mediterranean's annual tuna catch being restricted to 32,000 tons by the International Commission for the Conservation of Atlantic Tunas, the WWF reports yearly catches exceeding that figure by 50 percent as a result of "illegal, unregulated, and unreported production." The United Nations Food and Agriculture Organization considers that many edible fish stocks in the Mediterranean and its extensions have sharply declined in the past decade. This is due to pollution, excessive fishing, and taking fish illegally by exceeding internationally established quotas. However, creativity and modern technology have added another component for tuna population depletion in spite of environmental limitations and quota restrictions.

Tuna are in great demand for Japanese sushi and sashimi, at lucrative prices. In January 2009, a 282-pound Oma bluefin tuna was sold in Tokyo for $104,700, or $370 per pound, to provide gourmet sushi. Such commercial opportunity has resulted in Europeans establishing a network of fishing and fish farming companies to meet the financial opportunities of the Japanese marketplace. However, the solution extends an "almost extinct" status for the bluefin tuna, closer to complete extinction. Tuna, even the small juvenile fish, are caught in many areas of the Mediterranean and in the Atlantic Ocean. High-tech tuna boats from France, Italy, and Spain use sonar and airplane spotters to locate tuna off the coasts of Cyprus, Egypt, and Libya. The tuna are transported to fattening farms in Croatia, Spain, and Turkey and placed in underwater cages. The Japanese market is so profitable that 100 percent of Croatia's farmed tuna is exported as fatty raw tuna. The Croatian coastline provides an excellent site for tuna fattening farms, which have provided employment for the local out-of-work former fishermen, who have few native fish available to catch off their shores. In 2007, scientists estimated that bluefin tuna in the Eastern Atlantic and Mediterranean waters were being harvested at three times the sustainable rate. The U.S. fishermen caught about 10 percent of their allotted quota for the year.[27]

Such modern approaches to creating economic opportunities that ignore massive and irreversible depletion of the environment and natural resources are a stark reminder that the pursuit of wealth is a culture's primary motivator regardless of long-term cultural and environmental consequences. Also, it exemplifies the historic behavior of mature civilizations approaching cultural stagnation, which resort to dramatic, reckless, and short-sighted pathways to greater profits. The economic, political, and social processes implemented over the last century, producing these consequences for the Chesapeake Bay and the rivers and oceans of the world, utilize the same generic societal values, attitudes, and behavior as described by sociologists as normal human attributes in pursuit of survival and prosperity. Similar elements of Sorokin's sensate mentality, represented by self-interest in financial profits, power, and the luxuries of life attributed to cultural disintegration, are responsible for global ecological damage during the last century. These sensate values and the related ethics, motives, and general behavior normally associated with cultural growth and productivity are endemic to mature civilizations in a state of decline.

Sociological concepts, economic incentives, and scientific principles combine to explain human behavior and the consequences of the implementation of chemical, physical, and biological processes associated with the degradation of the Bay as well as with the more general reckless consumption of national wealth. In the final analysis, Mother Nature's principles control environmental systems as well as human manipulative efforts to gain greater benefits from her resources.

In his 2003 book *Chesapeake Bay Blues*, Howard Ernst takes the position that almost nothing of an ecological nature in the Bay is getting better and most historic problems are getting progressively worse. This applies to the global environment where a given marginal, regional improvement dwarfs the overall condition and the rate of degradation.

> **The overall effort to improve the water quality of the Bay, so as to achieve a corresponding improvement in the abundance of the Bay's living resources, has not succeeded.... The issue today does not represent a scientific problem, but a political problem.**[28]

Most of the proposed solutions to restore the fisheries industry in the Bay have ignored the basic fact that grossly inferior water quality for marine species and vegetation is equivalent to contaminated air for the human environment. The quality of the medium of existence is the primary variable for the quality of animal and plant life. Tom Horton in a *National Geographic* article states, "The Bay scene is changing and there's an air of finality to it now.... The Bay today has become the ecological equivalent to a morbidly obese person, force-fed nitrogen and phosphorous."[29] This description illustrates the debilitating effects of societal entropy production by a large, prosperous regional socioeconomic system. The historic sociopolitical aspects of promoting economic benefits from the Bay's resources, while giving the political appearance of governmental action to address the Bay's environmental problems, will continue. That is a political reality of the times at the local, state, national, and international levels.

The continuous net loss of environmental resources for future potential benefits to society may be described in nautical terms as "rowing ahead at four knots when the current is moving against us at five knots."[30] A more global statement is that society continues to produce ominous, increasingly more stagnant environmental conditions as a result of exponentially increasing consumption of the Earth's resources.

The history of the environmental degradation of the Chesapeake Bay represents a microcosm of the sociocultural transitions from prosperity to deterioration and stagnation that has plagued, on a wider scale, all civilizations since ancient Greece. This is a consequence of imperfect human priorities, behavior, and management of increasing quantities of wealth-energy resources. The same cultural mechanisms and properties that determine the fate of the Earth's ecosystems are applicable to the broader phenomena of cultural maturation. When combined with the limiting principles of energy-based socioeconomic models, the dynamic factors influencing the etiologies of deteriorating ecosystems and civilizations are recognized as mechanistically and energetically interrelated. The mechanistic-thermodynamic paradigm is equally applicable to the collapse of the Spanish Empire as to the Chesapeake Bay ecosystem.

Chapter 10
References and Notes

1. Diaz, Robert J. et al. "Spreading Dead Zones and Consequences for Marine Ecosystems." *Science* 321 (2008): 926-929. Also, Venkataraman, Bina. "Rapid Growth Found in Oxygen-Starved Ocean 'Dead Zones.'" *The New York Times*. 15 August 2008.
2. Ernst, Howard R. *Chesapeake Bay Blues: Science, Politics, and the Struggle to Save the Bay*. Oxford: Rowan & Littlefield. 2003. 9-10.
3. In 1988, Roger Newell of the University of Maryland Center for Environmental Science (UMCES) Horn Point Laboratory estimated the effect of the oyster's filtering capacity and consumption of algae on water quality and dissolved oxygen. See North, E. W., J. Xu, R. R. Hood, R. I. E. Newell, D. F. Boesch, M. W. Luckenbach, and K. T. Paynter. "Quantifying the Ability of Eastern Oysters to Filter the Waters of Chesapeake Bay." *Proceedings National Academy of Sciences*. Unpublished.
3. Lieutenant Francis Winslow of the U.S. Coast and Geodetic Survey carried out an intensive study of oyster beds in Pocomoke Sound and Tangier Island. See Merritt, Don and Leffler, Merrill. "Oyster Restoration in the Chesapeake Bay: Looking Back, Looking Forward." *Maryland Aquafarmer* Fall 2000. Also, Brooks, William K. *The Oyster. 1891*. Baltimore: The Johns Hopkins Press. 1999.
4. Maryland Department of Natural Resources data as reported by Marcy Mullins and Robert Ahrens. *USA Today*.
5. Ernst, Howard R. *Chesapeake Bay Blues*. 94-95.
6. Ibid., 96.
7. Harper, Scott. "Catch Data Suggest Sustained Dip." *The Virginian-Pilot*. 25 October 2007.
8. Chandler, Alision Michael. "Some Say Fish Deaths Are a Wake-up Call." *The Washington Post*. As reported in *The Virginian-Pilot*. 25 July 2005.
9. Bashara, Fred. "35 Years Later, the Magic—and the Fish—

Are Gone From the Bay." *The Virginian-Pilot.* 7 October 2007.
10. Merritt, Don and Leffler, Merrill. "Oyster Restoration in the Chesapeake Bay: Looking Back, Looking Forward." *Maryland Aquafarmer* Fall 2000.
11. Ernst, Howard R. *Chesapeake Bay Blues.* 13-17.
12. Ibid., 17.
13. Rowan, Jacobsen. "Restoration on the Half Shell." *The New York Times.* 9 April 2007.
14. Harper, Scott. "Elizabeth River Is Still a Polluted Mess—But It's Slowly Improving." *The Virginian-Pilot.* 6 November 2007.
15. Ernst, Howard R. *Chesapeake Bay Blues.* 54.
16. Chandler, Alision Michael. "Some Say Fish Deaths Are a Wake-up Call."
17. Vicky Blazer, Fish Pathologist, USGS National Fish Health Laboratory, Leetown, WV. NPR *Talk of the Nation*, September 15, 2006.
18. Dorabawils, Nelum and Gupta, Gian. "Edocrine Disrupter—Estradiol—In Chesapeake Bay Tributaries." *Journal of Hazardous Materials* A120 (2005), 67-71.
19. Williamson, Elizabeth. "Chesapeake Bay's Rockfish Overrun by Disease." *The Washington Post.* 13 March 2006.
20. "If Only Menhaden Wrote Campaign Checks." Editorial. *The Virginian-Pilot.* 20 April 2006: B8.
21. Worm, Boris. "Impacts of Biodiversity Loss on Ocean Ecosystem Services." *Science* 314 (2006): 787-790. Also, Dean, Cornelia. "Study Sees 'Global Collapse' of Fish Species." *The New York Times.* 3 November 2006. Also, Weise, Elizabeth. "Study: 90% of the Ocean's Edible Species May be Gone by 2048." *USA Today.* 3 November 2006.
22. Heilprin, John. "U.N. Study of Oceans Shows 200 So-called Dead Zones." *The Virginian-Pilot.* 21 October 2006.
23. Natural Resources Defense Council Report, "Contamination Forces More Beach Closing Nationwide." July 28, 2005. Also, Wheeler, Larry. "A High-Water Mark for Contamination?" *USA Today.* 7 August 2007.

24. U.S. Geological Survey, Lake Michigan Ecological Research Station Report. Richard Whitman. 2003.
25. Rosenthal, Elisabeth. "Fishing Depletes Mediterranean Tuna, Conservationists Say." *The New York Times.* 16 July 2006.
26. Kozak, Catherine. "U.S.-Proposed Limits on Bluefin Rejected." *The Virginian-Pilot.* 19 November 2007.
27. Ernst, Howard R. *Chesapeake Bay Blues.* 67-68.
28. Horton, Tom. "Saving the Bay." *National Geographic.* June 2005.
29. Horton, Tom, and Eichbaum, William M. *Turning the Tide: Saving the Chesapeake Bay.* Washington, D.C.: Island Press. 1991.

Chapter 11
Sociopolitical Evolution of Economic Power: The Stagnation of Western Civilization

> "The West is increasingly concerned with its internal problems and needs, as it confronts slow economic growth, stagnating populations, unemployment, huge government deficits, a declining work ethic, low savings rates, and in many countries, including the United States, social disintegration, drugs, and crime. Economic power is rapidly shifting to East Asia, and military power and political influence are starting to follow. India is on the verge of economic takeoff and the Islamic world is increasingly hostile toward the West. The willingness of other societies to accept the West dictates or abide its sermons is rapidly evaporating, and so are the West's self-confidence and will to dominate."
>
> **Samuel Huntington**[1]

The history of Graeco-Roman Western civilizations documents the nature, intensity, and scope of materialistic human instincts that have been fundamental to the basic survival of primitive social orders as well as to the acquisition of more abundant rewards and pleasures of more mature cultures. The provision of adequate food, clothing, shelter, and security constitutes the primary economic goals for the genesis phase of a social order, whereas enhanced material comforts and privileged circumstances of more luxurious products and services become the overriding priority for more mature cultures. Consequently, a culture's primary driving force has been to satisfy human needs and wants reflective of the economic opportunities and prosperity of a particular stage of societal maturation. Accordingly, sociopolitical strategies and tactics are typically formulated to satisfy economic objectives that reflect adopted *operational* human values, attitudes, and behavior, often incompatible with *professed spiritual, ethical, and cultural values.* Such conflicting human characteristics influence the design of a society's economic policies and practices as well as the personal and professional conduct of individuals, thus dictating the behavior observed within small businesses, major commercial enterprises, and governmental agencies. Consequently, the practice of political economics, intensely imbedded within more mature societies, has been a prominent expression of such human frailties and imperfections that have contributed to the inevitable cultural degradation of major civilizations. The political economics practiced within the United States during the first decade of the twenty-first-century will serve as an historic example of such human shortcomings.

A given level of socioeconomic maturity, whether that of an altruistic primitive culture or the more intense, sociopolitical, and self-serving twenty-first-century Western culture, provides a particular vision of the prevailing individual and cultural expectations of a social order. The *economic expectations of survival and contentment* constitute a psychological driving force and establish a rationale for specific sociopolitical structures and functions considered capable of achieving defined economic goals and objectives. Thus, economic priorities and expectations dominate the establishment and implementation of a society's socioeconomic game plan, which becomes the basis for the creation of a particular political culture intending to support the realization of the economic vision of success.

It is of interest to note that nation-states over the last century, whether espousing a form of political and economic socialism or a form of political democracy and economic capitalism, have evolved into hybridized mixtures of all of the above philosophical elements. Strategies and tactics depend on how the political, military, and business interests control, interpret, and react to the realities of the times and their potential economic and sociopolitical consequences. It is as if all such societies will embrace similar diffuse and often wanton social, economic, and political philosophies, dictated, in large measure, by random consequences of events and arbitrary management of the social order. The methods to achieve a desired endpoint are created and ultimately justified regardless of costs, issues of philosophy and values, collateral societal damage, and actual outcomes and their consequences. Many Americans of the twenty-first century revere an obtuse, unreal political ideology that generally has few fundamental linkages to major socioeconomic issues threatening to impose further degradation of the American culture. Perhaps, a lesson from history is that inevitable cultural change should be expected to include pragmatic modifications of ideological fundamentals as dictated by wisdom and unpredictable realities affecting a society of inherently increasing historic complexity.

The 2008 American financial crisis was pragmatically addressed by a conservative Republican administration by purchasing partial government ownership in a time-honored privatized financial system and by "bailing out" private corporations that were arbitrarily deemed "too large to allow to fail." This and other such socialistic responses were deemed necessary and desirable by members of both political parties and most of the nation's respected economists, while describing such actions as unfortunate and desperate but less damaging to the "free market economy." than not taking the actions. In frantic times, the fear for survival and the pragmatics of economic self-interest trump political and economic ideologies. If the capitalistic economy is not freely and profitably functioning on its own merits, government will adapt to even socialist principles while decrying the evils of socialism.

Reality is that a society's fear of potential negative physical, financial, and/or spiritual consequences associated with human choices in life will be psychologically pitted against the enticing opportunities and potential rewards of pursuing self-sufficiency and self-satisfying outcomes. Resolution of associated conflicting cultural values will ultimately determine dominant individual behavior and the social order's chosen plans of action. Individuals choose among those values, judgments, and actions that will predominantly benefit all members of society versus providing incentives and rewards for independent self-interest in a highly competitive and potentially rewarding socioeconomic environment. The balance is between a conservative, communal attitude of minimizing individual risk and the fear of society's failure and the more aggressive, perilous individualistic pursuit of economic success regardless of long-term consequences for future generations. As with past mature civilizations, the dominant motivation of twenty-first-century Western civilization is the abundant potential individual rewards of aggressive economic self-sufficiency, consumption, and competition utilizing the sword of politics.

Peter Steinfels reviewing Duncan K. Foley's *Adam's Fallacy: A Guide to Economic Theology* portrays the economist Adam Smith's "fallacy" as "the idea that the economic sphere of life constitutes a separate realm 'in which the pursuit of self-interest is guided by objective laws to a socially beneficent outcome,' a realm unlike all the rest of social life, 'in which the pursuit of self-interest is morally problematic and has to be weighed against other ends.'" Foley states, "At its most abstract and interesting level, economics is a speculative philosophical discourse, not a deductive or inductive science.... These are discussions, above all, of faith and belief, not of fact, and hence theological ... to justify the ways of the market to men." Thus, contemporary capitalism may be viewed, according to Foley, as a system for creating material wealth ... not a stable, self-regulating one.... It will "not 'solve the problems of poverty and inequality... Economics functions in a theological role in our society." [2]

In reality, "poverty and inequality" are inherent by-products of a successful economy, recognized, anticipated, and accepted by the political and economic leadership, as public and corporate policies and strategies are designed and implemented to pursue economic development and a vision of continuous prosperity.

Historically, a society's economic practices have been akin to those of formal religions, both utilizing similar psychological tactics focused on individual incentives and rewards. Theology offers individuals the option of peace of mind, salvation, and eternal life based on the negative incentives of eternal damnation and the fires of hell. Economics offers positive materialistic incentives utilizing aggressive marketing tactics, attractive (and sometimes dubious and unethical) financial strategies, and an abundance of luxury goods and services for a feel-good society.

Foley's perspective of sociopolitical economics is a twenty-first-century view that reflects Sorokin's concept that shifting cultural values from a predominantly altruistic to sensate mentality is an inherent outcome of a civilization's increasing economic prosperity. Additionally, prevalent sensate values and behavior inherently and ultimately undermine sound economies and destroy traditional cultural values. As a civilization's economic profitability increases, the human instinct to adjust and adapt one's values and behavior toward a more commonly accepted sensate mentality of self-interest, ruthlessness, and greed increases in direct proportion to the magnitude of the potential wealth to be acquired. This is the attraction of modifying one's values and ethics and aggressively and ruthlessly pursuing the rainbow to the proverbial pot of gold.

In primitive societies, political systems are created out of necessity for the common good, while for mature civilizations, political functions continuously expand to accommodate and create opportunities for acquiring additional socioeconomic control and benefits for specific targeted segments of society. Consequently, with increasing cultural maturity, socioeconomic processes continuously undergo increasing political and governmental interventions for planning, financing, coordination, and regulation. The objective is to create and refine operational structures and functions so as to achieve a healthy national economy and social and financial stability. Thus, a particular political philosophy will reflect and support the adopted economic priorities, strategies, and objectives of the prevailing socioeconomic system. Unfortunately, a society's interest in sound political ethics and good government does not rise to the level of the inherent preoccupation with economic profitability. Durant states, "Man is not willingly a political animal. The human male associates with his fellows less by desire than by habit, imitation, and the compulsion of circumstances.... He fears solitude, ... isolation endangers him."[3] In modern times, political motivations and actions are based on individual and group financial interests; politics is viewed as a vehicle for protecting one's economic turf, ambitions, and prerogatives.

Primitive hunters were among the earliest organized human associations to be formed based on shared economic interests and an obvious need to collaborate in order to realize a more orderly and disciplined approach to mutually beneficial outcomes. Such associations, beyond family members, typically exhibit voluntary cooperation with a minimum of emphasis on a designated leader, formal regulations, or arbitrary authority while maintaining, if possible, the familiar family-type customs and rules. As the nature and focus of these associations of mutual economic interest evolved over time from family to clan to tribe, and to state, the initially informal association progressed toward more formal regulations and centralized authority. Such sociopolitical transitions are motivated by the primary objectives of economic and physical security, as measured by individual and collective power and by the related prospects of securing more favorable socioeconomic outcomes. That is, physical and financial security, the universal motivator, involves continuously evolving interrelated social functions and structures, all dependent on the effective practice of political economics. When the politics of protecting economic interests and security is insufficient to successfully resolve conflict, words are replaced "in crises by the sword."

This sociopolitical evolution to more progressive and competitive economic strategies and tactics parallel the sequence of domination by primitive tribes, clans, infant states, and empires, which eliminate the weak, reward violence, and develop more effective instruments of combat and military tactics. Armies that prevail increasingly improve their military organization, discipline, and tactics while enslaving the conquered. Durant summarizes the model of transitional conflict and warfare:

> **Property was the mother, war was the father, of the state.... The state is the product of force, and exists by force.... Societies are ruled by two powers: in peace by the word, in crises by the sword; force is used only when indoctrination fails.**[4]

While the usual practices of political economics rely heavily on "indoctrination," the intensity of economic self-interest and the level of ruthlessness applied are determined by the magnitude of the potential rewards, which also determines the intensity of the force applied to the sword.

Societal forces may be an official or subtle component of a culture's political tactics but tend to become more formalized, aggressive, and tactically less subtle as a society matures. The evolution of primitive socioeconomic associations reflects increasing expansion and integration of communication, commerce, transportation, security, formal regulations, education, and governance. The twenty-first-century environment of Western civilization is a dramatic illustration of maximizing complexity to achieve progress and of the dysfunctional nature of mature social institutions. While primitive associations rely on the simplicity of kinship, it is pragmatically and gradually replaced by individualism, including the opportunities and prospects of wealth from a more advanced and competitive socioeconomic environment. Ultimately, regulation, more autocratic rule, a body of laws, and village and urban communities became realities.

Laws come with property, marriage, and government.[5]

As a context for appreciating the cultural degradation of twenty-first-century Western civilization, the history of the transition from cultural growth to stagnation of the Greek and Roman civilizations will be reviewed from a political economic perspective based on wealth-energy economics. Their historical growth and decline patterns are illustrated in Figure 8-2.

All economic history is the slow heart-beat of the social organism, a vast systole and diastole of naturally concentrating wealth and naturally explosive revolution.[6]

About 300 B.C., ancient Greece reached an intolerable level of chaos and disorder as its socioeconomic and political systems became increasingly dysfunctional. Athens is described as having a population of about 21,000 citizens, 10,000 aliens, and remarkably, 400,000 slaves. Estates were being expanded, with absentee owners relying on managers to work a slave labor force. Rain and flooding conditions, coupled with man's aggressive deforestation, resulted in significant soil erosion, which contributed to substantially reduced productivity. Farms were abandoned, and it became necessary to import food. Mines were becoming depleted, and some materials could be imported more cheaply than produced internally. Commerce declined in the cities, as the number of citizens capable of being productive diminished while, in villages, the economic trend was toward new types of commerce.

Employment was irregular, wages were low compared to inflated costs, and thousands of men abandoned the mainland cities for mercenary soldering abroad or to survive their poverty in rural isolation. As the gulf between the poor and the wealthy broadened, bitter class warfare proliferated, with loss of life and property. Durant describes the Greek wealthy class:

> The wealth of the wealthy grew beyond any precedent in Greek history. Homes became palaces, furniture and carriages more sumptuous, servants more numerous; dinners became orgies and women became show windows of their husbands' prosperity.... Every city, young or old, echoed with the hatred of class for class, with uprisings, massacres, suppressions, banishments, and the destruction of property and life. When one faction won it exiled the other and confiscated its goods; when the exiles returned to power they revenged themselves in kind, and slaughtered their enemies; imagine the stability of an economic system subject to such decerebrations and disturbances.[7]

Thus, the stage was set for Rome to easily overcome or, more precisely, to gradually absorb Greece's territory. Durant's assessment: "The essential cause of the Roman conquest of Greece was the disintegration of Greek civilization from within. No great nation is ever conquered until it has destroyed itself."[8] The conditions cited include the following:

- Deforestation and the abuse of the soil
- Depletion of precious metals
- Migration of trade routes
- Disruption of economic life by political disorder
- Corruption of democracy
- Decay of morals and patriotism
- Declining population
- Replacement of citizen armies by mercenary troops
- Human and financial waste of war
- Loss of capable people via the revolutionary environment

It is noteworthy that the conditions described by Durant as leading to Greek self-destruction are the direct or indirect consequences of human behavior, motivated by materialistic attitudes, that unwisely and wastefully consumed wealth and natural resources. Such behavioral outcomes, typical of mature civilizations, become increasingly prominent as depleted wealth-energy resources sooner or later insidiously corrupt and eventually cripple the socioeconomic and political systems. The social order runs out of time and resources while lacking the moral discipline and the commitment and support of the masses to resolve accumulating societal chaos.

The same Greek political and socioeconomic characteristics of cultural deterioration are also attributed to the later disintegration of the Roman civilization.

> **A great civilization is not conquered from without until it has destroyed itself within. The essential causes of Rome's decline lay in her people, her morals, her class struggle, her failing trade, her bureaucratic despotism, her stifling taxes, her consuming wars.**[9]

Authors of the civilization studies literature agree that a society's processes of existence create self-imposed elements of sociocultural degradation, thus creating a pathway for long-term, intense cultural deterioration. The Roman civilization entered its initial collapse phase about A.D. 193, but factors responsible for its ultimate final collapse originated about four centuries earlier.

Our main thesis focuses on human interactions and manipulation of a society's wealth-energy-based economic system, responsible for creating the conditions and consequences leading to cultural collapse. Therefore, it is instructive to identify the underlying Roman economic origins and the resulting human behavior responsible for cultural collapse. The Agrarian Revolt (145 to 78 B.C.) is an appropriate historical period from which to gain insights into the socioeconomic characteristics that contributed to the final stages of the Roman civilization's collapse. In this way, cultural conditions may be identified and associated with the processes, driven by wealth and influenced by political economics, which, over centuries, were responsible for modifying cultural values, attitudes, and behavior and thus contributing to societal degradation.

The Roman agrarian economy was undergoing fundamental changes during this period that clearly were destined to generate social unrest, poor economic returns, and a general deterioration of the social order. Slave labor was rapidly replacing free workers at a time when imported grain was cheaper than domestic production prices. The elite business and political figures were able to buy large quantities of farmland by establishing financial and political procedures that were favorable to the wealthy, but that constituted impossible hurdles for the general population. Inappropriate and unjust economic conditions doomed many free workers into slavery, accounting for an estimated half of the total slave population. Some leaders raised the issue of land reform and redistributions, but the wealthy class stifled such efforts.

As a result, the Roman socioeconomic labor force was transformed from free workers to a combination of internal slaves and an external military force. Durant states, "In the city all domestic service, many handicrafts, most trade, much banking, nearly all factory labor, and labor on public works, were performed by slaves, reducing the wages of free workers to a point where it was almost as profitable to be idle as to toil."[10] This description of the Roman civilization as it approached the time of Christ is an excellent illustration of a society creating and accumulating a fatal dose of accumulated societal entropy during a period that produces great prosperity for a small segment of society. Shortly beyond the time of Christ, from A.D. 50 to 100, and approximately a century prior to the Roman civilization entering its final collapse phase, the Roman Empire was considered to have achieved the greatest Mediterranean prosperity in history, "the material zenith for the ancient world."[11] However, much of Italy was threatened by starvation and was required to import food. The state owned the natural resources, factories, and much of the cultivated land. The government produced building materials and the weapons of wars and operated the banking business. Clearly, the economic and political systems produced benefits for a small segment of the population, who were also successful in minimizing business opportunities and land ownership for the masses. Consequently, internal production and commerce were severely restricted and limited to local distribution.

By A.D. 190, the economic system principally benefited a ruthless and greedy minority, who were rapidly consuming and exhausting the empire's wealth. In the process, the general population was losing hope and faith as social unrest approached a breaking point. The familiar socioeconomic transitions from "stagnation" to "disintegration" and to "collapse" had been rooted in Roman cultural traits extending back to the Republic (500 B.C.). The accepted period of the Roman Empire collapse, from A.D. 193 to 305, represents the exacerbation of pre-existing and intensifying elements of cultural degradation that had existed for centuries.

Rome's societal decay followed the familiar mechanism of socioeconomic development that also generates an inevitable, continuous flow of costly problems, issues, and opportunists. As the Roman civilization matured, politics increasingly became the vehicle by which the socioeconomic elite functioned and manipulated the social order in the pursuit of material rewards. Increasingly, the disappearance of ethics and integrity from socioeconomics closely paralleled their absence from the conduct of political affairs. During the latter stage of cultural stagnation, the political and military systems became major contributors to the cultural dysfunction as Rome approached collapse. In such desperate circumstances, leadership ultimately resorts to despotism and political anarchy, which, in turn, only accelerates socioeconomic deterioration and destroys any remaining sense of common purpose within the population. Inherently, this sets the stage for a next phase of greater societal disorder consisting of mass violence; class warfare; and evasion of taxes, laws, and military service. People begin to seek new avenues of psychological and material security necessary to survive the time of troubles. The Roman leadership response was to create additional bureaucratic controls (including the removal of local autonomy), administer harsh punishment for opposing views and activities, and create villains as the source of society's problems.

At this point, a stagnant Roman civilization lacked a source of much-needed wealth-energy capital to address its rising financial problems, as the previous sources of new wealth from conquests were no longer available. Attempts to raise money internally through taxation were unsuccessful due to the failed economy and a lack of the population's commitment to assist the failing social, economic, and political systems and the abusive leadership. Currency depreciation and the general economic conditions discouraged any significant long-term investment by the wealthy class, as the empire was viewed as a sinking ship.

Durant describes the labor force:

Artisans deserted their trades, peasant proprietors left their overtaxed holdings to become hired men, many villages and some towns were abandoned because of high assessments, and citizens fled over the border to seek refuge among the barbarians.[12]

Consequently, by A.D. 250, the population declined by almost half in parts of the empire. Government regulated any attempts by the labor force to change employers and imposed price controls to replace a free market economy. Such government regulation greatly expanded the complexity of everyday life as well as the bureaucracy and cost of government. The traditional, unquestioned Roman authority and the fear of government soon vanished among the people and the military, as well as the enemies of the empire. This is illustrated by the unique occurrences during the period of Roman emperors being defeated in battle, captured by enemies, and assassinated by the military. In less than four decades, thirty-seven men were established as emperor.

The failure of the empire is traceable to economic collapse, resulting from adopted cultural values and activities that created chaos that exceeded the society's management capability. Specific factors include the failure of the agricultural estates and food production; abuse of the land; deterioration of transportation routes and commercial exports; insufficient supplies of precious metals; conflict between the rich and poor; expanding costs of armies; ineffective public works and bureaucracies; currency depreciation; high taxes; and loss of investment capital and labor.

Entering the third century A.D., Roman cultural instability attained an intolerable level for the population, reflecting massive disorder, a collapsing economy, and ineffective governance. Consequently, the new religion of Christianity found a population desperately seeking a pathway to a more personally tranquil and economically sound way of life. The characteristics of these sought-after changes in the cultural environment coincide with the constructive attributes of Sorokin's *Ethics* of Principles, *discipline* of spirituality, and *truths* of faith. Consequently, the pendulum of the adopted Roman cultural values began reversing its long-established sensate behavior as people increasingly came to possess great concerns for their physical and economic survival and a fear of spiritual damnation. Thus, the population increasingly embraced a new spirituality and gradually rejected the established excesses of materialistic happiness and luxury. The primitive fears of inadequate food, shelter, and security and the avoidance of spiritual damnation and punishment increasingly became the primary concerns of the general Roman population. Societal motivations were redefined and redirected toward self-preservation and more primitive, basic human needs, more appropriately met by the new, spiritually uplifting religion of Christianity. Durant states:

The growth of Christianity was more an effect than a cause of Rome's decay.... It was because Rome was already dying that Christianity grew so rapidly. Men lost faith in the state not because Christianity held them aloof, but because the state defended wealth against poverty, fought to capture slaves, taxed toil to support luxury, and failed to protect its people from famine, pestilence, invasion, and destitution; forgivably they turned from Caesar preaching war to Christ preaching peace, from incredible brutality to unprecedented charity, from a life without hope or dignity to a faith that consoled their poverty and honored their humanity. Rome was not destroyed by Christianity, any more than by barbarian invasion; it was an empty shell when Christianity rose to influence and invasion came.[13]

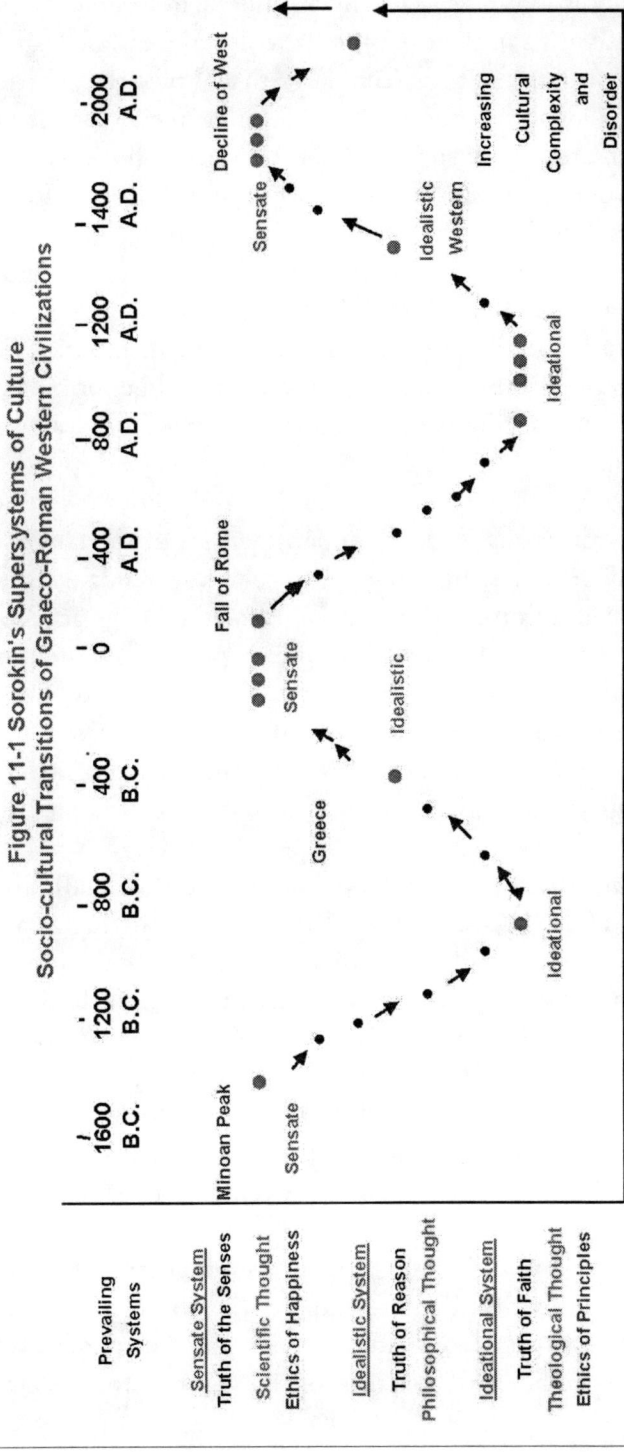

Sorokin's analysis indicates that sensate values had declined about the time of Christianity's initial expansion as cultural ideational values became increasingly dominant between the sixth and twelfth centuries. More broadly, he identifies fluctuations in his defined "Ideational, Idealistic, and Sensate Supersystems of Culture" during the 580 B.C. to A.D. 1920 period of the Graeco-Roman and European cultures. Recall that each "supersystem" consists of subsystems of truth, ethics, and knowledge that are based on the disciplines of thought.[14] See Figure 11-1. Such cultural development normally incorporates the integration of such subsystems into dominant cultural themes and mentalities. For example, the prevailing value system of a given era will be formulated by the integration of varying degrees of religious, philosophical, and scientific thought.

Sorokin's enormous research undertaking indirectly sought to identify cultural transitions in the dominant systems of the truths of faith, reason, and the senses utilizing epistemological trends detected in Graeco-Roman and European cultures. This methodology attempts to gauge the influence of the prevailing philosophies of empiricism, religious and idealistic rationalism, mysticism, skepticism, and fideism as expressed by the "leading thinkers in the field." A *dominant system of truth* is fundamentally characterized as primarily a truth of faith (religious or magical), a truth of reason, logic, or the senses of the mind.[15]

The *ideational system* of truth is based on the assumed infallibility and superiority of faith over those of human reason, logic, and judgment as interpreted by the human senses. *The sensate system* of truth assumes that the senses can be the only judge of truth as well as of falsehood. A middle ground is the *idealistic system* of truth, blending various degrees of faith, reason, and the senses. *Ideational rationalism* is mainly faith-based, while *idealistic rationalism* relies heavily on reason and logic (e.g., the idealistically rationalistic system of truth of the medieval scholastics of the twelfth to the fourteenth century).[16]

Consequently, Sorokin's categorizations of truth, thought, and ethics possess broadly differing capabilities to stimulate and guide a broad spectrum of human attitudes and behavior. Over many centuries, specific characteristics of these dominant systems, reflecting adopted values, attitudes, and behavior, fluctuate due to natural and human events and thereby alter socioeconomic priorities, policies, and practices. Thus, Sorokin's concept of fluctuating "systems" influences the nature and wisdom of human behavior and accordingly the degree of socioeconomic advancement derived from cultural priorities and the efficiency of the consumption of national wealth and energy resources. This provides an appropriate framework upon which to review and evaluate the nature of Graeco-Roman and European cultural development and the evolving sociocultural transitions of the past 2,500 years.

Basically, the periods from 800 B.C. to 500 B.C. and from A.D. 1000 to 1500 (illustrated in Figure 11-1) both represent eras of increasing sensate influence, each followed by a period of gradual resurgence of the ideational mentality. The twentieth century was distinguished by excessive sensate materialism as ideational truth, discipline, and ethics approached a minimum. Sorokin recognizes that the elements of truth that best serve the broad interests of a culture to sustain socioeconomic advancement appear to be a mixture of ideational (altruistic) and sensate (materialism) characteristics.

> **All in all, a certain predominance of Idealism over Materialism is necessary for the continued existence of human culture and society and that a balance of the Mixed-Idealistic-Materialistic systems is still more indispensable than pure, otherworldly Ideationalism and Idealism.... These facts make one ponder whether any culture can subsist with only materialism, and especially with mechanistic materialism! It seems that a considerable proportion of idealism is a prime requisite for the durable existence of society.... [*Materialism*] almost always occurred before or during crises, hard times, social disintegration, demoralization, and other phenomena of this kind.[17]**

Ethics plays an integral role in organized societies, thus Sorokin also analyzes societal interrelationships and implications of ethical fluctuations, specifically long-term sociocultural transitions related to his defined "Ethics of Happiness" and the "Ethics of Principles." The Sensate system embodies the Ethics of Happiness, the achievement of the comforts and pleasures reflecting "relativism, expediency, and changeability of the rules." The Ideational system of ethics is absolute, "bringing its followers into unity with the supreme and absolute value.... Its principles are considered as emanating from God or some other supersensory absolute value."[18] Thus, dynamic societal equilibrium shifts over time between a dominant Ethics of Happiness and the Ethics of Principles. Sorokin concludes that the Ethics of Happiness and Principles were relatively balanced during the period of 500 B.C. to A.D. 100, while from 100 to 1500 the Ethics of Principles prevailed to a significant degree. However, from 1500 to 1900, the two once again came into close balance. Sorokin concludes that:

> **all in all, social existence of man requires that in the relationship between the Ethics of Principles and that of Happiness, the former must be generally much stronger than the latter.... Some degree of sacrifice and altruism is always necessary, and that the Ethics of Principles stimulates these forms of relationship much more than the Ethics of Happiness.**[19]

Sorokin's analyses of Graeco-Roman and Western cultural transitions reveal fluctuations in the prevalence and influence of the systems of truth, thought, and ethics, which parallel fluctuations of cultural growth and degradation. This periodicity and strong stages of productivity also coincides with wise investment of national wealth, effective consumption of energy, and the adoption and application of idealistic truths and ethics.

Figure 11-1 illustrates and summarizes the fluctuations of the behavioral components of Sorokin's "Supersystems of Culture": Sensate, Idealistic and Ideational. Each is associated with the subsystems of truth, thought, and ethics that contribute to and account for sociocultural transitions. This summary represents Sorokin's methodology applied to the cultural development and integration processes experienced by Graeco-Roman and European cultures. These derived from his "*causal-functional formulas*" that "sum up briefly a prodigious number of separate relationships" and his "*logico-meaningful method*" of "ordering chaos.... Uniformity of relationship is the common denominator of causally united phenomena."[20]

The causal-functional and the logico-meaningful methods of integration both act as the means of ordering into comprehensible systems the infinitely numerous and complex phenomena of the sociocultural world, ... [which] consists of endless millions of individual objects, events, processes, fragments, having an infinite number of forms, properties, and relationships.... They give us the patterns of uniformity that are to be found in the relationships of a vast number of individual components of this infinite chaos.[21]

Sorokin's generalized characteristics of sociocultural transitions may be summarized using his data of "indices" for each period.[22] From this data, the *Ideational to Materialism truth ratio* and the *absolute Principles to Happiness Ethics ratio* has been determined for each period. The Happiness Ethic is considered a sensate trait, while the Principles and Love Ethics are considered ideational traits. The major transitions and ratios are summarized:

- **500 B.C. to A.D. 100:** While ideational values decline, a relative balance is ultimately achieved as materialism increases (1.08). The Ethics of Happiness and Principles are also approximately balanced (1.49). This period corresponds to the final stage of the Roman civilization prior to its collapse.
- **A.D. 100 to 600:** Ideational values increase significantly relative to materialism (7.12). The Ethics of Principles increases relative to the Ethics of Happiness (15.2). This constructive period represents the initial transition to a more intense Ideational system and the formative years of Christianity.
- **A.D. 600 to 1500:** Initially, ideational values increase but materialism sharply increases in the latter stages of the period (33.8). The Ethics of Principles to Happiness ratio continues to increase (126) due to a significant increase in the Ethics of Love, as other contributing factors of the Ethics of Principles and the Ethics of Happiness significantly decline. This period initially coincides with the embryonic stage of European cultures, culminating in the mature stage.
- **1500 to 1900:** Ideational values decline relative to increasing materialism (5.43). The Ethics of Happiness and the Ethics of Principles once again come into balance (1.06) as in the 500 B.C. to A.D. 100 period. This corresponds to the

zenith of Western civilization, the final phase of the Growth stage as it enters Stage III, Breakdown.
- **1900 to 1920:** Ideational values continue to decline relative to materialism (1.73), and the characteristics of the truth of the senses prevail, continuing through 2008 as Western civilization enters the cultural stagnation stage.

The economic and sociological concepts that have been cited and applied to the rise and fall of both the Greek and Roman civilizations are equally applicable to the period since the collapse of Rome. Carroll Quigley interprets the history of the Western cultures as consisting of three "expansion" periods: 970 to 1270, 1420 to 1650, and 1730 to 1929, with each expansion being followed by a period of "conflict" and "decay." He adopts sequencing labels that are conceptually identical to Toynbee's four periods of genesis, growth, breakdown, and disintegration but are more mechanistically and descriptively defined as *mixture, gestation, expansion, conflict, universal empire, decay, and invasion*.[23] Prior to considering each of Quigley's three expansion periods in the context of the mechanistic-thermodynamic paradigm, it is appropriate to recall the various viewpoints of fluctuations in the rate of societal development adopted by other authors and previously presented.

It is suggested that Quigley's representation of three cyclic periods is the same phenomena that Toynbee describes as a "rout" followed by a "rally" and that Kroeber depicts as societal "pulses" and "lulls." Additionally, Huntington describes the fading of the West, noting, "Decline does not proceed in a straight line. It is highly irregular with pauses, reversals, and reassertions of Western power following manifestations of Western weakness."[24] These authors consider such rhythms as normal for a developing culture. Quigley considers "the history of Western civilization to the middle of the twentieth century … a series of at least three successive pulsating movements of expansion," each followed by an "Age of Crisis." This may be interpreted as normally fluctuating societal growth phenomena, an outcome resulting from an exceedingly large number of socioeconomic and political processes and variables.[25]

These socioeconomic fluctuations of Western cultural advancement and retreat are the consequence of decisions and actions inherently associated with vital processes of maintaining human existence, whether in an economic environment of extravagance or poverty. The mechanisms of these processes are normally subjected to random variations in the conditions and factors that affect the rate of cultural growth and, in turn, alter the dynamic equilibrium of a particular process. Such variables are related to shifting ethics, philosophies, attitudes, behavior, and available resources, and Mother Nature's restrictions and influences. This oscillating dynamic equilibrium may be envisioned as a normal cultural phenomenon existing between the extremes of Sorokin's Ideational and Sensate mentalities or between the extreme cultural motivations of primitive fear for survival and a self-centered, economic greed typical of more advanced cultures. In Quigley's system, this dynamic equilibrium may be expressed as the relative balance at any given time between all positive contributing factors related to his *mixture, gestation, and expansion processes* and the net negative variables of his *conflict, universal empire, decay, and invasion* processes. See Equation 8-1 for an analogous summation expression in formula form. Additionally, an associated fluctuating dynamic equilibrium may be expressed between the rate of society's creation of order and simplicity and its rate of generating disorder and complexity. The outcomes of Quigley's phases and related processes as well as the dynamic equilibrium state at any given time rely on the generalized variables associated with wealth, energy, and human behavior.

Quigley identifies his first expansion period of Western cultural development as lasting about three centuries, from 970 to 1270. This represents a prosperous era when a feudal system provided the masses with the necessities of life and also began to produce a surplus of goods. The economy may be characterized as opportunistic, simple, creative, and being responsive to a broader constituency. Consequently, by the twelfth century, a new socioeconomic segment of middle class town dwellers emerged. By the end of this period, 1300, increases in economic prosperity, population, territory, and the knowledge base have occurred. However, rapidly increasing socioeconomic prosperity also creates an inevitable accumulation of societal issues and conflict. For example, economic prosperity was based on a high priority of providing luxury goods for the upper class.

Consequently, by 1274, according to Quigley, the feudal organization (i.e., feudal lords) had become "institutionalized into an obsolescent structure with few functions and a powerful determination to resist further change and to defend its own social position."[26] Socioeconomic progress brought greater organization (that is, "institutionalized feudalism") and altered the military system (mercenary men-at-arms); political systems (royal and princely rulers served by clerical officials); and the social system (feudal nobility challenged by officials and the middle class). Furthermore, the evolution of the economic system into a money economy where money became the focus of exchange produced significant socioeconomic structural changes. As a result, the fixed incomes of the nobles based on in-kind transactions were converted to a fixed money-based income. Thus, the nobles found little financial incentive to invest in the marketplace and instead invested in warfare in order to potentially achieve greater economic rewards.[27]

Consequently, an economic crisis emerged, characterized by declining investments, falling prices, scarcity of labor, and, due to primitive metallurgical technologies, declining bullion production. Additionally, the cost of producing goods did not decrease as rapidly as prices declined, resulting in lower or nonexistent profit margins. Thus, the stage was set for Quigley's Age of Conflict in 1380, which produced mass class hostility in England, the Low Countries, France, and Germany. The signs and symptoms of socioeconomic stagnation and societal breakdown proliferated and intensified as the rate of cultural deterioration exceeded the rate of advancement, thereby exacerbating any potential immediate economic revitalization.

Quigley's second Western expansion period, from 1440 to the end of the seventeenth century, the Renaissance period, is credited to economic reform and a new economic model assisted by more sophisticated sociopolitical structures and functions. Commercial capitalism evolved, as the former craft economy was replaced with elementary forms of manufacturing. In what ultimately became "economic globalization" in the twentieth century, the former Mediterranean economy expanded into a more comprehensive Atlantic-centered market. The economy of the period, not unexpectedly, produced new problems and issues that inherently accompany socioeconomic success.

The second expansion period completed the transition from "commercial capitalism" to "state mercantilism"; that is, a municipal-centered economy became more of a statewide function with all of its inherent controls and regulations. Once more, greater centralization and control were created out of necessity and opportunism, intended to provide additional economic prerogatives and advantages for the upper class. However, this also created an environment that contributed to the socioeconomic stagnation during the second Age of Conflict between 1650 and 1750. This period witnessed the formation of great military powers as well as significant international economic and military conflicts that confiscated and redistributed abundant wealth from the New World. This era also produced major class conflict and an inability to create supportive socioeconomic structures to wisely and appropriately invest the newly acquired wealth. Unfortunately, human aspirations, based upon rapidly expanding economic opportunities, were undisciplined and greatly exceeded administrative skills and techniques of the era. Thus, significant, broad-based problems arose from expanded commercial markets, particularly involving transportation and communication issues. In reaction, political strategies and tactics were devised to deal with these new opportunities and issues and to provide regulation and control of socioeconomic progress. Dynastic monarchies and state mercantilism emerged as the controlling influence of the period.

Quigley's third Age of Expansion, from 1730 to 1929, is highlighted by the impact of the Industrial Revolution, refined capitalist economics, and innovation in financial and administrative systems. This era, perhaps more than any other in history, clearly illustrates the diversity of mechanisms by which human creativity and opportunism simultaneously produce both rapid socioeconomic progress and conversely a high degree of societal complexity and disorder. A rapidly expanding and prosperous technologically oriented economy responds to unique opportunities, and out of necessity, society creates social and political entities that expand sociopolitical structures, functions, responsibilities, and challenges. Technological inventiveness and related commercial applications of the industrialization of transportation, communication, and manufacturing required more advanced and effective concepts of capital investment, banking, corporate structures, and business practices. These fundamental changes of a radically altered and demanding socioeconomic system had significant cultural ramifications, rapidly escalating the pace of cultural evolution.

The Industrial Revolution provided innovations in science and technology, which catalyzed the creation of new materials, energy sources, machines, and work structures, and the skills and methods of business management, transportation, and communication. Innovation generates rapid and prosperous economic growth, but also requires an enormous increase in the consumption of resources. During this third Age of Expansion, agricultural improvements provided abundant and more widely distributed food sources for an expanding population. Also, prosperous economic conditions provided for a wider distribution of individual wealth and expanded international commerce. Social changes were significant and included the urban development phenomena, a more complex work environment, broader cultural attitudes and practices, and a more intensely regulated and controlled existence. More highly integrated political and governmental structures, functions, rules, and bureaucracies were implemented as the socioeconomic system became increasingly energized and affected by sensate values, materialism, and the prospect of acquiring abundant wealth and status.

Society's message was, and remains, that the acquisition of wealth and social standing provides access to political favor, thus enhancing one's chance of realizing additional riches and influence. However, the accumulation of substantial political and economic leverage by a small segment of society while the majority endures significantly less desirable socioeconomic benefits set the stage for Quigley's designated third Age of Conflict, beginning in the latter stages of the nineteenth century. Business and other financial interests, assisted by the political and judicial establishments, created more effective, self-serving capital investment and corporate control strategies and tactics. These included the business practices of cartels and trade associations, tariff protection, common stock, and government subsidies. Financial capitalism evolved into monopolistic capitalism as banking secured control of the industrial economy during the twentieth century. Since the money supply was controlled by the banking industry, industrialists were at the mercy of financial institutions for the necessary investment capital needed by an expanding industrialized economy.[28] By the middle of the twentieth century, the American government, reacting to political pressure from organized groups representing industry, civil service workers, commercial services, farmers, transportation, and labor, established appropriate monetary policies and controls. However, this was perhaps the initial stage of modern political lobbying efforts, which were to secure enormous, detrimental influence by the late twentieth century.

The history of global economic expansion resulting from the Industrial Revolution and capitalism illustrates the primacy of economics and human aspirations for wealth, social status, and political power and influence. The Industrial Revolution is recognized as having its earliest major success in Britain during the period of 1760 to 1830. Despite the English government's attempts to restrict the export of new machines, techniques, and skilled labor to other countries, Belgium began its economic transformation in 1807, and by 1848 France had become industrialized. In contrast to other European countries, the British, Belgians, and French possessed the necessary resources, political conditions, and technical knowledge for successful economic transformation very early in the global industrial movement. Although Germany possessed the necessary assets for industrial production, its political climate prevented it from achieving its industrial expansion until the 1870s, but its efforts were rapidly successful. American industrialization during the nineteenth and twentieth centuries achieved world dominance and was followed by Japan and the Soviet Union. In the latter stages of the twentieth century, China and India initiated impressive industrial growth. In the past two decades, China's economy has grown at an average annual rate of 9 percent.

Sorokin, writing in 1930s and 1940s, describes "the present status of Western culture and society, ... the twilight of our sensate culture, ... a tragic spectrum of the beginning of the disintegration of their Sensate supersystem."[29] His work, covering the period since 1600 B.C., describes the Graeco-Roman Western civilizations as undergoing two periods of cultural deterioration, each followed by two new periods of cultural growth, the last of which led to twenty-first-century Western cultural stagnation. Sorokin's view of Western civilization appears to have been an accurate forecast: "that the existing order has passed its creative phase and is on the verge of bankruptcy; that it spells bullets rather than bread, destruction rather than construction, misery rather than prosperity, regimentation rather than freedom, confusion rather than order, death rather than life. Its decline is not due to the murderous assaults of barbarians, revolutionaries, or plotters, but to its own senility, the exhaustion of its creative forces."[30] Current signs indicate an increasing rate of Western civilization's Stage IV Deterioration… a victim of squandering wealth, energy, and human resources. Its decline was inevitable, only one component of the triple rhythm of sociocultural transitions during the last 4000 years.

> **The dominant system prepares its own downfall and paves the way for the ascendance and domination of one of the rival systems of truth and reality, which is, under the circumstances, more true and valid than the outworn and degenerated dominant system.**[31]

In the foreword to his 1957 abridged version of *Social and Cultural Dynamics*, Sorokin states:

> **As to the changes in the original text of the work, practically no change is made because none is needed. Since the original publication historical events have been unfolding according to its diagnosis and prognosis, and its main forecastings have been coming to pass during the last twenty years. There is no need for correction of any of its significant propositions, since up to this moment, the historical processes have been proceeding as outlined.**[32]

At the time of this statement, his original 1941 forecast of wars, conflict, and general cultural disintegration of Western civilization had proven to be painfully and explicitly accurate. Sorokin believed that in the twentieth century, the post-medieval Western sensate cultures were already in a last phase approaching destruction, described as a pathological, decadent, and self-destructive way of life. He accurately predicts the declining political and cultural influence of Europe in international affairs and foresees the "shift of creativity" toward North and South America, Australia, Russia, and Asia. Based on his 1930s investigations, he predicts the late twentieth-century growth of the then-undeveloped nations.

Perhaps as significant, he identifies and predicts the evolution of Western values and the behavioral characteristics of the late twentieth-century cultural trends:

> The boundary line between the true and false, the right and wrong, the beautiful and ugly, positive and negative values, will be obliterated increasingly until mental, moral, aesthetic, and social anarchy reign supreme.... With all values atomized, any genuine, authoritative, and binding "public opinion" and "world's conscience" will disappear. Their place will be taken by a multitude of opposite "opinions" of unscrupulous factions and by the "pseudo consciences" of pressure groups.[33]

Pertinent to his accurate vision are the cultural, religious, ideological, and economic origins for human incivility, discrimination, warfare, atrocities, and mass cruelties of the modern era exemplified by Nazi Germany, Northern Ireland, the Middle East, Eastern Europe, and Africa. Additionally, the international situation of the twenty-first century provides ample examples of governments investing more in "bombs" and "bullets rather than bread" while restricting freedoms and opportunities and being unable to manage lawlessness and destruction. Any effective effort by nations to moderate and rejuvenate the decaying Western civilization would require that the world's more progressive and wealthy nations exhibit more intense ideational values rather than the prevailing, overwhelming sensate values and behavior. Sorokin's 1941 predictions of governments' leadership in the latter part of the twentieth century have also been accurate:

> Governments will become more and more hoary, fraudulent, and tyrannical, giving bombs instead of bread; death instead of freedom; violence instead of law; destruction instead of creation.[34]

In closing this chapter, it is emphasized that Graeco-Roman Western cultures have exhibited typical behavioral characteristics and properties during their growth phase as well as decay phases. The commonalities among cultures reflect the self-destructive nature of civilizations: "A society does not ever die 'from natural causes,' but always does from suicide or murder,"[35] the direct and indirect consequences of inappropriate investment of wealth-energy capital, misuse of natural resources, and sensate human behavior driven by materialistic values of greed and ruthlessness. Such behavioral outcomes proliferate with increasing cultural advancement and insidiously corrupt and cripple socioeconomic and political systems. Ultimately, the social order arrives at a point when it no longer possesses the appropriate and indispensable values, behavior, resources, commitments, and human capabilities to resolve the mounting and accumulating societal complexity, chaos, and deterioration. In the first decade of the twenty-first century, the societal entropy accumulation within Western civilization has reached and exceeded the population's maximum level of tolerance and endurance.

> **Western people were finding their own culture hopelessly perplexing. Its ethics and morality were steadily eroding with nothing replacing abandoned codes but the threat of ultimate chaos or order by force. Equally perplexing was the steady atrophy of many of the culture's major institutions. Stable family life was diminished.... Religions continued losing influence. Schools failed.... Business suffered from increasing dishonesty and lessening efficiency. Governments lost the trust of the governed as they reeled from mismanagement and scandal.[36]**

Chapter 11
References and Notes

1. Huntington, Samuel P. *The Clash of Civilizations and the Remaking of World Order*. New York: Simon and Schuster, 1996. 81.
2. Steinfels, Peter. "Economics: The Invisible Hand of the Market." *The New York Times*. 25 November 2006. Also *Adam's Fallacy: A Guide to Economic Theology* by Duncan K. Foley.
3. Durant, Will. *The Story of Civilization: Part I: Our Oriental Heritage*. New York: Simon and Schuster. 1954. 21.
4. Ibid., 21.
5. Ibid., 25.
6. Ibid., 18-19.
7. Durant, Will. *The Story of Civilization: Part II: The Life of Greece*. New York: Simon and Schuster. 1954. 563-564.
8. Ibid., 659.
9. Durant, Will. *The Story of Civilization: Part III: Caesar and Christ*. New York: Simon and Schuster. 1944. 665-666.
10. Ibid., 111.
11. Ibid., 336-337.
12. Ibid., 644.
13. Ibid., 667-668.
14. Sorokin, Pitirim A. *Social and Cultural Dynamics*. Boston: Porter Sargent. 1970. 414-417, 676-679.
15. Ibid., 226-227.
16. Ibid., 227.
17. Ibid., 296.
18. Ibid., 415.
19. Ibid., 429.
20. Ibid., 9-10.
21. Ibid., 9.
22. Ibid., 429.
23. Quigley, Carroll. *The Evolution of Civilizations*. Indianapolis: Liberty Fund. 1979. 146-166.
24. Huntington, Samuel P. *The Clash of Civilizations*. 83.
25. Quigley, Carroll. *The Evolution of Civilizations*. 334.
26. Ibid., 361.

27. Ibid., 362-367.
28. Ibid., 390-412.
29. Sorokin, Pitirim A. *Social and Cultural Dynamics.* 699.
30. Brander, B. G. *Staring into Chaos: Explorations in the Decline of Western Civilization.* Dallas: Spence. 1998. 353-354.
31. Ibid. 265.
32. Sorokin, Pitirim A. *Social and Cultural Dynamics.* Foreword.
33. Ibid. 699.
34. Ibid. 700.
35. Toynbee, Arnold J. *A Study of History*, ab. Sommervell, D. C. New York: Oxford University. 1957. I-VI: 273.
36. Brander, B.G. *Staring into Chaos.* 16.

Chapter 12
Degradation of the American Culture: Abuse of Founding Principles

"The people who created this country built a moral structure around money. The Puritan legacy inhibited luxury and self-indulgence. Benjamin Franklin spread a practical gospel that emphasized hard work, temperance, and frugality. Millions of parents, preachers, newspaper editors, and teachers expounded the message. The result was quite remarkable. The United States has been an affluent nation since its founding. But the country was, by and large, not corrupted by wealth. For centuries, it remained industrious, ambitious, and frugal. Over the past 30 years, much of that has been shredded. The social norms and institutions that encouraged frugality and spending what you earn have been undermined. The institutions that encourage debt and living for the moment have been strengthened. The country's moral guardians are forever looking for decadence out of Hollywood and reality TV. But the most rampant decadence today is financial decadence, the trampling of decent norms about how to use and harness money."

David Brooks[1]

At the beginning of the twenty-first century, America continues to consume record levels of wealth-energy resources, continue its global interventions, and create an increasingly debilitating level of internal and external complexity and disorder. The transition of American cultural values, integrity, and ethics from Colonial times to the twenty-first century is striking. This history represents a textbook illustration of Pitirim Sorokin's sociocultural transitions of truth, thought, and ethics associated with his "Supersystems of Culture" as presented in earlier chapters. America's 400-year evolutionary period reflects recognizable signs and symptoms of a sociocultural transition from rapid, prosperous cultural growth to stagnation and deterioration as experienced by previously successful civilizations.

The socioeconomic consequences of the American industrialized, capitalist system have been spectacularly successful as is evident by the nation's unparalleled high standard of living, research and scientific leadership, financial and business innovations, military power, and general worldwide influence during the last half century. However, the evolution of America's profitable capitalist free market economy has also been consistent with the outcomes of the dark side of socioeconomic success, as depicted by Joseph Schumpeter's concept of creative destruction. The American pattern of rapid socioeconomic achievements with the inherent accumulation of societal degradation leading to stagnation repeats the ultimate fates of ancient Greece and Rome, Imperial Spain, and the British Empire. Ultimately, the positive and negative ramifications of internal and external economic competitiveness in the sociopolitical environment of materialistic self-interest generate rudiments of accumulating cultural deterioration. History documents that successful socioeconomic systems inevitably breed increasing poverty; crime; political divisiveness; urban decay; financial insolvency; and declining ethics, values, and morality.

There exists a broad array of modern illustrations of the American culture exhibiting a significant degree of cultural deterioration and having reached Toynbee's designated condition of "breakdown" and "stagnation." These include (1) the massive invasion of illegal immigrants; (2) the inept management of the Gulf Coast hurricanes; (3) massive financial, corporate, and governmental complexity, corruption, inefficiency, and financial insolvency; (3) increasing wealth distribution inequities; (4) lack of adequate health care for all citizens; (5) inferior public education; (6) escalating incivility, violence, crime, and poverty; and (7) the debilitating socioeconomic costs of ideological and imperialistic wars.

However, perhaps the most revealing and significant evidence of American society's self-destructive pathway of cultural degradation is the massive failure of the American financial and economic systems in the latter part of 2008. The nation's most serious financial crisis since the 1930s shook the world's financial markets. Wall Street's financial giants

Bear Stearns, Merrill Lynch, Lehman Brothers, Goldman Sachs, and Morgan Stanley suddenly became financially insolvent as government deregulation allowed the world's leading financial banks to fail an unproctored test of human values, integrity, and ethics. This financial disaster, reflective of the American cultural environment dominated by greed, misplaced priorities, sensate human values, and ruthlessness, has massive and unprecedented long-term worldwide ramifications.

The collapse of the U.S. home mortgage market resulted when a modern financial instrument known as a *derivative* was tactically and unethically utilized by Wall Street financial institutions to provide mortgages in vast numbers for clearly unqualified, high-risk homebuyers. As the predictable massive foreclosures rapidly accumulated, the five major Wall Street financial banks collapsed. In March 2008, J. P. Morgan, funded by a $29 billion federal government taxpayer dowry, hastily acquired and salvaged Bear Stearns. During the next six months, Bank of America, with government funds, purchased Merrill Lynch, Lehman Brothers filed for bankruptcy, and Goldman Sachs and Morgan Stanley converted their banking status to become regulated commercial banks.

This global crisis has restricted individuals and businesses with good credit histories from engaging in normal business practices, as fear and uncertainty has driven liquidity from the financial system worldwide. Entering 2009, increases in unemployment, home foreclosures, business failures, and skepticism in the economy and in government's ability to address the financial instability have paralyzed the nation's economy.

During 2008, the federal government, attempting to revive the financial system, infused the American economy with hundreds of billions of dollars while also guaranteeing investments, loans, and deposits estimated at $8 trillion. These financial commitments exceed the combined total government expenditures of the post-WWII Marshall Plan, the Louisiana Purchase, the New Deal, and the savings and loan bailout.

The scope and scale of this global financial meltdown may be appreciated by the chronology of events involving the world's most prestigious, successful, and wealthiest financial institutions, selling large quantities of their products worldwide. At the end of 2006, Merrill Lynch celebrated record revenues and earnings with its share price for the year up by 40 percent. In December 2006, it paid $1.3 billion to purchase a major lender specializing in risky mortgages. Suddenly, in 2008, Merrill Lynch collapsed as a result of record mortgage defaults, thereby acquiring $71 billion of bad debts while losing many more billions. The insurance giant American International Group, having provided insurance to protect against defaults of sub-prime mortgages, survived only because of a $150 billion federal government bailout.

During 2005, Wall Street financial banks issued $178 billion in risky mortgages and other assets backed by "synthetic" *collateralized debt obligations* (CDOs), a combination of bundled sub-prime loans plus insurance against loan defaults (i.e., *credit default swaps*). This figure increased to $316 billion in 2006 with estimated collected fees, based on a variable scale, ranging from $1.3 billion to $8 billion.[2] Ironically, the multimillion-dollar annual salaries of Wall Street employees represented illusory profits that never existed for processing the paperwork of debt, which Frederick Soddy appropriately labels as "virtual wealth," as opposed to real wealth.

The severity of these global consequences, perhaps like none before in American history, best illustrate the self-destructive nature of mature civilizations and more specifically the abuse of national wealth and disrespect for the responsibility to one's social order by unethical, greedy, and ruthless human behavior. In the aftermath of the failure of these financial institutions, it is noteworthy that highly respected and successful economists and financial professionals acknowledge that few appeared to comprehend the high level of complexity or the ramifications of these modern financial instruments. Regrettably, this saga is a fitting, illustrative capstone of the sociological concepts and predictions of Sorokin and is relevant to Toynbee's comment that "a society does not ever die 'from natural causes,' but always does from suicide or murder—and nearly always from the former."[3] The individuals representing these Wall Street firms and their associates, pushing risk and greed to historic heights, reflect the sensate mentality of a degenerating, materialistic American culture, pursuing quick profits regardless of ethical and long-term social consequences.

The early 2009 response to America's deepening economic recession by Congress and the nation's leading financial and business institutions illustrates the ultimate and inevitable cultural stagnation and breakdown of mature or "overripe" civilizations, as depicted by Toynbee and Sorokin. Logic indicates that the approach to the America's worsening socioeconomic condition involves increasing productivity through job creation and addressing home foreclosures without disadvantaging those mortgage holders who have been ethically and financially responsible. Unfortunately, the business and financial services sectors continue past self-serving attitudes and practices that initially contributed to the country's problem. Meanwhile, the political nature, excessive complexity, and limited resources of the federal government continue to be ineffective in creating mechanisms to resolve the most pressing national issue in many decades. Effective solutions appear beyond America's limited and imperfect ethical standards, cultural values, and intellectual capabilities. This national condition is also exemplified by the national debate as to whether such torture tactics as water-boarding and the indefinite incarceration of terrorist suspects without legal representation and timely trials are justified by the potential of avoiding crimes being perpetrated against the nation. That is, long-held cultural values may be expediently modified or disregarded and the law of the land ignored when government decides that the means justify the ends (a slippery slope taken in the past by many cultures and nation-states, adding another element to their societal degradation).

Clearly, America has entered Toynbee's final transitional stages having evolved from the ideational culture of Colonial times, reflecting theological thought, the truth of faith, and the Ethics of Principles, to the twenty-first century sensate culture dominated by the truth of senses, scientific thought, and the Ethics of Happiness. The prevalent "currents of philosophical inquiry"—criticism, skepticism, and empiricism—characterize America's current cultural ambiance.

How did the United States get to this situation? Clearly, American preoccupation with materialism, consumerism, debt, and maintaining global economic and military dominance at the expense of its internal social and economic stability has corrupted and diminished American society. These overriding priorities have resulted in decades of inadequate national wealth being invested in important socioeconomic segments of American society. Specifically, the large accumulating national debt based, in large part, on post-WWII military expenditures to achieve ideological conversion of the world will become a historic example of wasted resources that significantly contributed to the decay of the American culture. Poor choices severely abridged America's prerogatives and vitality for constructive innovation by virtue of self-imposed limitations being placed on resources, creativity, and time devoted to health care, economic opportunities, basic research, education, and social well-being of less fortunate citizens. Consequently, creeping cultural paralysis has resulted from abundant misplaced priorities and missed opportunities. A point of economic nonsustainability is realized when the investment of national wealth becomes extensively diverted for military activity, nonessential priorities, and satisfying sensate values and the Ethics of Happiness. It has become impossible for the American economy to generate an adequate level of new wealth-energy capital to maintain a positive rate of cultural growth for its bloated, unrealistic economic agenda and national debt. Toynbee observed: "The ultimate criterion and the fundamental cause of the breakdowns which precede disintegrations is an outbreak of internal discords through which societies forfeit their self-determination."[4]

Post-World War II American cultural afflictions and gradual deterioration have been addressed by a new generation of authors whose work complements and updates the pioneering efforts contained in the earlier twentieth-century literature. This modern view reflects human behavioral patterns based on innate human nature and instincts as modified over generations by acquired cultural values, resulting in the modern creation and pursuit of excessive materialistic and self-serving objectives. These modern-day authors, while recognizing historic American socioeconomic successes, identify and focus on the mitigating circumstances and by-products of accumulating cultural deterioration, appropriately labeled "internal barbarism" by Morris Berman.

Berman's two volumes, *Dark Ages America: The Final Phase of Empire*[5] and *The Twilight of American Culture*,[6] describe characteristics and properties of modern America's cultural deterioration. His focus is:

> structural factors endemic to American society that are ... bringing about its decline---factors such as the widening gap between rich and poor; the growing climate of apathy, cynicism, and corruption; and the dramatic drop in levels of literacy and overall intellectual awareness ... what might be collectively termed "internal barbarism" ... crucial to the collapse of Rome and ... at the heart of the American crisis as well.[7]

Berman references literature of the last few decades "that document[s] the obvious decline of this country in nearly all of its aspects—loss of community, steady erosion of an intelligent middle class, progressive degradation of our cities and our culture."[8] This literature depicts a deteriorating twenty-first-century American society, illustrating a broad pattern of cultural degradation (i.e., societal entropy production), consistent with the characteristics of the last cultural phases of ancient Greece and Rome. These properties and characteristics found in this literature will be incorporated into this chapter.

The term "liquid modernity" is used by Zygmunt Bauman[9] to characterize a society unable to maintain a cultural orientation and stability due to a lack of generally adopted and observed standards, values, and cultural norms. Essentially, a self-orientation gradually replaces an individual's sense of societal orientation as stability becomes defined in terms of one's self-dependency and achievements rather than on societal success and advancement.

Berman cites Lewis Lapham's *Waiting for the Barbarians* and Robert Kaplan's *An Empire Wilderness,* as examples of books that describe widespread deterioration within the twenty-first-century American culture. He summarizes their views: "Crumbling school systems and widespread functional illiteracy, of violent crime and gross economic inequality, of apathy, cynicism, and what might be called 'spiritual death.'... The system has lost its moorings and, like ancient Rome, is drifting into an increasingly dysfunctional situation."[10] Berman quotes Lapham:

> Sallust's description of Rome in 80 B.C.—a government controlled by wealth, a ruling-class numb to the repetition of political scandal, a public diverted by chariot races and gladiatorial shows—stands as a fair summary of some of our own circumstances.[11]

Richard Sennett in *The Corrosion of Character* describes America's evolving cultural state as the disappearance of "narrative coherence" and emphasizes that maximizing consumerism and profitability requires an economic "strategy of permanent innovation."[12] However, in reality, such *innovation* may pertain only to modern marketing strategies applied to insignificant existing technologies that may or may not increase national productivity. For example, Gillette replaced its "obsolete three-blade razor technology" with a five-blade razor to compete with Schick's four-blade design.[13] This illustrates an investment in a supposed technological improvement that exceeds an existing *good enough technology*. In essence, it is a *marketing innovation* requiring an additional investment of resources that does not perceivably improve technical capabilities. Such marketing innovations of good enough technology contribute little to the net production of national wealth or any tangible improvements in the significant aspects of the quality of life. From a national economic perspective, how advanced does the shaving technology really need to be, and what are the criteria for balancing the need, cost, and return to society of a given level of investment of national wealth? Viewing a much broader perspective, today's electronic products associated with computing, automotive, and entertainment devices continue to add new features, often superfluous and expensive for most purchasers. Surveys indicate that two thirds of current users do not take full advantage of the technology they already own.

Meanwhile, the modern baker who operates the programmed machinery and technological devices that require no specific knowledge or skill for baking bread illustrates the potential social ramifications of "innovation" for many workers. There is no artistry and little pride in such commercial baking, but there are monetary profits for stockholders. As a result, there increasingly exists within the American workforce a void of work-related knowledge, skills, and pride in the accomplishments of being an essential contributor to a highly successful project or thriving business. Increasingly, the personal rewards of employment have become relegated to the satisfaction of stagnating wages and perhaps health and retirement benefits, which are becoming less available for all but the more highly compensated. Such a working environment breeds worker indifference toward an employer's operational systems, fellow employees, product quality, potential profits, and the company's long-term financial viability. The implications for societal coherence are described by Sennett: "A regime which provides human beings no deep reasons to care about one another [or, one might add, about their work] cannot long preserve its legitimacy.... The primacy of self-loyalty becomes a necessary survival strategy—what marketing guru Tom Peters approvingly calls 'The Brand Called You.'" This attitude is the essence of unabashed self-promotion; the individual's everyday microcosm and the macrocosm of the world become one and the same or, as concluded by Robert Reich, "the economy is us."[14]

Clearly, such cultural traits may empower some individuals to become self-actualized, more materialistic, and economically self-reliant and successful. Alternatively, this employment environment may discourage, alienate, and create fears of inadequacy and failure within a society's work force. However, a net cultural effect of a society's loss of "narrative coherence" is to encourage individualism and economic competition, reduce community spirit and civic responsibility, and place less worth on previously cherished altruistic standards of civility, integrity, courtesy, and cooperative participation in social and political endeavors. Social values that promote cultural coherence, other than to promote common economic and political interests, have come to be considered deterrents to the achievement of individual and corporate success.

Fundamentally, the modern American's experience with bureaucratic and impersonal governmental, corporate, and nonprofit institutions promotes a profound sense of *cultural incoherence and insecurity*, the result of an increasingly complex, materialistic, competitive, and chaotic social order. Consequently, American cultural values have shifted since the Colonial era as socioeconomic conditions have been transformed by the "strategy of permanent innovation" that has continuously altered the realities and tactics of economic competition and the pragmatic definition of human needs and survival. Unfortunately, such self-serving strategies and tactics detract from broadly defined and generally beneficial socioeconomic growth and productivity as the cultural priorities and human judgments influencing the investment of national wealth become increasingly misplaced and imprudent.

Civility by Stephen Carter describes the linkages among America's declining sense of national purpose, prevailing attitude of incivility, loss of community spirit, and a self-serving materialistic economy, all of which dominate its twenty-first-century social order.[15] This illustrates how intense market forces have triumphed over basic altruistic values, as societal complexity, stress, and competition have become acceptable by-products of modern socioeconomic advancement. These behavioral aspects of the mature American civilization are manifestations of the twentieth century's successful scientific and technologically based capitalist economy that was able to secure major global markets for appealing goods and services. This is the basis for the unprecedented American socioeconomic prosperity that catalyzed a cultural values shift toward extreme materialistic and self-serving consumerism.

Thus, Sennett's depiction of a gradual loss of "narrative coherence" may be traced to rapid advancements in knowledge, technology, and sophisticated financial and marketing strategies and tactics, all important elements of increasing economic productivity. These factors have broadly and significantly affected the American economy and, by implication, the nature of its social order, particularly the sociopolitical environment. Robert Putnam in *Bowling Alone* and Todd Gitlin in *Media Unlimited* relate, and place in perspective, political thought, sense of community, consumerism, and the declining intellectual pursuits of twenty-first-century America. Berman, referencing Putnam, states, "the evaporation of interest groups, neighborhood alliances, and all forms of civic life—the very stuff of democracy, ... the life of the mind has been drowned in a huge consumerist fantasy."[16]

Gitlin addresses the dominant role of a broad array of media in unproductively consuming enormous amounts of time: "The ceaseless quest for disposable feeling and pleasure hollows out public life altogether."[17] This includes watching television for an average of four hours per day, corresponding to about 40 percent of free time. American television programming has increasingly become "meaningless," mindless entertainment with commercial ads often being the more creative and intellectual content.

The Numbing of the American Mind by Thomas de Zengotita describes an American addiction of participating in the "busyness" of a "flow of events they have conspired with their fellows to create" that is devoid of intellectual substance. This culture is characterized as "vast goo of meaningless stimulation."[18] Americans increasingly spend more time entertaining themselves with an expanding array of media technologies that "numb the mind." This begins at an early age. Gary Cross reports that children's toy sales increased from $4.2 billion in 1978 to $17.5 billion in 1993, which does not include $4 billion for video games, whose sales reached $30.6 billion by 2002. In 1983, $221 million was spent for toy advertising, which by 1993 reached $790 million.[19] Neil Postman told us two decades ago that we were "amusing ourselves to death":

> **When a population becomes distracted by trivia, when cultural life is redefined as a perpetual round of entertainments, when serious public conversation becomes a form of baby-talk, when, in short, a people become an audience and their public business a vaudeville act, then a nation finds itself at risk; culture-death is a clear possibility.**[20]

Nicholas Kristof notes that a thirty-four-nation study "found Americans less likely to believe in evolution than citizens of any of the countries polled except Turkey."[21] The president of the United States, asked for his reaction to this finding, remarked that "the jury is still out." The survey noted that only one American in ten understands radiation, one in three has an idea of what DNA does, and one in five thinks that the sun orbits the Earth. Susan Jacoby in *The Age of American Unreason* blames the culture of "infotainment," sound bites, fundamentalist religion, and ideological rigidity for a "strain of intertwined ignorance, anti-rationalism, and anti-intellectualism."[22] She questions how, given this culture, Americans could participate in an informed and reasoned debate on such important issues as embryonic stem cell research. Kristof concludes that the "dumbing-down of discourse" in America has been particularly striking since the 1970s: "Think of the devolution of the emblematic conservative voice from William Buckley to Bill O'Reilly. It is enough to make one doubt Darwin."

Berman asks, "How did the United States arrive at the point that money, power, speed, and mindless entertainment came to be the defining characteristics of American civilization … to reduce the country to a cultural and emotional wasteland?"[23] The answer may be derived from the collective thoughts and cultural examples of these modern authors, which interestingly are consistent with Sorokin's thesis that as civilizations become more mature and economically prosperous, cultural mentalities shift from his defined ideational to sensate values. America's early history evolved from its more primitive culture, reflective of altruistic values and inspired by fears of physical, spiritual, and economic survival, to twenty-first-century America absorbed and driven by self-centered materialism. This is an inevitable consequence or by-product of its socioeconomic aspirations and achievements. The spontaneous directionality of America's sociocultural transition is also reflective of the cultural evolution of Sorokin's Ethics of Principles toward the adoption of the Ethics of Happiness.

The Ethics of Principles associated with an ideational culture embraces the self-discipline of a spiritually based value system and a communal-oriented work ethic that was characteristic of seventeenth- and eighteenth-century America. Since this period, this mentality has been modified and transformed into a twenty-first-century version of the Ethics of Happiness and sensate values that focus on the selfish pursuit of wealth, trivial entertainment, personal entitlements, and extravagant consumerism.

America's *consumption*, defined as inflation-adjusted personal consumption expenditures, increased on the average by 4 percent annually over the past fourteen years. In 2007, consumer spending reached 72 percent of gross domestic product (GDP), a modern world record. However, for only the first time since 1950, consumption declined in the second half of 2008 by more than 3 percent, only the fourth time of back-to-back quarters of decline. For decades, increasing consumer consumption was, to a large extent, financed by debt and a reduction in personal savings. As a share of disposable income, personal savings declined from about 6 percent in the early 1990s to 0 percent during the period 2005 to 2007. Additionally, homeowners removed accumulated equity from their homes to accommodate increased spending. According to the Federal Reserve, the net equity extracted from homes rose from 3 percent of disposable personal income in 2000 to 9 percent in 2006. Household debt reached a record 133 percent of disposable personal income by 2008, compared to 90 percent a decade earlier. America's consumption of a record 72 percent of its GDP in 2007 exemplifies an extreme level of expenditure for nonessential goods and services, characteristic of the cultural deterioration of previous civilizations. Excessive consumption and debt does not financially permit the regeneration of national wealth sufficient to maintain a positive rate of cultural growth. [24]

Prior to entering the credit cards era fifty years ago, the American consumer utilized layaway plans, Christmas saving clubs, and a philosophy of *paying for it before securing it*. As of 2008, Americans have a credit card shopping debt of $822 billion. This translates into an average household debt of over $8,000 compared to less than $2,000 in 1980 (adjusted for inflation). In 1980, credit card debt per dollar of after-tax household income was 30 percent, compared to 110 percent in 2008. Beginning in 1996, American revolving credit card debt exceeded the nation's after-tax income.

The tactics of credit card promotion utilized by lending institutions during the last fifty years parallel and provide insights into the evolution of intensifying greed and declining ethics exhibited by the major Wall Street financial institutions, culminating in the 2008 synthetic collateralized debt obligation scam and the associated sub-prime home mortgage collapse.

During the early 1980s recession, lenders targeted senor citizens, students, and other low-income customers with high credit card interest rates, ultimately profiting from defaults. During the early 1990s recession, credit was extended to a wider range of borrowers, particularly those with existing multiple cards. Also, sub-prime lenders sought out unemployed workers with bridge loans until they found employment. After the technology bubble burst of 2001, lenders loosened lending standards and sought indebted borrowers with zero interest rates, expecting to be repaid from the equity in homeownership. In 2008, these strategies were no longer profitable as monthly unemployment rates reached record thirty-year levels and home equity from the era of continuously increasing home values had disappeared. Consequently, credit card lenders are expected to write off about $45 billion of bad debts in 2008 and an estimated $60 billion in 2009. The credit card business is no longer as profitable as in the early years.[25]

The economics inherent in the modern subculture of big-time college athletics is another illustration of the self-oriented materialism and the corruption of early American values, ethics, and principles. The massive media revenues and corporate and individual donations supporting college athletics have become a driving force supporting a major lucrative and popular entertainment business of the American culture that mirrors the generally dominant twenty-first-century culture. Oklahoma University's football coach earned $6 million in 2007, which included a $3 million "stay bonus" for successfully completing ten seasons. The university's athletic director enunciates the principle that "the people who are generating the revenue, why on Earth wouldn't they share in it? ... Because without them, we wouldn't have any of it." In his words, the coach "is worth every penny and always has been and always will be."[26] The annual revenue from the University's athletic programs increased from $26.1 million to $66.3 million during this time period, with football accounting for $28.5 million. It is instructive to philosophically compare and contrast seventeenth- and eighteenth-century American values, ethics, and principles to the accepted twenty-first-century cultural priorities for education, entertainment, commercialism, celebrity worship, and the distribution of wealth. Recall that the more primitive societies generally exhibit altruistic values, with private property and personal possessions being nonexistent; all members share equally in the bounty of the community's assets and good fortune. Thus, the answer to Berman's question of how American society evolved to such an "emotional and cultural wasteland" is that economic success has gradually perverted Colonial American values, integrity, and ethics. This is illustrated by the general population's acceptance and financial support for universities, routinely compensating athletic coaches with multimillion-dollar annual contracts. Many such institutions are public universities whose student cost is unaffordable for a large segment of the American population.

As with past successful cultures, twenty-first-century American society is the product of insidious, centuries-long shifting of sociocultural values and principles. These transitions are motivated by inherited human survival and competitive instincts coupled with learned cultural traits and modified standards of social, ethical, and intellectual behavior. America's cultural advancement and prosperity since Colonial times is generally viewed as the success of democracy and capitalist economics. The dominant human values and characteristics of a particular time, place, and culture will guide mankind's decision-making processes and methodologies, configuring pragmatic priorities likely to maximize the probability of achieving socioeconomic objectives.

Colonial America's initial success was due in large measure to the strict adherence to altruistic values and principles by highly motivated people who acquired the necessary resources and attempted to create a better society for the majority. The concern for the survival of the social order within the Colonial American population was clearly the first priority; in contrast to the twenty-first-century emphasis, the primacy of self-loyalty becomes a necessary survival strategy. The genesis and early growth stages of American society, as with past civilizations, while being highly successful also produced and cultivated the foundation for its own ultimate cultural degradation by virtue of the misuse and abuse of wealth, energy, and human resources.

The sociocultural transitions from the collapse of the Roman Empire through the genesis, affluence, and degradation of Western civilization, including the American experience, serves as an illustration of the continuous *refinement of a culture's survival instincts and methodologies*, which determine human behavior. The spiritual elements of the new religion of Christianity provided a new, more conducive cultural environment of ideational values that contributed to a revitalized population. Cultural renewal was gradually achieved, as the dominant Roman sensate mentality, incapable of providing the hope of survival for the discouraged masses, was displaced. Thus the initial genesis phase of Western civilization (A.D. 400) evolved into a fully developed ideational cultural system (A.D. 1000), followed by the transition to an idealistic culture (1300), and ultimately to the dominant sensate American culture during the twentieth century.

Thus, America was the by-product of a cultural rebirth process, based on a redefined social order emerging from the collapse of the Roman civilization. Over a millennium, the successful genesis and growth phases of Western civilization emerged from inspired, disciplined, and creative people who successfully established a new social order of increasing economy and social stability. The advent of the Industrial Revolution witnessed America contributing substantially to new knowledge and advanced technologies that, combined with innovative economic and production strategies, provided vast commercial opportunities. American industrialization generated a progressively greater variety of attractive products and services to an expanding global marketplace, thereby maintaining the momentum of a new, prospering socioeconomic system through the twentieth century. America's active participation in the scientific, technological, and economic development was handsomely rewarded.

Continuing economic success breeds additional economic opportunities, thereby inspiring more intense human desires, expectations, and competition for boundless wealth, influence, and social position. Consequently, the twentieth-century American culture increasingly embraced a more dominant sensate mentality as creative marketing and financial strategies made available an abundance of nonessential goods and services to seduce the population into historic consumer debt. In the process, formerly dominant ideational values and ethics associated with America's founding principles continue to gradually disappear.

The accelerated momentum of post-WWII American socioeconomic productivity could not be indefinitely sustained. The country's increasingly superficial priorities applied to its consumption of national wealth-energy resources and its more intense sensate values have had profound and debilitating effects on significant segments of the economy and its productivity. Consequently, America, while generating great wealth in the latter part of the twentieth century, accumulated major systemic cultural deficiencies that have progressively reduced its net rate of productivity by insidiously increasing its rate of societal degradation. This dynamic cultural equilibrium continues to shift with time, inflicting a more infectious and debilitating level of cultural degradation on the nation. The American system that had effectively fashioned and sustained historic socioeconomic advancements and a worldwide competitive advantage has gradually undergone significant cultural deterioration during the latter part of the twentieth century.

The financial rise and fall of General Motors, an American icon of unprecedented industrial success, provides an excellent case study that parallels the collective transition of the nation's socioeconomic system during the twentieth century. This general phenomenon, typically exhibited by mature civilizations and, in modern times, by large corporations and governments, has been described in a previous chapter as resulting from a depletion of resources and the inherent human limitations and capabilities to deal wisely with inevitable cultural complexities, competition, and human opportunities afforded by an advancing civilization. Whether applied to twenty-first-century American business, industry, or government, a paralyzing organizational and functional atrophy has been established that is unable to sustain the increasingly more complex, expensive, and wasteful nation. The American accumulation of substantial personal and governmental debt, a zero individual savings rate, escalating national debt, and currency devaluation are also typical characteristics of cultural degradation.

The period during which Western civilization's creativity, self-determination, self-discipline, inventiveness, and other ideational characteristics were dominant predates the era of its most profitable commercial applications and exploitation. The early, intensively creative scientific period of America was the foundation for subsequent historic socioeconomic advances throughout the twentieth century. Thus, it should be noted that the seeds of America's economic successes as well as it stagnation, as exemplified by General Motors, were sown during the initial success phase of industrial and commercial creativity and development of new scientific, technological, and commercial production processes that evolved from the Industrial Revolution. However, the effects of the cumulative negative cultural by-products of economic success, identified by Berman and other modern commentators, were masked for many decades by their relatively limited existence, overshadowed by higher rates of more visible economic productivity and social benefits. Additionally, the less-developed nations that did not share in the economic prosperity of the first wave of industrialized capitalism gradually became modernized and increasingly competitive with Western nations. This was the beginning of America's gradual loss of its global market share of manufactured goods and services, specifically automobiles to Japan, low-tech manufacturing to China, and business services to India.

Global diffusion of knowledge, technology, and economic theories and practices has been instrumental in the natural progression of mankind's quest for human survival and socioeconomic advancement since the beginning of recorded history. This is illustrated by the technological evolution of weaponry from gunpowder to nuclear weapons and by the economic transitions from Middle Age villages to Coke bottled and distributed in China. Such diffusion of more advanced knowledge and technological capabilities is the mechanism for worldwide transference of enhanced human opportunities, survival strategies, and economic achievement. Thus, today's expanding global economy is the natural consequence of the regionally restrictive twentieth century's industrially based capitalism, but now practiced in a more highly advanced, intense, and understood worldwide context of twenty-first-century opportunistic, competitive commercialism. Fundamentally, the global diffusion of socioeconomic principles and practices is an unavoidable and continuous phenomenon, a natural progression representing numerous systems incessantly undergoing change, but continuously seeking an evasive dynamic global equilibrium.

The economic challenge facing twenty-first-century America, or any declining culture, is the creation of new opportunities for more highly developed and profitable global economic prospects, as well as greater economic and societal stability in an atmosphere of greater altruistic values. This first requires the continuous creation of the next generation of advancements of human knowledge, applied technologies, and commercial production systems, which requires a spirit of innovation.

William A. Wulf, past president of the National Academy of Engineering, notes the importance of "the ecology of innovation" to America's future economic productivity, in addition to the usual elements of a more highly educated workforce and increased support for basic research. This "ecology" also pertains to updating and improving intellectual property law, tax codes, patent procedures, export controls, and immigration regulations. He contends that many of these important modern elements affecting the success of innovation are currently unworkable, irrelevant, outdated, or simply broken and must be repaired if the United States is to advance economically, particularly to the next level of manufacturing technology. America's opportunity is to create "a knowledge-intensive kind of manufacturing.... It will not be low-wage labor."[27] However, Wulf also laments the general lack of science and technology literacy in America: "Here we are with 90 percent of the population incapable of intelligent conversation about some important policy issues of the day." His perspective and recommendations address America's outdated infrastructure as well as the necessity of creating the next evolutional phase of scientific, technological, and engineering knowledge base, applicable to an increasingly rapid rate of change and the more challenging twenty-first-century economic environment.

The history of past civilizations, as interpreted by Sorokin, Toynbee, and other historians and sociologists, casts severe doubt on America's capability to rise to its twenty-first-century challenges. Such cultural rejuvenation has proven to be possible for failing mature civilizations only if the dominant sensate cultural environment is reversed and society increasingly embraces more altruistic values and ethical principles. Clearly, the current American population does not possess the collective discipline, integrity, initiative, ethics, and values necessary to begin re-creating a significantly more ideational cultural environment and the related behavior necessary to affect societal rejuvenation in the fashion outlined by Toynbee. Additionally, such cultural rejuvenation also requires an economic reorientation of the priorities for the consumption of limited wealth-energy resources. Berman addresses the "rebuilding" of the American social order:

> **The rebuilding of social capital cannot occur in a context in which power, money, celebrity, and the like have become the key values of the dominant culture ... a kind of mass pathology ... the ethos of globalization and late–capitalist corporation hegemony.**[28]

He continues by referring to "Putnam's observation" that America has destroyed the physical basis of community so as to reorient the landscape to maximize commercial opportunity, or as Vincent Scully professes, society will build what society values (i.e., financial success).

Meanwhile, the American culture continues to engage the challenges of twenty-first-century socioeconomic stagnation as the formerly less-developed Asian nations establish and cultivate their particular economic growth phases, learning much from the experiences of Western industrialized capitalism. These nations have increasingly become more economically competitive with the West, resulting in the historic flow of wealth into America gradually declining and being reversed, as America becomes a debtor nation. In earlier decades, imperialistic expansion and capitalistic economic advantages provided the necessary additional wealth-energy resources for America to more fully develop and maintain its unprecedented rate of increasing prosperity. However, this advantage has proven to be of limited duration. Berman, citing Herman Schwartz (*States Versus Markets*) and Barry Eichengreen (*Globalizing Capital*), provides a commentary that recognizes our theme that inadequate wealth and energy resources ultimately propel a society on a course of cultural exhaustion: "In the late phase of capitalist development, the relentless pressure for markets leads to imperialism, or what we now euphemistically call globalization. This process is the norm"[29] in a laissez-faire economy. America now must compete in a global marketplace while handicapped by significant accumulating financial liabilities and cultural disadvantages, making it significantly less competitive worldwide.

An additional troubling American liability, commonly found in historically deteriorating cultures facing stagnation, is the continuous inequitable broadening of the distribution of income and the massive accumulations of wealth by a very small minority. Frequently, this great wealth is derived from activities that contribute little to the production of national wealth or long-term societal improvement. This has been the general experience in America and, even more so, for the people of less-developed nation-states during the 2,000-year Western civilization cycle extending from the collapse of Rome to the cultural degradation of modern times. Data, representing the distribution of wealth in America and worldwide, provide a rationale for those who are fearful and demoralized by what appears to be the threat of an inevitable domination of modern economic globalization. The worldwide fear is that the more-developed nations, as occurred in the past, will continue to perpetuate the long history of the inequitable and harsh wealth accumulation by a small minority of people and nations.

In a commentary regarding "lavishly rewarded" major banking executives, Paul Krugman comments: "Around twenty-five years ago, American business—and the American political system—bought into the idea that greed is good."[30] He notes that the CEOs of Merrill Lynch and Citigroup were paid $48 million and $25.6 million, respectively, in 2006. Recent data support the view that such greed and inequity have been generally adopted within the U.S. culture as an acceptable and expected consequence of competitive capitalism. This is illustrated by the accelerating disproportionate sharing of America's productivity with those who provide the labor. In 2005, the two highest paid hedge fund managers received $1.5 billion and $1 billion in compensation. The average take-home pay for the top twenty-five hedge fund managers was $363 million. In 2007, the two top hedge fund managers earned

$3.7 billion and just under $3 billion. Even with the financial crisis, in 2008 the top twenty-five received $11.6 billion. In first place, a former math professor and now successful computer-driven trading strategist, continued his billion-dollar annual compensation pace with $2.5 billion. [31]

According to Northeastern University's Center for Labor Market Studies, the top five Wall Street firms (Bear Stearns, Goldman Sachs, Lehman Brothers, Merrill Lynch, and Morgan Stanley) were expected in 2006 to award $36 billion to $44 billion in bonuses to 173,000 employees, for an average bonus exceeding $200,000. Ironically, by October 2008, all five financial institutions had declared financial insolvency and ceased to exist as investment banks.

The center estimates that between 2000 and 2006, the non-farm labor productivity segment of the economy increased by 18 percent while the inflation-adjusted weekly wages of workers increased by 1 percent. These 93 million production and nonsupervisory workers in the United States had a combined real annual earnings increase from 2000 to 2006 of $15.4 billion, which is less than half of the combined bonuses awarded by the five Wall Street firms in one year.[32]

According to a survey by Pearl Meyers & Partners, the average total pay for chief executives of 200 large American companies rose 27 percent to $11.3 million in 2005. For those individuals who were surveyed each of the last three years, compensation increased by an average of 15 percent per year.[33] An analysis of the compensation of top executives at 102 large companies from 1936 to 2003 reveals executive compensation beginning to accelerate in 1980 relative to worker wages. In 1980, the average compensation was $1.6 million in 2006 dollars, $2.7 million in 1990, and $7.6 million in 2004, for an average increase of 6.8 percent per year. From 1980 to 2004, the average worker's wages increased by 0.8 percent per year, from an average of $36,000 to $43,000.[34]

In 1995, John Cassidy reports the existence since 1973 of American's "unprecedented redistribution of income toward the rich." Between 1947 and 1973, actual American income rose at the same rate for all income quintiles, after which the highest income quintile rose significantly. From 1977 to 1989, the top 1 percent of the population had an average income increase of 78 percent and, by 1989, owned 40 percent of American wealth. The two lowest quintiles, 40 percent of the population, suffered a decline in income equivalent to $275 billion per year being shifted from the middle class to the richest Americans. Also, Robert Reich notes that in 1995, 1 percent of the American population owned $4 trillion in assets, or 47 percent of the nation's wealth, with the upper quintile owning 93 percent.[35] Meanwhile, between 1979 and 1990, the number of American children living below the poverty line rose by 22 percent.[36] The lack of ethical behavior highlighted by the nation's 2008 financial crisis and the inequity of sharing national gains in American economic productivity are sad commentaries on the values and ethics of the American culture.

While America's productivity has steadily risen, wages and salaries now constitute the lowest share of the nation's gross domestic product since such data have been kept (1947). However, by 2006, corporate profits reached their highest share since the 1960s. One investment bank, UBS, describes the current times as "the golden era of profitability." Goldman Sachs economists noted, "The most important contributor to higher profit margins over the past five years has been a decline in labor's share of the national income."[37] From 1980 to 2004, the top 20 percent of American earners increased real income by 59 percent, and the top 10 percent were expected in 2006 to obtain 45 percent of the nation's household income, an increase from 40.6 percent in 2000.

The increasing stratification of wealth or net worth in America, distinct from income, constitutes a troubling long-term inequality. About 85 percent of the nation's wealth is owned by the richest 15 percent of American families, while the bottom 50 percent possess less than 3 percent of household net worth. The average net worth of the richest 10 percent of American families rose to $860,000 in 2005, a 6.5 percent increase over 2001, with the bottom 25 percent declining by almost 2 percent. Additionally, taxes for the upper income are lower than any time during the last sixty years.[38]

Continuing a quarter-century trend, American's rich are increasing their wealth. In 2004, the wealthiest 1 percent of American households had an average annual income of $326,720, which is 19.8 percent of the nation's gross income before taxes. According to the study by University of California economist Emmanuel Saez, this percentage was an increase of 2 percent from the previous year. Meanwhile, the richest 0.1 percent of individual Americans acquired 9.5 percent of the country's total gross income for the year. However, the median family income of Americans, adjusted for inflation, rose by only 1.6 percent between 2001 and 2004. This is in stark contrast to the income of corporate CEOs, professional athletes, and entertainers.[39]

The wages and benefits of the workers in other industrialized nations have also been eroded. The U.S. Bureau of Economic Analysis reports that from 2000 to 2005, wages, health insurance, and pension benefits for U.S. workers, as a percentage of gross domestic product, declined by 2.5 percent. The Organization for Economic Cooperation and Development reports that such compensation as a percent of GDP fell in Germany by 3.1 percent and Japan by 3.0 percent, while Britain and France increased by 0.8 and 0.2 percent, respectively. Such declines and extremely modest increases in the workers' share of productivity among the most industrialized countries have been the trend since the 1970s.[40]

The pervasive unfairness in the way the great wealth of the United States is distributed should be seen for what it is, an insidious disease eating away at the structure of the society and undermining its future.[31]

In 1996, the 447 richest people on the planet had assets equal to that of the poorest 2.5 billion, 42 percent of the world's population.[41] Robert Kaplan reports that in 1997, of the world's 100 largest economies, 51 were corporations and exceeded the economies of many countries. The 500 largest corporations accounted for 70 percent of world trade. "Corporations are like the feudal domains that evolve into nation-states; they are nothing less than the vanguard of a new Darwinian organization of politics, ... the forefront of real globalization."[42]

Not surprising, the global diffusion of Western knowledge and technology during the late twentieth century has expanded worldwide economic opportunity and advancement to the historically less-developed nations. Accordingly, successful economic principles and practices of industrially based capitalism have been widely adopted, modified, and improved worldwide. Consequently, the elements of modern production have catalyzed a broader and more competitive worldwide market for consumer goods and professional services. Unfortunately for the West, the former undeveloped nations of Asia can provide lower labor and material processing costs, thus placing the economies of Western nations at a substantial disadvantage. This phenomenon has increasingly broadened international competition, wealth distribution, and composition of the twenty-first-century individuals, corporations, and nations dominating the global economy. This rejuvenation of Asian cultures as Western cultures decline will be addressed in the next chapter.

The twenty-first-century authors cited in this chapter report the same characteristics and properties of cultural deterioration within the American culture that early twentieth-century authors found to be exhibited by Greek and Roman civilizations in their declining stages. Huntington describes such elements of modern American cultural degradation as "slow economic growth, stagnating populations, unemployment, huge government deficits, a declining work ethic, low savings rates, ... social disintegration, drugs, and crime."[43] In recent decades, the escalating intensity and scope of cultural deterioration have produced remarkable abuses of America's political, governmental, and corporate institutions. The complexity, disorder, and ineffectiveness of federal, state, and urban governing bureaucracies is as discouraging for the American people as the associated financial burden is for a deeply flawed governance system.

> **The grim truth is that the political leadership of the country, especially in Washington, is almost dysfunctional in grappling with the big issues.... From energy to education, climate change to health care, budget deficits to trade deficits, progress is perilously slow. And time is definitely not on our side.**[44]

Negative reviews of government service agencies parallel those of political figures whose interests appear to be focused primarily on serving those representing sources of wealth and secondarily, and occasionally, the general population. This atmosphere reflects to a great extent the amalgamation of political and business self-interests that successfully manipulate governance processes in a culture dominated by materialism, greed, and a ruthlessly competitive socioeconomic system. The political environment of twenty-first-century America is not conducive to substantive, constructive public discussion and debate, or to definitive and timely governmental action to address important national issues. The modern proliferation of well-organized and-funded special interest groups aggressively utilize professional media strategies and tactics and, in the process, successfully mute political discourse in America. It is widely accepted that successful political careers depend on a Hollywood-style charisma and the ability to manipulate public opinion and the media with messages that lack integrity, placate major interest groups, and confuse the public with ambiguous, shallow, and inaccurate rhetoric. The politician's objective is to avoid taking sides on crucial but potentially unpopular issues while not offending any segment of the voter population or the nation's power structure. This normally entails embracing the nebulous topics of hope, change, and spiritual righteousness while promising additional spending on education and health care, combating poverty, and increasing employment, at the same time unabashedly supporting lower taxes. Controversy is avoided at all costs, usually by accomplishing little of substance that is capable of making significant, long-term improvements in the social order. Time and effort is devoted to raising campaign funds from wealthy constituents and commercial and nonprofit organizations while protecting the economic interests and philosophies of supporters.

These modern-day political attitudes and practices have come to reflect society's general lack of respect for and commitment to intellectual, financial, and moral discipline and to a sense of integrity. Meanwhile, major policy issues facing America's future generations have become increasingly complex, often structurally, functionally, and politically irresolvable. This is illustrated by the federal income tax code, illegal immigration, annual budget deficits, the national debt, lack of universal health care, and the pending financial insolvency of Social Security (often referred to as the "third rail of politics").

Ethics, values, and integrity in the practices of twenty-first-century American business, politics, and governance are blatantly absent relative to America's early history. The accepted political agenda is to win ideological battles and to secure and maintain increasing levels of power, public recognition, and campaign funds. In every civilization, the conduit to achieving this objective has involved the collusion of business, financial institutions, and politics. American financial investors in association with business and corporate entities have utilized and manipulated government and politics to create mechanisms whereby the affluent are able to increase their assets to the detriment of the general society, as materialism and commercialism flourish. The methodologies of dominating and controlling economic interests is analogous to that found within the Roman Empire, whereby the wealthy elite successfully restricted land ownership to those of wealth and position by dominating and restricting the mechanics of political and financial systems.

We return to Berman's question of how the United States arrived at the point in time where "money, power, speed, and mindless entertainment came to be the defining characteristics of American civilization." The early American cultural environment of deference toward a mutually beneficial social order, embracing altruistic cultural norms, and a respect for a socioeconomic system that benefits all members of society has been replaced by increasing individualism that promotes and rewards self-oriented materialism and competitiveness. Meanwhile, American politics and government have increasingly become more complicated, centralized, costly, controlling, ineffective, and greatly influenced, if not directed, by well-funded and powerful special interest groups and large corporations.

A number of major American self-inflicted, dysfunctional conditions of complexity, disorder, and faulty judgments have been selected to illustrate the misuse of wealth, energy, and human resources, which has become instrumental in creating an escalating deterioration of the American culture.

Perhaps one of the best examples of government systematically creating incoherent complexity is the annual federal income tax preparation process that, on the average, requires thirty-four hours for the completion of a 1040 long form. Those who use the standard Form 1040 must complete seventy-six lines, some with multiple parts that utilize 142 pages of instructions. The 2007 version of the American federal income tax code consists of 66,498 pages, twice the size of the *Encyclopedia Britannica*, with more than 20,000 pages being added within the last six years. During the previous four years, the number of federal tax regulations has increased by over 40 percent.[45] Since 2006, the Internal Revenue Service has used three private contractors in an attempt to collect over $1 billion in unpaid taxes. In 2008, it was revealed that the program collected only $49 million from delinquent citizens, representing more than a $37 million loss for the government's program intended to increase revenue.[46]

In recent years, changes to the tax code designed to produce general economic benefits have assisted wealthier citizens to a greater extent than the lower and middle economic groups, thereby reducing the historic progressivism of the tax code. The tax rate reductions on income in 2001 and on capital gains and dividends in 2003 lowered the average tax rate for the richest 0.1 percent of Americans by 3.8 percent. However, the poorest 20 percent of the population receive a tax reduction of just 0.03 percent. Of the savings on investments, 70 percent went to the richest 2 percent of the taxpayers, while 90 percent of taxpayers making less than $100,000 benefited by only 14 percent from the dividend tax reduction and by 5 percent from the capital gains tax cut. Effectively, taxes for those earning over $10 million have been reduced to a lower rate than those with incomes between $500,000 and $1 million.[44] Consequently, the share of the nation's income for the wealthiest 1 percent of taxpayers has increased from 9 percent to 14 percent in this decade.[47]

In the late 1960s, a "stealth tax increase," the Alternative Minimum Tax (AMT), was adopted, intending to ensure that an identified 155 of the wealthiest citizens who hadn't paid any federal tax in 1967 would be required to do so. Over three decades later, the AMT unintentionally affects about 19 million taxpayers and is projected to include married couples with incomes over $100,000 by 2010. Essentially, the AMT is a second system of taxation that requires individuals and families to calculate their taxes twice, using two different methods, paying the higher amount. This process has its own fifty-five-line form that is required to be completed, if only to determine that this method doesn't apply to the individual.[48]

The irony of this government policy is reflected by its implication for taxpayers in 2006. It is estimated that Americans who earn $50,000 to $200,000 will pay an additional $13 billion in taxes, which approximates the total amount saved by those earning more than $1 million. Approximately 30 million taxpayers will benefit from such savings, while 19 million will be expected to pay the AMT. One in four families with children, an increase from one in twenty-two from the previous year, will owe up to an additional $3,640 in additional federal tax. According to the Tax Policy Center, those with incomes of $1 million are paying only 0.1 percent of the alternative tax revenue or more, even though this is the group initially targeted by the 1967 legislation.[49]

The AMT is an illustration of governmental patchwork applied to fundamental but historically incomprehensible and complex legislation that, in this case, regulates one of the more bureaucratic agencies in the federal system. The design flaws in the AMT legislation have resulted in its unintended application to numerous taxpayers. The many structural flaws or errors include the lack of an adjustment for inflation that automatically yields a yearly tax increase, the loss of personal exemptions (which penalizes families), and the loss of deductions for state and local taxes in high-tax states. All of these outcomes are unintended and will become more unjust and economically ineffective with the passage of time. Of a political interest, the impact of the AMT has reduced, if not eliminated, much-heralded tax reduction programs that were intended to stimulate the American economy and ultimately reduce the national debt.

The federal tax code illustrates a mature social order continuously creating new forms of structural and functional complexities that inherently generate new cultural problems, whose solutions also provide new economic opportunities for nonproductive consumption of wealth. For example, the continuous propagation of federal income tax laws and regulations has spawned a large and profitable tax preparation business that is utilized by 84 million people, or about 60 percent of taxpayers. Americans spend $65 billion annually for professional income tax preparation assistance and related services, estimated to add an additional 20 percent levy on top of the $1.5 trillion paid in taxes.[50] The increasing reliance on certified public accountants and lawyers to comply with governmental complexity is just one of many imposed requirements that consume significant national wealth, leaving less for investment in the regeneration of national wealth and for long-term beneficial social programs.

Creeping confusion and complexity in the overall system of taxation is also illustrated by the surcharges, taxes, and fees added to the monthly telephone bill. Unnecessary multiple doses of repetitive governmental regulations and related infinitesimal, annoying consumer usage charges have accumulated over the last century. The list includes the Local Number Portability Charge (23 cents), Cost Recovery Surcharge (0.4

percent), Local Utility User Tax (20 percent maximum), Relay Center Surcharge (16 cents), Gross Receipts Tax (3 percent), Emergency 911 Surcharge ($3), End User Access Fee ($2), Carrier Cost Recovery Fee ($1.49), Federal Subscriber Line Charge ($6.50), Universal Service Fee (10.2 percent), and a Federal Excise Tax of 3 percent. The latter excise tax was first imposed in 1898 to finance the Spanish-American War but now goes to the government's general fund.[51]

Governments are constantly creating additional laws, some designated as reforms, which inherently expand societal complexity and bureaucratic agencies. This is illustrated by expanding number of pages in *The Federal Register*, which records new and existing government regulations. The 1936 version of 2,620 pages has steadily increased to 74,402 in 2007. However, few new regulations result in greater cultural efficiency, lower cost, less stress, and fewer pressing issues and conflicts. The American culture, as with former deteriorating social orders, has characteristically suffered severe consequences from this self-induced chaos that includes inequities of wealth distribution and taxation, expensive and ineffective bureaucracies, financial debt and insolvency, and disproportionate military spending. Many social thinkers have identified such cultural characteristics of civilizations deteriorating prior to their total collapse. Durant expresses this view: "No great nation is ever conquered until it has destroyed itself."[52]

America is being "conquered" not by a military or actual barbarian invasion, as occurred in the eras of ancient Greece and Rome, but by its own individual, corporate, and governmental priorities, values, and practices. These include the invasion of foreign capital to finance America's escalating national debt and the invasion of illegal immigrants to provide pseudo-slave labor. The massive legal and illegal Latin American immigration is an illustration of a successful, mature civilization willing to compromise its values, integrity, and rule of law to achieve the never-ending quest for greater wealth. This explosive issue is just one example of America's unwillingness to manage its socioeconomic affairs based on high ethical, moral, and legal principles that it once held sacred. By a variety of legislative, budgetary, and political tactics, elected officials have neutralized law enforcement's efforts to curb a dramatic influx of illegal immigration. In addition, local and state governments have been saddled with significant expenditures for health care, law enforcement, education, and other public services to support an estimated population in excess of 11 million (mostly low-skilled and impoverished illegal immigrants and their families).

This illegal immigration is in sharp contrast to America's historic European and modern Asian legal immigration of people with sufficient and appropriate education and skills to support themselves and to contribute to socioeconomic growth. Furthermore, such immigrants have a history of possessing the desire to fully assimilate into American society and support the fundamental principles and ideals of the American culture. Teddy Roosevelt's statement regarding immigration captures this spirit and the historic meaning of cultural assimilation in the American culture:

> **In the first place we should insist that if the immigrant who comes here in good faith becomes an American and assimilates himself to us, he shall be treated on an exact equality with everyone else, for it is an outrage to discriminate against any such man because of creed, or birthplace, or origin. But this is predicated upon the man's becoming in very fact an American, and nothing but an American.... There can be no divided allegiance here. Any man who says he is an American, but something else also, isn't an American at all. We have room for but one flag, the American flag, and this excludes the red flag, which symbolizes all wars against liberty and civilization, just as much as it excludes any foreign flag of a nation to which we are hostile.... We have room for but one language here, and that is the English language, ... and we have room for but one sole loyalty, and that is a loyalty to the American people.[53]**

Support for the illegal immigrant population by elements of the business community and its political base is founded on the proposition that such labor is a necessity for the nation's long-term economic vitality. In actuality, this is a faulty, self-serving rationale that enables the business community to acquire inexpensive labor to improve profits while ignoring the imposed, unavoidable state and local government burden for providing necessary public services. Additionally, the laws of the land are disregarded and immigration violations are excused with the justification that illegal immigrant labor is essential to a vibrant economy. This issue has resulted in numerous negative, long-term cultural effects that will be difficult, if not impossible, to rectify to any perceivable degree.

Since the time of ancient Greece, ambitious civilizations, motivated by human aspirations for great wealth and power, have ruthlessly sought cheap labor including slave trading, conquest with virtual enslavement, and incentives for immigration. Modern economic and political strategies employed by the U.S. government to permit, if not encourage, massive immigration of cheap labor are based on the same economic objectives of the slave trade during the prosperous eras of Imperial Spain, the British Empire, and ancient Rome and Greece.

The immigrant population since 1970 has tripled and currently represents more than one third of all people ever to enter America. The annual arrival of approximately 1.5 million legal and illegal immigrants plus the annual birth rate for immigrant women represent three fourths of the total American population growth. In contrast to early twentieth-century immigrants, many of whom eventually returned to their country of origin, a much larger proportion of today's entering immigrants remain in the United States. It is estimated that by 2060, current immigration policies and practices will result in about 105 million persons being added to the U.S. population.[54]

Census Bureau data reveal that the U.S. immigration population reached a record 37.9 million in 2007, more than double the 1910 record peak. Immigrants account for one in eight U.S. residents, the highest level in eighty years; one in three immigrants is an illegal alien. A second record was achieved when, for the seven-year period since 2000, 10.3 million new immigrants entered the country, half of which were illegal immigrants.[55] Since 2000, approximately 1.5 million immigrants have entered the United States annually, adding an enormous financial burden to a social order already encountering functional, structural, and cultural stress and dysfunction. Immigrant families represent 20 percent of the school-age population, 24 percent of those live in poverty, and 30 percent are without health insurance. Given that only about one third of the adults have completed a high school education and arrive in the United States from a much less developed economic environment, the resulting dependence on government welfare is not unexpected.[56]

Immigration has resulted in a large increase in the low-wage, low-skill labor market, resulting in fewer opportunities for the native population. Immigrants are 60 percent more likely to be employed in low-skilled occupations than native-born workers. The average American wage may exceed the average wage of the immigrant's native country by a factor of ten; thus, the lowest-wage American jobs become very attractive to immigrants, who are often paid informally in cash, avoiding tax payments. In addition, the arrangement is attractive to employers who pay employees "off the books" and avoid responsibilities for related employer-paid taxes and benefits.

While immigrants received a disproportionate share of new employment between 2000 and 2004, the unemployment rate for adult immigrants increased from 4.4 to 6.1 percent, a 43 percent increase of 400,000 persons. During this period, there was an increase of more than 1 million immigrants per year; thus, the number of immigrants employed and the number unemployed both increased. Importantly, the net increase in the size of the immigrant population showed no indication of being closely linked to the rate of employment fluctuations for native-born Americans.[57]

It is noteworthy that the countries of origin of recent legal immigrants, such as China, Latin America, the Philippines, and India, are significantly poorer and lacking technical skills relative to the U.S. standard, compared with the nineteenth-century immigration from Great Britain, Italy, Germany, Ireland, and Russia.[58] Consequently, in recent decades local, state, and federal government agencies, out of necessity, have assumed responsibility for providing a broad range of basic social services for immigrants at significant government expense. The National Research Council has estimated that U.S. immigration in recent years has saddled the nation with a net annual expense ranging from $11 billion to $22 billion, with most for public assistance paid from state and local government tax revenues. This figure represents the net deficit between the total taxes collected from individuals and total government expenditures for social services. The deficit reflects lower tax levels paid by the mostly low-skill, less-educated immigrants during recent decades and their high rate of use of government services.

The National Academy of Sciences estimated that the average immigrant without a high school education imposes a net fiscal burden on public coffers of $89,000 during the course of his or her lifetime.[59] The average adult Mexican immigrant's deficit was estimated in 2000 to be $55,200. "Mexican immigration becomes, in effect, a subsidy for employers of unskilled labor, with taxpayers providing services such as education, health insurance and medical care, and income-transfer programs such as the Earned Income Tax Credit to workers, who because of their low incomes, pay nowhere near enough in taxes to cover their consumption of services."[60]

American immigration has been significantly altered by the large influx in Latin American immigration, particularly from the leading country of origin, Mexico. In 2004, Mexico accounted for 31 percent of America's foreign-born, compared to 28 percent in 2000, 22 percent in 1990, and 16 percent in 1980. The previous leading country of origin was Italy in 1970, with only 10 percent of the total immigrant population.[61] Obviously, such dominance by a single culture as has existed during the last twenty-five years raises serious issues regarding the future diversity of the American culture and more specifically the impact of integration and assimilation, as reflected by Teddy Roosevelt and more recently by Samuel Huntington.

The net effect of America's current immigration profile, laws, and enforcement policies and practices is that immigration issues have become a major source of societal instability. During the past twenty-five years, immigration has emerged to catalyze appreciable socioeconomic discord within some segments of the population, while providing profits for some elements of the business and corporate sector. This coincides with the executive branch of the federal government allowing financial burdens to befall local and state taxpayers who have been required, by default, to provide public services for an unexpected and uncontrollable expansion of illegal immigration. The profile of modern U.S. immigrants reveals extreme poverty, inadequate education, and a lack of high-level employment skills and cultural experience within a developed country. For example, in 2005, 31 percent of immigrants age twenty-five to forty-four lacked a high school diploma, compared to 9 percent of natives, and since 1990, immigration has increased this segment of the labor pool by 25 percent.[62] Immigrants constituted about 16 percent of the U.S. adult workforce, which constitutes more than 40 percent of adults in the workforce without a high school education. Since 2000, this segment of the workforce without a high school degree has increased by 14.4 percent.[63] These characteristics constitute a significant handicap and a formidable challenge for these immigrants to achieve socioeconomic success and to assimilate into the modern American culture. Thus, it is not surprising that substantial public assistance has been necessary to provide food, shelter, health care, and education. There are 8.6 million school-age children of immigrants in the United States, with Mexican immigrants accounting for 3.2 million, or more than one third of the total and more than one third of the national increase in the school-age population since the early 1980s. On the average, Mexican households have twice the number of school-age children compared to the native population.

From 1979 to 1997, immigrant households increased in the United States by 68 percent, while the poor population increased by 123 percent and the gap between immigrant and native poverty rates tripled during this time period. For the most part, local and state governments have provided humanitarian aid to cope with an avalanche of illegal immigrants.

Legal and illegal immigration has created major long-term negative socioeconomic consequences for the United States. Faulty human judgment and ineffective public policy has generated a self-induced divisive social climate in America. Past immigration to the United States brought valuable assets into the country with relatively few liabilities while meeting the labor market's needs; it also contributed immeasurably to the building of a great early twentieth-century society. Unfortunately, the last few decades of uncontrollable Latin American immigration has brought into the country an extremely large number of people with low employment assets and high financial liabilities from a vastly dissimilar culture. "Setting aside the lower socioeconomic status of immigrants, no nation has ever attempted to incorporate more than 35 million newcomers into its society."[64]

In 1996, Samuel Huntington placed the immigration issue in a global twenty-first-century context:

> Western culture is challenged by groups within Western societies. One such challenge comes from immigrants from other civilizations who reject assimilation and continue to adhere to and to propagate the values, customs, and cultures of their home societies. This phenomenon is most notable among Muslims in Europe, who are, however, a small minority. It is also manifest, in lesser degree, among Hispanics in the United States, who are a large minority. If assimilation fails in this case, the United States will become a cleft country with all the potentials for internal strife and disunion that entails, ... a country not belonging to any civilization and lacking a cultural core.... No country so constituted can long endure as a coherent society. A multicivilizational United States will not be the United States; it will be the United Nations.[65]

Unfortunately, America's twenty-first-century illegal immigration issue is driven by the nation's unwillingness, and perhaps inability, to create sound socioeconomic policies and practices based on a high sense of integrity and moral values and to enforce existing laws for the general, long-term benefit of the nation as a whole. This cultural limitation is indicative of a more general cultural deterioration, the broader symptoms for which have been associated over the ages with failing civilizations. The pre-eminent objectives of wealth and power appear to ultimately reach an opportunistic, greedy stage that capitalizes on a low-cost, disadvantaged labor pool, which inevitably contributes to the erosion of the moral core of the social order.

> A civilization which has become the victim of a successful intrusion has already in fact broken down internally and is no longer in a state of growth. For our present purpose it is enough to observe that of the living civilizations every one has already broken down and is in the process of disintegration except our own.[66]

Historically, mature civilizations, exemplified by the American culture, suffering from "successful intrusion" have already endured prolonged dysfunctional bureaucracies; a deteriorating infrastructure; severe social and economic instability; and debilitating materialistic values, motivations, attitudes, and behavior. Concurrently with these characteristics of cultural degradation, one also normally detects a critical and deepening depletion of wealth-energy resources due to the increasing costs of escalating functional and structural integration and stratification (i.e., the gradual creation of greater sophistication and complexity of cultural processes). These characteristics also describe the twenty-first-century profile of the American culture. The underlining foundation responsible for this ultimate cultural stagnation phase has been extensively developed in previous chapters. It is derived first from a *declining marginal return* of society's consumption of its national wealth-energy resources. And second, from *declining energy returns on investment (E.R.O.I.)*; that is, the *declining returns, in energy units, per unit of energy invested* in a nation's energy production processes of acquiring, refining, and distributing new energy capital to consumers. Also recall that the restrictive *science-based definition of wealth requires physical matter possessing value,* as illustrated by steel and coal, in contrast to Wall Street's twenty-first-century financially engineered paper products of debt.

On a broader scale, the human machine requires a continuous supply of energy from agricultural products. Thus, the twenty-first century's increasing global population has created a dramatic increase in the demand for new sources of food energy as well as for fossil fuels. Worldwide advancements in general knowledge, communication, technology, economic markets, and opportunities, and the attractiveness of the Western standard of living, have created greater global expectations, often perceived as individual entitlements. One consequence is a continuously increasing sophistication and redefinition of "modernity" with concurrent higher aspirations for a materialistic life style, a continuous progression from earlier centuries. This will continue to require a significantly larger and expanding worldwide supply of both food products and fuels, as the global consumption for both rapidly narrows the gap between the demand and limitations of the Earth's resources.

The mature industrialized countries, historically the largest consumers of fuels and food per capita, are gradually losing their long-held economic advantage to the energy-rich and newly industrialized nations as the projected total worldwide demand for wealth-energy resources rapidly multiplies. The less-developed countries, in early stages of rapid economic growth, are realizing new opportunities, wealth, and global economic empowerment but, in the process, are also dramatically increasing the world's consumption of energy at an unprecedented rate. As a result, the American economy, losing its former vitality and competitiveness, faces the challenges of its energy consumption greatly exceeding its available internal energy sources as energy costs dramatically increase and the dollar declines in value.

In 2008, as oil prices widely fluctuated between $50 and $150 per barrel, America endured its third energy crisis in a generation, but unlike the sudden disruptions in Middle East exports that produced the first two, this occurrence resulted from increasing global demand. In the period of a decade, the price of crude oil increased by over 350 percent.[67] For the first time, Asia is consuming more oil than North America, and China has become the world's second-largest oil importer. China has tripled its oil consumption since 1980, and since the 1990s, its economy has experienced an average 10 percent per year increase, lifting about 300 million people out of poverty. However, per capita, China consumes only 10 percent of the energy used by Americans, 20 percent of that used by the Japanese, but twice the average of India. Given the increases in China and India's energy consumption, it is projected that worldwide demand is growing by an average of 2 million to 3 million barrels of oil a day. Within twenty-five years, China and India together will consume three times the energy that the United States currently uses. Jointly, they are expected to add 80 million cars within a generation, and general socioeconomic progress can be expected to sharply increase the demand for consumer goods and services.[68]

America's importation of oil costs the nation about $1 billion a day, and from 2001 to 2005, oil producers' revenues increased from $300 billion a year to $800 billion. The current U.S. consumption of about 12 million barrels a day is expected to reach 20 million barrels by 2025. Two thirds of the U.S. oil consumption is used for transportation, with three out of four Americans commuting by car to work, as compared to one in five European workers. America is the only industrialized nation less energy efficient in 2008 than it was two decades ago.[69]

The worldwide trend of new oil discoveries relative to oil consumption is sobering. In 1930, new deposits of 10 billion barrels were discovered and 1.5 billion barrels were consumed; in 1964, 48 billion were discovered with 12 billion used; in 1988, 23 billion were found to 23 billion used; and in 2005, 6 billion were found and 30 billion consumed. The International Energy Agency estimates that during the next forty years, 90 percent of new energy supplies will come from developing countries, in contrast to three decades ago when 40 percent originated from industrialized nations.[68]

Given the current and expected global increases in economic activity and the demand for consumer goods, some oil-rich nations, which have historically been net oil exporters, are expected to become net importers. This has recently occurred in Indonesia and could occur in Mexico within five years. Iran, Algeria, and Malaysia are projected to be in this same category within a decade. From 2005 to 2006, internal oil consumption by the five biggest oil exporters grew 5.9 percent: Saudi Arabia (7.0 percent), Russia (5.8 percent), Norway (6.3 percent), Iran (5.7 percent), and United Arab Emirates (2.4 percent). This consumption is consistent with a World Bank report that economic growth in the Middle East, Russia, and North Africa doubled during the 1990s.[70]

The historic early twenty-first-century dramatic increases in the demand and price for oil parallels a similar scenario for agricultural products that constitutes a serious broad-based global food shortage (i.e., the energy or fuel supply for the human machine). In both cases, current and anticipated inadequate supplies of fuel and food will increasingly influence human behavior, politics, government policies, and the events of history.

With the emerging late twentieth-century socioeconomic advancements and profitability of newly industrialized developing countries, a so-called *dietary globalization* has emerged, a by-product of economic globalization and population growth. Acquisition of new wealth has permitted less-developed countries to modernize agricultural methods and to import food, including commodities, reflective of affluent Western nations. The resulting global demand for a more satisfying diet has resulted in the consumption of more protein and calories, specifically meat, dairy products, and wheat. This has resulted in an enhanced quality of life for some that includes a more energy-rich diet and greater joy in dining, while others simply acquire a healthier survival diet. Meanwhile, these successfully developing countries anticipate continuing population increases averaging 7 percent per year.

Thus, projections call for severe pressure on the world capability to produce adequate food to meet future demands.

The world's grain production is expected to increase by 8 percent in 2008. However, the world's wheat surplus is at its lowest level since 1947, as during seven of the last eight years, the world's wheat consumption has outpaced production. In 1967, worldwide wheat consumption was 300 million metric tons but by 2007 had grown to 600 million metric tons. Given the current demand-to-supply imbalance, it is not surprising that this year's wheat price is the highest, adjusted for inflation, since 1975. American agriculture exports are expected to increase 23 percent in 2008 to a record $101 billion, 50 percent greater than the average of the last ten years.[71]

Additionally, as a result of global industrial development, worldwide coal consumption has increased in recent years by more than 4 percent annually. The U.S. coal price increased by over 60 percent as American exports increased from 49 million tons in 2006 to 59 million tons in 2007, with an estimated 80 million tons expected for 2008. Given that America possesses 27 percent of the world's coal supply, it is projected that this export business should flourish in the future, as America becomes a major worldwide exporter. Additionally, as with oil, developing countries that have historically been coal exporters are consuming more coal and are in the process of becoming net importers. These countries include China, South Africa, Indonesia, and Vietnam, while India and Russia are significantly increasing coal imports. The dramatic increase in the global demand and price for agriculture products and coal, coupled with the declining value of the dollar, provides a bright spot for America's export revenues.

It is instructive to view twenty-first-century global fuel and food production/consumption data relative to Great Britain's exhaustion of its rich coal resources of the Newcastle-upon-Tyne region during the eighteenth century. The cultural consequences of this Industrial Revolution-era experience of a nation exhausting its supply of wealth-energy resources in a competitive global environment may be worth relearning.[72]

Historically, when inadequate wealth-energy resources severely limit socieal aspirations in achieving economic advancement or sociopolitical dominance, human values, ethics, and behavior adjust accordingly to justify the pursuit of desired objectives. Such a deficiency also applies to the availability of human energy resources (i.e., food commodities) necessary to nourish the population as it conducts the social, economic, and political affairs of the social order. This energy supply of food is as necessary to cultural survival and prosperity as the energy requirements for fueling mechanical, chemical, and electrical processes and their associated machines and devices. Over the millennia, insufficient wealth-energy resources have initiated abundant conflicts, suffering, and chaos, contributing to social and economic instability and ultimately cultural degradation. Such circumstances have also provided the background for one civilization fading from global prominence while another more primitive, vital culture begins its passage from a genesis stage to rapid economic growth and cultural prosperity by virtue of its more favorable profile of wealth, energy, and human behavior. The twenty-first century is witness to this passage of Western civilization and the resurgence of Asian cultures.

Thus America, having consumed an enormous and disproportionate quantity of the planet's natural resources, has achieved enormous prosperity in its 400-year history as well as creating an unparalleled degree of cultural complexity and disorder while evolving through Toynbee's four stages of cultural development. This journey completes Sorokin's sociocultural transformation from the prevailing ideational culture of America's Colonial times to twenty-first-century America dominated by sensate values. America's available wealth-energy resources are approaching inadequacy to support its current socioeconomic system and the population's expectations for a continuously escalating standard of affluence and entitlements. Additionally, the current cultural characteristics, traits, and behavior make improbable any near-term achievement of cultural "rejuvenation" as defined by Toynbee. Meanwhile, "ossified" Asian cultures become revitalized and reborn.

America's insufficient national wealth, relative to expenditures and its associated cultural deterioration, is reflective of government's budget priorities, budget and trade deficits, and a mounting national debt. In 2005, America's trading imbalance for high-tech manufactured goods was $24 billion, while in 1990 it had been a surplus of $33 billion, constituting a $57 billion turnaround. In 2004, America's trade deficit in financial services was $6.82 billion, while Britain had a $32.56 billion surplus.[73] As of January 2008, according to the U.S. Census Bureau, the combined goods and services trade deficit reached $58.2 billion. Foreign investments in American properties in 2005 totaled $130 billion, an increase of 20 percent as foreign investments and ownership of U.S. assets dramatically increased.[74]

U.S. federal spending in 2006 exceeded economic growth at a rate unseen in more than half a century, as government spent 20.8 cents of every dollar the American economy generated, an increase from 18.5 cents in 2001. This trend is expected to continue, as Congress remains unwilling to offset new expenses by reducing current spending or by increasing taxes. The common-sense rationale for government budget preparation should be to avoid inflated spending. However, the federal budget is projected to add $3 trillion of debt during the next five years due to increases in record spending for the military; domestic programs; and benefits for the elderly, poor, and disabled.[75]

Prior to the late 2008 financial crisis, the federal budget deficit for 2008 was expected to reach $400 billion, with the first economic slowdown stimulus package expected to add another $150 billion. The proposed 2009 $3.1 trillion federal budget would increase military spending and curb the growth of Medicare and Medicaid, with a budget deficit of about $410 billion. The proposed $515.4 billion budget for the Defense Department would, adjusted for inflation, be more than at any time since World War II. The budget includes "savings" of $208 billion over five years including Medicare ($178 billion), Medicaid ($17 billion), and student loans ($6 billion). Federal debt held by the public, which was $3.3 trillion in 2001, is expected to reach $5.9 trillion in 2009.[76] The sudden Wall Street financial collapse and consequences of late 2008 are projected to add another $800 billion to $1 trillion to the national debt, with additional billions expected in subsequent years.

From 2000 through 2005, the enrollment in federally funded social programs increased by 17 percent from 263 million to 307 million recipients, the largest increase since the program was created in the 1960s. In 2005, this represented an expenditure of $1.3 trillion, which is an increase adjusted for inflation of 22 percent since 2000 and accounts for half of all federal spending. Enrollment growth is responsible for three fourths of the social program spending increase, which, in part, is due to the increase in the poverty rate from 11.3 percent in 2000 to 12.7

percent in 2004. An additional major expenditure increase is anticipated in 2008, when 79 million baby boomers first become eligible for Social Security, and in 2011, when they become eligible for Medicare. Also, over the next decade, Medicare's new prescription-drug coverage is projected to cost an average of $80 billion a year, adding nearly 20 percent to the annual cost.[77] Clearly, projected future resources are less than the anticipated future costs of social services for an expanding, aging American population and the nation's decaying infrastructure.

The American people have come to expect little progress from the federal government's dealing with the broad scope of critical issues facing the nation. Year after year, the U.S. Congress appropriates more of the taxpayers' money, often for illogical, political, and wasteful purposes, accomplishing little to resolve critical national issues.

In 2007, according to the Taxpayers for Common Sense, over 500 members of Congress arranged for more than $18 billion in "pork barrel projects," or so-called earmarks, to be appropriated for their constituents. A National Drug Intelligence Center in Johnstown, Pennsylvania, received $39 million to provide drug war information to police departments across the nation, a function already served by nineteen other government agencies. Also, while the Air Force requested that the production of C-17 cargo planes be halted after producing two more aircraft, Congress "earmarked" $2.4 billion for ten additional planes.[78] Recently, these pet projects that historically have been quietly added to appropriation bills were recommended to receive the new title of "legislatively directed spending." House Minority Leader John Boehner, one of fewer than twenty members of Congress who refused to seek home-state earmarks, stated, "The earmarking process in Congress has become a symbol of a broken Washington."[79] The long history of earmarks is illustrated by the record of Senator Ted Stevens, a forty-year veteran from Alaska, who has led that state's delegation to acquire record levels of pork barrel projects over the years. These include a 2005 Senate highway transportation bill containing funds for the now-infamous bridge dubbed the "Bridge to Nowhere." The bridge was to connect the town of Ketchikan, Alaska (population 8,900), with its airport on the island of Gravina (population 50) at a cost to federal taxpayers of $320 million. At the time, a ferry provided the service to the island, but required a fifteen- to thirty-minute wait and a fee of $6 per car. In addition, the bill included $230 million for the "Don Young Way," a bridge between Anchorage and a swampy undeveloped port, which is named for Alaska's Congressman Young.[80]

As of 2008, the American gross national debt stood at $9.3 trillion and is currently increasing at a rate of $600 billion a year. Politicians often speak of the "debt held by the public" rather than the "gross national debt" owed to the general fund. This enables those in political office to subtract the annual surpluses of the Social Security Trust and other government trust funds such as Medicare and highway construction tax revenues, which the federal government borrows and owes the public's trust funds. From 1945 through the 1970s, the American national debt increased slightly, but never exceeded $1 trillion despite the Korean and Vietnam wars. However, beginning in the late 1970s, the debt increased dramatically to the 2006 level of $8.3 trillion, at which time Congress raised the national debt ceiling for the fourth time in a five-year period, to $9 trillion. In 2006, the national debt amounted to $150,000 per citizen. In fiscal year 2005, the American government spent $352 billion in interest on the national debt, compared to $61 billion on education and $56 billion on transportation. The interest expense paid on the national debt is the third-largest expense of the federal budget, exceeded only by defense and by health and human services costs.[81]

In 2005, spending for the Afghanistan and Iraq wars was $277 million a day or over $100 billion a year. By mid-2006, the Iraq war cost was reported by the Congressional Budget Office (CBO) to be $300 billion and was expected to reach half a trillion dollars by 2009. However, in 2008, the cost reached $12 billion a month, and the Congressional Budget Office projected a cumulative cost by 2017 of $1.2 trillion to $1.7 trillion.[82] The CBO estimate includes $705 billion in interest expense from the war's deficit spending. Joseph Stiglitz, a Nobel prize economist, has projected an ultimate full cost of $1.7 trillion to $2.7 trillion. He notes that the national debt has increased by about $2.5 trillion since the beginning of the war; about $1 trillion is due to the war directly and will increase to $2 trillion by 2017. Robert Hormats, vice chairman of Goldman Sachs International, points out the "opportunities lost" as a result of military spending; for example, the *daily cost of the war* was sufficient to enroll an additional 58,000 children in Head Start for one year.[83]

In recent decades, the deepening ineffectiveness and inefficiency of large government and corporate structures and functions have had dramatic effects on the American population's confidence in the nation's social and economic stability. One of the most large-scale and traumatic illustrations has been the erosion and complete failure of retirement and health care benefits and programs as a result of corporate insolvencies and the lack of assurances that a government safety net will provide commensurate financial resources. Nearly two thirds of Fortune 1000 companies have defined pension benefit plans. In 2004, seventy-one companies, 11 percent of those that had such plans, "froze" them (i.e., halted future benefits that would have been earned by active employees) or "terminated" their plans. This represents an increase from forty-five companies in 2003 and thirty-nine companies in 2002. Employees of such failing companies expect the federal government to provide a safety net for retirement benefits.[84]

The U.S. Pension Benefit Guaranty Corporation (PBGC) is a federal corporation created by the Employee Retirement Income Security Act of 1974. The PBGC plays the role of insuring private pension funds, analogous to the Federal Deposit Insurance Corporation, which insures bank deposits. The PBGC currently protects the pensions of 44 million American workers and retirees of over 30,000 private, single-employer, and multiemployer benefit pension plans. The PBGC receives no funds from general tax revenues. Operations are financed by insurance premiums established by Congress and paid for by sponsors of defined benefit plans, investment income, pension plan assets managed by PBGC, and recoveries from companies that formerly were responsible for their own plans.

In 2003, the PBGC had exhausted its entire $8 billion surplus as a result of a series of major corporate bankruptcies during the previous year. Three steelmakers accounted for most of the drain on PBGC's resources: Bethlehem Steel required $3.7 billion, National Steel $1.1 billion, and LTV Steel $1.6 billion, its second pension bailout in sixteen years. All three pension plans suffered bankruptcy due to the collapse of the U.S. steel industry. Bethlehem's 67,000 retirees receiving benefits and 15,000 laid-off workers were eligible to receive future pension benefits supported by only 13,000 active workers producing company revenues.[85]

Meanwhile, beginning in 2006, PBGC assumed liability for more than $3 billion in pension contributions over a five-year period for United Airlines. Additionally, Delta Airlines' pension plan is underfunded by $5 billion and Northwest's by $3.8 billion. Thus, as of 2006, the PBGC was underfunded by $108 billion, an increase of $96 billion from the previous year, and had a $22.8 billion operational deficit for fiscal 2005. Thus, currently the PBGC is responsible for the pensions of over 600,000 current and future retirees, has no assets, and projected annual revenues of only $800 million a year. *The Wall Street Journal* describes PBGC as a "slow-motion train wreck."

It is estimated that as of 2006, the pension funds of American companies are underfunded by $450 billion. Credit Suisse First Boston reports that 326 companies that are listed in the Standard & Poor's 500 have underfunded benefit plans, with a total underfunding of up to $185 billion. As Standard & Poor's downgrades GM and Ford stock to junk status, a potential $50 billion in benefits responsibility could be assumed by the PBGC.[86] The current trend of corporate pension funds defaulting on their obligations to workers is anticipated to affect additional millions of future retirees.

The failure of corporate pension funds is symptomatic of the degradation experienced by America's large, aging segments of former economic success, such as Detroit's Big Three automobile manufacturers, the airlines, and the steel industry. The mid-twentieth-century corporate culture of General Motors provides valuable insights into the institutional values and economic principles that guided an unwise utilization of resources and the squandering of opportunities for continuing profitability. GM accumulated enormous long-term financial obligations due to overly generous wages as well as retirement and health care benefits designed to rely on the faulty assumption of the company's unwavering prosperity. Unfortunately for some American corporate sectors, a very competitive global business environment evolved in the late twentieth century, which rendered America and other mature industrialized countries dramatically less competitive. Specifically, these American industries have either stagnated or collapsed in accordance with Schumpeter's postulate of "creative destruction" (i.e., creative innovation of capitalism ultimately leads to internal corporate demise). The inability of an organization or society to constructively adapt during its success phase may result in its original ideas, technologies, skills, and tools becoming obsolete and its products becoming less competitive in the dynamics of the marketplace.

It has become more apparent in recent times that regardless of current profitability, business and industry must incessantly respect the "ecology of innovation" and continuously adopt attitudes, behavior, and "strategies of permanent innovation." The corporate culture of General Motors did not embrace such a perspective in the latter part of the twentieth century. Worldwide diffusion of industrial technologies and engineering methodologies, particularly in computer-assisted design and manufacturing, have significantly reduced the historic American competitive advantage. A major American disadvantage is higher production costs, relative to underdeveloped countries, related to its high standard of living and more generous wages, pension plan provisions, and health care benefits. America's global twenty-first-century economic standing in the financial services, computing, and telecommunications businesses is following the same pathway as the steel and automotive industries during the late twentieth century (i.e., becoming less competitive in a more broadly industrialized, commercialized, and progressive world market).

The stages and specific aspects of General Motors' industrial growth and decline provide exemplary insights into the decline of the American culture and illustrate Schumpeter's hypothesis of the inevitable, creative destruction of capitalist economic systems. During periods of corporate growth and prosperity, GM, under heavy sociopolitical pressure, provided unrealistically high salaries and benefits for its unionized employees. During the 1950s and 1960s, GM, facing relatively insignificant worldwide competition, was able to expand a costly, bureaucratic enterprise as a result of its high profitability. The overly generous wage and benefit concessions provided to the workforce were unnecessary from the perspective of the competitiveness of the existing national labor market. Excessive nonessential expenditures and overly generous profit sharing was the rule at the expense of the need for a continuous *strategy of permanent innovation*. Faulty logic and misplaced priorities, as well as insufficient thought and investment resources, created a comfortable status quo attitude at a time when global expansion of technology, engineering innovation, and modern business practices was inevitable.

GM failed to utilize a sufficient portion of its generated wealth, created in its early, highly profitable growth stage, to effectively respond to the predictable worldwide challenge to its market share. More effective and timely investments in technological innovation, competitive product design, and automated manufacturing capable of achieving lower labor costs per vehicle could have lessened the impact of the unavoidable foreign competition.

Instead, the GM organizational dynamics, on a smaller scale, followed an institutional pattern leading to "stagnation" analogous to the progression of the cultural stages from genesis through stagnation followed by all civilizations since the time of the ancient Greeks. Consequently, the by-products of GM's early economic success of creating large profits were its own particular varieties of organizational complexity, bureaucracy, and disorder that eventually produced irresolvable human and financial issues. This *corporate entropy production*, a component of America's accumulated *societal entropy production*, contributed to a gradual, inevitable corporate deterioration.

In 2005, GM posted an annual loss of $4 billion. The company had attained an enormous competitive disadvantage of about $2,500 per vehicle with its foreign competitors as a result of its accumulated financial obligations to generations of retired and laid-off workers. GM's 1960s partial, but noteworthy, worldwide monopoly based on style, function, quality, and price permitted it to hire unskilled employees at rates and benefits that have been estimated to be twice the appropriate national marketplace value at that time. In 2006, it was estimated that GM was responsible for health care insurance for 145,000 employees and 1.1 million retirees and dependents, adding $1,500 to the price of each vehicle produced. Other legacy issues include employees retiring after thirty years at full pay, regardless of age, and workers not contributing to the GM health program, while retirees pay only a small amount. Retirement benefits cost the company the equivalent of $1,000 to $1,500 per vehicle. To contrast, in 2005 GM paid $5.4 billion in health care costs for its 600,000 retirees, current workers, and dependents, while Toyota's annual report referred to its health care expenses for its retirees as "not material." Toyota employees pay the major portion of their health care premiums and are eliminated from company support at age sixty-five.[87]

In 1984, at GM's suggestion, the United Auto Workers agreed to a contract with GM, Ford, and Chrysler that established a worker layoff program, the Jobs Bank. This enabled those workers laid off to perform charity work or go to school while drawing full pay as a laid-off employee. At that time, the companies erroneously assumed that any future worker layoffs would be temporary. In 1969, the UAW had a membership of over 1.5 million, compared with less than half of that number in 2006. Clearly, the 1984 assumption that future layoffs would always be temporary was faulty, resulting in a major continuing financial liability for the companies. In 2006, GM had about 7,500 employees in the Jobs Bank, each receiving wages and benefits valued at between $100,000 and $130,000 a year, for an annual cost to GM of approximately $850 million.[88] These former workers are not required to contribute any value to the company. Such private sector welfare programs are founded on the reckless assumption that continuing company profitability would indefinitely supply the necessary future revenues for these contractual employee benefits.

In essence, the concept of generously sharing large profits during the early years of prosperity with employees in the form of permanent, multiyear commitments rather than providing one-time bonuses and holding wages and benefits to more reasonable long-term levels stifled wise long-term company investments. Consequently, GM has been unable to constructively adapt to the evolution of a global automotive industry and to generate sufficient new capital to ensure long-term financial solvency. The company did not adopt appropriate financial priorities and strategies based on a sound economic model. As a result, it was unable to effectively invest in a corporate culture of continuous innovation and create sufficient new ideas, products, and opportunities; thus, production problems and enormous financial loses ensued as its market share declined.

In 2006, while Ford and General Motors announced plans to close thirty factories and eliminate 60,000 jobs through 2012, Toyota was finalizing plans to construct its eighth North American assembly plant. Perhaps, Toyota learned from GM's history to avoide unionized states and build a national political power base, seeking lucrative state incentive packages as it moves toward its objective of increasing its share of the American car market from 14 to 20 percent. Toyota has become the richest car company in the world, earning $16 billion worldwide in 2005, while GM lost $10 billion.[89] During its initial period of prosperity, Toyota is successfully avoiding financial commitments over which it has no long-term control by placing health and pension burdens on their American plant workers rather than on the company. In 2006, Toyota's American enterprise has 258 retired production workers (compared to GM's 400,000 retirees), which will increase to only 1,700 by 2012. Toyota expects to pay about $700 million in pension benefits worldwide in 2006, which is one tenth of GM's payments. Toyota insists that workers make contributions to a company matching pension plan that is dependent on the financial markets, as in a 401(k) plan.[87]

Currently, the number of retirees per active worker for the U.S. Honda, Toyota, and Nissan automobile plants is 0.02, 0.05, and 0.11, respectively, while Detroit's Big Three range from 1.25 to 3.10. The Japanese have avoided providing overly generous benefits during the early stages of financial prosperity in favor of a more sensible long-term financial strategy. This was the lesson to be learned from GM's 2006 unfunded pension and health care liability of $85 billion and from observing GM's payroll shrinking by two thirds from 1990 to 2005.[87] The Japanese approach provides for a future during which a larger portion of its profits will be available for reinvestment in innovative new products. This philosophy recognizes the principle that continuously maximizing the reinvestment of profits into new wealth production endeavors will enhance the probability of long-term profitability (i.e., minimizing unproductive consumption of wealth-energy capital maximizes long-term profitability). This principle could be considered as a thermodynamic efficiency perspective and representative of Galbraith's view of social and economic stability, discussed in previous chapters. Sufficient resources must be available to continuously support the private sector's mission of continuous profitability that also provides, in a balanced manner, for employee benefits and social services, whether bestowed by the private or public sector. While the General Motors history violates this principle, other American corporations have also undergone the same experience.

Globally, industrially mature countries face the same socioeconomic challenges characteristic of America's twenty-first-century cultural deterioration. Their citizens and newly arrived legal and illegal immigrants expect the benefits, protection, and entitlements that have come to be associated with successful industrialized nations. Such countries, whether dominated by a capitalist or socialist philosophy, have created generous but expensive public and private social service environments of entitlements that are costly and ultimately capable of becoming economically debilitating, as prophesied by Schumpeter and Galbraith.

The cultural and political realities of the stagnating Western civilization are, in part, a consequence of corporate and governmental commitments that have grossly exceeded long-term economic capabilities. The resultant social and economic instability within Western civilization, and specifically the American culture, is increasingly sapping its vitality and ability to work toward common societal objectives. Fundamentally, unwise cultural priorities and limited national wealth relative to expenditures severely limit America's ability to effectively compete in the twenty-first century's global economy.

Galbraith addresses the folly of underestimating the economic consequences of providing a society with desirable but excessive public services by either the public or private sector:

> **Public services ... are an incubus. They are necessary, and they may be necessary in considerable volume. But they are a burden which must, in effect, be carried by the private production. If that burden is too great, private production will stagger and fall. At best, public services are a necessary evil; at worst, they are a malign tendency against which an alert community must exercise eternal vigilance. Even when they serve the most important ends, such services are sterile.[90]**

Galbraith recognizes the necessity of a higher priority being assigned to investments that are capable of regenerating national wealth in lieu of the production of excessive frivolous goods and well-meaning public services. He warns against underestimating a disproportionate humanitarian attack on social issues through the "public services" mechanism. While public services may be necessary in "considerable volume," the cost may become a "burden" (i.e., "private production will stagger and fall"). This dilemma is framed and illustrated by debates associated with the 2006 U.S. rallies for "rights" by illegal immigrant workers. The emotional and humanitarian rhetoric generally ignored the realities and long-term effects associated with legal issues, border security, and governmental costs of social services and law enforcement, as well as the impact of attracting additional waves of illegal immigrants.

Galbraith's warning of debilitating private and public expenditures for public services that restrict necessary long-term socioeconomic investments is exemplified by General Motors' unnecessarily generous wage and benefits program and by the social services expenditures of governmental agencies in recent years for illegal immigrants. As a result, fewer resources are available for investments capable of generating sufficient new capital to maintain a positive rate of economic growth. This is representative of America's large-scale cultural deterioration of corporate and governmental institutions, consequences of the culture's adoption of an increasingly sensate, materialist mentality. It would appear that America's consumption of its national wealth exemplified by such corporate and government fiscal management has consistently breached Galbraith's warnings and advice.

Additionally, during the last half of the twentieth century, America's excessive expenditures to support a world-dominating military force and its continuous global engagements is another example of the misuse of national wealth that has contributed to cultural deterioration. According to the Defense Department's annual "Base Structure Report" for fiscal year 2003, the Pentagon supports over 700 overseas bases in about 130 countries and another 6,000 bases in the United States. Overseas bases support about 250,000 uniformed personnel and over 200,000 additional Defense Department personnel, contractors, and local foreign workers, utilizing almost 50,000 U.S.-owned structures valued at over $100 billion.

Since World War II, America has utilized its military presence and power as an economic strategy to promote its global economic and political influence and, more specifically, to expand and protect its commercial interests. A strong and direct correlation has existed since the time of ancient Greece between a civilization's militarism and its objective of securing dominance, wealth, and energy such as gold, slaves, territory, and fuels, often disguised as missions of religious righteousness, moral conversion, or the imposition of a political ideology. The costs of such conquests or imposed liberations extend beyond a financial balance sheet of expenditures and confiscated wealth-energy resources to the sociological, psychological, and physical trauma for the invader and the conquered populations. America's post-WWII emphasis on military power and global interventions in promoting its political, ideological, and economic interests is such an example of the excessive and misguided consumption of human and wealth-energy capital that has degraded the American culture.

The destructive cost of a half century of militarism to the American moral fiber, national wealth, and international relationships has been a major contributor to the deterioration of its culture. Toynbee refers to militarism as being the "commonest cause of the breakdown of civilizations during the last four or five millennia,"[91] and America is no exception. World War II may be considered the last American war based on honorable rationale involving the actual defense of the nation and its democratic values and ideals. Since that period, American wars have been foreign invasions for attempted ideological conversions to supposedly defend democracy against differing social and political thoughts of foreign entities lacking the military power to seriously challenge America's homeland. The Korean War cost $456 billion in 2008 dollars, Vietnam was even more costly, and current estimates for the Afghanistan and Iraq wars extend to $2 trillion.

Decades of large military expenditures and related accrued interest, contributing to national debt and direct war costs, have had major long-term financial ramifications. America has waged wars in foreign lands to stabilize geographical areas of economic interest. Meanwhile, issues of housing, health care, public schools, transportation, and scientific development and applied technologies have received an insufficient proportion of the national wealth resources. The nation's inability to adequately respond to the 2005 hurricane that devastated the Gulf Coast is an illustration of the general failure of government and the American society's misplaced humanitarian and financial priorities. In 1948, President Truman signed legislation establishing the Marshall Plan, which was a major contributing factor in the rebuilding of post-WWII Europe. More than a half century later, the U.S. government was unable to rescue and rebuild in a timely manner its own Gulf Coast region devastated by hurricanes.

Despite enormous federal government resources, the national tragedy and embarrassment of the inept bureaucratic governmental response to Hurricanes Rita and Katrina was a major failure resulting from the unmanageable complexity and disorganization of a mature civilization's governance system. The federal government's priorities and financial resources were overwhelmingly absent, primarily devoted to its military agenda in foreign lands. Hurricane Katrina struck the American Gulf Coast in August 2005 and is considered to be the worst natural disaster in U.S. history. The subsequent disjointed attempts by local, state, and federal government agencies to adequately provide immediate and long-term aid to the Gulf population will be recorded as a colossal failure of epic proportions. The inadequate effort to rescue citizens from life-threatening conditions and to restore some semblance of order and normalcy in a reasonable time frame after the storm is inexcusable.

The U.S. House of Representatives report entitled *A Failure of Initiative* states, "Katrina was a national failure, an abdication of the most solemn obligation to provide for the common welfare." Federal disaster relief is described as late, ineffective, or nonexistent. Government agencies exhibited failures of agility, readiness, coordination, responsiveness, planning, training, command and control, efficiency, and effectiveness. The report concludes: "The preparation for and response to Hurricane Katrina show we are still an analog government in a digital age."[92]

The Department of Homeland Security is responsible for protecting the country from terrorist attack and for responding to natural disasters. It is composed of twenty-two federal agencies of over 180,000 employees with a 2004 budget of $36.5 billion. DHS was formally established in 2002 and is the third-largest cabinet department in the U.S. federal government. The various governmental and nongovernmental reports of the DHS performance during the Gulf Coast rescue and restoration efforts have unanimously reflected failure at all levels of the federal government. The Government Accountability Office (GAO), the investigatory arm of Congress, concludes that the void of leadership "serves to underscore the immaturity and weaknesses related to the current national response framework." GAO Comptroller General David M. Walker stated, "There were multiple chains of command, a myriad of approaches and processes for requesting and providing assistance, and confusion about who should be advised of requests and what resources would be provided within specific time frames."[93]

The Institute for Southern Studies issued a report of conditions six months after the storm: "Despite promises from national leaders to 'do what it takes' to rebuild New Orleans, the devastated city has been mostly left to fend for itself—with tragic results." A few of the major failures from what should have been a well-organized national plan of preparation and response to natural disasters include the following:[94]

- After catastrophic flooding of New Orleans, 75,000 people of the 500,000 population were stranded. It was previously known that 100,000 city residents had no cars and relied on public transit and that Amtrak and Greyhound would stop service prior to the storm striking the city. The failure to execute an effective evacuation plan resulted in hundreds of deaths.
- Evacuations and medical care for special needs populations including hospitals and the Veterans Affairs Medical Center did not occur. As a result one hospital with 2,000 patients and staff was left with no electricity, water, food, flushing toilets, elevators, cell phones, computers, or life-saving electrical equipment, resulting in over forty deaths during the first two days after the storm. Citywide, about 12,000 patients were evacuated while 24,000 were left behind in twenty-two hospitals.
- A total of 215 patients died in nursing homes.
- Thousands of prison inmates were left behind, with 600 prisoners in one building being left locked in chest-deep water.
- The 40,000 people who took refuge in the Superdome experienced failed water supply and toilets, while about 20,000 people marooned at the Convention Center had no water, food, or medical care.

Recovery efforts were as inadequate as the rescue attempts. The most startling examples of the inadequacies and incompetence of the recovery efforts include the inability of government to simply remove the debris from neighborhoods that were completely destroyed and to deal with the public health threats of mold and soil pollution (e.g., high arsenic levels). Six months after the storm, homeowners were unable to rebuild without a permit, which was based on a nonexistent but required new building code. Assistance for renters, who compose two thirds of those displaced by the storm, had not been provided. Also, 65 percent of homes were without power, 73 percent of homes remained damaged, and only seven of twenty-two hospitals were open, with 15,000 hospital beds being available, compared to 53,000 beds prior to the storm. The city that had operated 117 public schools for 60,000 students had only 11 public and 8 charter schools functioning to support 13,000 students.[95]

More than a year after the storm, New Orleans and the other Gulf Coast areas had received little assistance from the federal government. Over 200,000 people remained displaced; aid for homeowners in Louisiana and Mississippi, approved ten months after the storm, had not been dispersed; and few resources had been allotted to rebuild rental units. Eighty percent of public housing in New Orleans remained closed despite suffering only minimal damage. Over $135 million in corporate fraud related to Katrina contracts had been reported, and government investigators cited contracts worth $430 million as being misappropriated or lacking appropriate oversight.[96] One year after the hurricane, about a third of the debris remained and reconstruction had barely begun. Of the $17 billion appropriated to the Department of Housing and Urban Development for cash assistance to homeowners, only $100 million had been spent.[97] Equally embarrassing was the nonexistence of a functioning legal system.

Eight months after Katrina, the New Orleans chief criminal court judge and his staff were still discovering people in jail who should not have been incarcerated. Some defendants had been kept in jail even after charges had been dropped, while others remained in jail awaiting trial longer than their potential maximum sentences. These people included indigent defendants arrested for misdemeanors just before or just after the storm, referred to at the courthouse as "doing Katrina time." The right to legal representation for all accused persons was nonexistent, and people remained in jail while cases remained lost in the system. During the eight-month period, no criminal jury trials were held. The underfunded public defender system, which is required by law to provide legal representation, was unable to function. One judge announced suspension of the prosecution of cases in which defendants were represented by the public defender's office until the state appropriated sufficient money to allow public defenders to provide a competent defense.[98] The Justice Department estimated that New Orleans would require $8.2 million a year to operate a public defender system capable of appropriately protecting clients. A man accused of taking $50 from his girlfriend's house was in jail for 400 days and had not yet been interviewed by his court-appointed lawyer.[99]

Clearly, the performance of DHS, which includes the Federal Emergency Management Agency (FEMA), constitutes a paralysis of governmental bureaucracies headed by political appointees of questionable experience and training. It is also representative of mature civilizations that have maximized their complexity, disorganization, and confusion beyond society's ability to constructively respond to the needs of its population. Its societal chaos has accumulated to the point of being unmanageable and ineffective. The general indifference of so many Americans outside the affected area in tolerating this plight of human suffering and to the government's gross negligence and ineffective response to the Gulf Coast calamity is reflective of America's general apathy to its deteriorating infrastructure and cultural values. Perhaps the most telling illustration of the American culture's loss of altruistic values, a sense of national purpose, and the Ethics of Principles, is that almost three years after the hurricane hit New Orleans, the number of homeless was one in twenty-five residents, unprecedented for a U.S. city. An estimated 12,000 homeless people, 4 percent of the city's population, four times the rate of most cities in America, remained stranded by the paralysis of their government.[100]

The Gulf Coast's infrastructure required rapid rebuilding; however, the failure of local, state, and federal governments to respond adequately in a timely manner reflects the absence of such a capability. As of January 2008, only one quarter of the $4.5 billion federal appropriation to replace schools, firehouses, and other public works had been spent. FEMA reports $4.5 billion of infrastructure projects have been funded for Louisiana and Mississippi, with only about $1 billion being spent; the remaining funds are in state accounts delayed by extensive, complex local and state rules and procedures.[101]

The inability to reconstruct the Gulf Coast infrastructure after the hurricanes parallels America's more generalized inability to adequately maintain the nation's infrastructure system over the last few decades.

The nation's infrastructure of transportation, communication, public education, health care, and governance has become structurally, functionally, and financially encumbered. Historically, a society's capability of applying creativity and innovation to develop sound, cutting-edge infrastructure systems is a major contributing element for a growing, evolving, productive economy and a stable social order. "A modern economy needs a modern platform, and that's the infrastructure.... It has been shown that the productivity of an economy is related to the quality of its infrastructure."[102]

An American Society of Civil Engineers' report in 2005 assigned an overall grade of D to the nation's infrastructure, projecting the investment cost for rehabilitation at $1.6 trillion. It noted that 27 percent of the country's 5,900 bridges were structurally deficient or functionally obsolete and that since 1998, the number of unsafe dams had increased by 33 percent to 3,500. Additionally, drinking water, hazardous waste, roads, schools, transit, and wastewater were all rated D.[103] The nation's massive, widespread deteriorating infrastructure is another symptom of cultural degradation that America has in common with past great civilizations that became stagnant or collapsed. Such infrastructure deterioration, combined with broader organizational and functional degradation, disproportionately affects lower socioeconomic populations, breeding an environment of escalating crime, poverty, societal inequity, and hopelessness.

By most statistical measures, the American economy has continued to be strong. However, the benefits of prosperity are increasingly restricted to the wealthier segment of the population, as the rate of poverty in America continues to increase, with about 37 million Americans living in poverty. The U.S. median household income decreased by 5.9 percent from 2000 to 2005 from $49,133 to $46,242, as more than half of the nation's income went to the top 20 percent of wage earners. The number of Americans not covered by health insurance increased by 1.3 million between 2004 and 2005, reaching 46.6 million, representing 15.9 percent of the population, with 11.2 percent of this group being under eighteen. In 2005, the average person living in poverty earned $3,236 less than the poverty line of $19,971 for a household of four, the largest such gap ever measured by the Census Bureau.[104] During the 1980s, 13 percent of Americans in their forties spent at least one year below the poverty line, with that figure increasing to 36 percent in the 1990s.

According to the National Urban League, based on government data, the median net worth for a typical African-American family is $6,166, compared to $67,000 for a white family. More than 70 percent of white families own their homes, compared to 50 percent of blacks, which is partially responsible for this large gap in wealth. Additionally, the unemployment rate for blacks is 10 percent, compared to 4.4 percent for whites.[105] In 1970, 85 percent of children lived with two parents and 11 percent with only a mother, while in 2004 these figures were 70 percent and 23 percent, respectively. The data for children living with two parents broken down by race and ethnicity are 87 percent of Asians, 78 percent of non-Hispanics, 68 percent of Hispanics, and 38 percent of blacks.[106] Disparities in life expectancy parallel America's widening gap in income and wealth accumulation. From 1980 to 1982, the most affluent segment of the population had an increase in life expectancy of 2.8 years relative to the most deprived segment of the population. In the period from 1998 to 2000, it increased by 4.5 years (79.2 vs. 74.7). The gap between poor black men and affluent white women was fourteen years.[107]

Since 1987, the U.S. prison population has nearly tripled, increasing state spending for corrections by almost 130 percent when adjusted for inflation. Between 2003 and 2004, the population of America's prisons increased by about 900 inmates each week, totaling 2.1 million people, or 1 in every 138 citizens. This data place America as the country with the highest rate of incarceration, followed by Britain, China, France, Japan, and Nigeria. In 2004, 61 percent of U.S. prison and jail inmates were racial or ethnic minorities: 12.6 percent were blacks in their late twenties, followed by 3.6 percent Hispanic men and 1.7 percent white men in that age group.[108]

By 2008, the incarcerated population reached a record of 1 out of 100 American adults, or more than 1.6 million people being in prison, with another 723,000 in local jails. This included one in fifteen adult black men and one in nine young black men ages twenty to thirty-four. Additionally, a record 5 million people were on parole or probation. According to the National Association of State Budget Officers, states spent $44 billion in tax money on corrections, a 127 percent increase adjusted for inflation from 1987. This figure is projected to increase by $25 billion by 2011. The current figure of $44 billion equals or exceeds the total amount spent on higher education in the states of Vermont, Connecticut, Delaware, Michigan, and Oregon. About one in nine state government employees works in corrections, with California spending more than $500 million in 2006 just for overtime payments.[109]

There exists a direct correlation between the socioeconomic unraveling of past great civilizations and increasing disparities of wealth, basic human support services, and economic opportunity. A resulting dispirited general population gradually abandons faith in the social order and increasingly relies on self-oriented principles and priorities. The consequence is the rejection of community and altruistic values and the creation of a higher level of social and economic instability. There are disturbing similarities between the past collapsed and ossified cultures and the current American profile of the widening distribution of acquired wealth, inequitable access to quality health care and housing, rising incarceration rates, and declining educational opportunity.

At the opposite extreme of acquired wealth, America's celebrity culture of entertainers and athletes, with their large financial rewards and adoring public, has become the envy and role models for both young and old. Additionally, there exists a segment of celebrities that are simply "famous for being famous," who have no accomplishments or talent other than the ability to gain media attention. The media, in the pursuit of identity and profit, chronicles the rumors, antics, and extremes of their fast-track lives of expensive clothes, cars, jewelry, chemical abuses, and self-stimulation. This twenty-first-century American culture of self-indulgence and entitlement is a modern version of the priorities and values of ancient Greece and Rome as they reached cultural stagnation and proceeded down the pathway of self-destruction.

The actual productivity of a New York Yankee baseball player having completed five years of his ten-year $250 million contract was analyzed and reported as $142,000 for each game played and $37,000 for each time at bat.[110] Meanwhile, five U.S. professional baseball players have contracts totaling $863 million for periods that vary for seven to ten years, while the average annual salary for all professional baseball players in 2006 was $2.9 million. Higher salaries exceed $20 million per year. After the first quarter of the 2008 baseball season, seven players were "released or designated for assignment" for inadequate performance. These actions cost the teams involved a total of $31 million in "guaranteed salaries to players set free." The individual payouts for not playing baseball ranged from $10 million to $550,000.[111]

In 2006, the National Football League (NFL) completed a six-year collective bargaining contract with its players for $21 billion, with the players entitled to about 60 percent of this amount. The media networks provide about $3.7 billion per year for NFL broadcasting rights by obtaining significant revenues from their advertisers.[112] Advertisers signed up to pay a record $2.6 million for thirty-second commercials during the 2006 Super Bowl XLI broadcast to reach over 100 million viewers. One new company that anticipated total company revenues of less than $1 million for its first year of operation spent $3.5 million to run an ad five times during the game.[113] In America's major universities, basketball and football coaches routinely command compensation packages significantly higher than university presidents and the most accomplished scholars and faculty members.

The top ten American celebrities, defined as entertainers and athletes, have annual incomes ranging from $32 million to $225 million. Talk show celebrity Howard Stern has a $500 million five-year contract with Sirius XM satellite radio, and Rush Limbaugh has an eight-year $400 million contract. The 2005 compensation rate for the top ten Hollywood superstars ranged from $12 million to $20 million per movie, while the yearly income for the more general category of the top twenty entertainers ranged from $20 million to $40 million. The list of the *Forbes World* richest people has Americans occupying eleven of the top twenty rankings, with net worth ranging from $15.6 billion to $46.5 billion.

The Wall Street Journal reported that in 2005 the CEOs of the 350 major U.S. companies were paid $4.1 billion in salary, bonuses, and realized long-term compensation (about $12 million each), with the median pay for these executives increasing by 23 percent from the previous year.[114] In the early 1980s, executive compensation including salaries, bonuses, and option grants began to rise faster than the earnings of most workers, in contrast to the previous forty-year period, when the pay of both groups grew at about the same rate of 1.3 percent. During this period of time, the median ratio between executive to worker compensation ranged from 25 to 34. However, since 1980, the ratio increased to 55 in 1990 and 119 in 2000, and has remained above 100 since. In 2005, the total average pay for chief executives rose 27 percent to $11.3 million, compared (in 2006 dollars) to $1.6 million in 1980, $2.7 million in 1990, and $7.6 million in 2004.[32]

The ever-widening gap in individual wealth and the associated accumulation of personal possessions is reflective of society's insatiable appetite for "collectables." A 2006 U.S. car auction included four vehicles that sold for a total of more than $10 million: a 1950 General Motors bus, a 1954 Pontiac Bonneville, a 1970 Plymouth, and a 1953 Chevrolet Corvette.[115] In the world of art, collectors paid $71.5 million in 1998 for Vincent van Gogh's *Self-Portrait Without Beard* and in 1990 paid $82.5 million for his *Portrait of Dr. Gachet*.[116] In 2006, a Picasso portrait, *Dora Maar with Cat,* sold for $95.2 million at action, the second-highest amount ever paid for a painting. His *Boy with a Pipe* sold for $104.2 million in 2004.[117]

The vast inequities of wealth accumulation are prevalent in many aspects of American life, following the familiar theme that the rich get richer and more privileged. This also applies to private, nonprofit institutions favored by the wealth elite that have acquired enormous wealth in recent decades. In 2007, donations to colleges and universities totaled $30 billion, of which $7.7 billion or 25 percent went to just twenty institutions. The top recipient, Stanford, raised $832 million in private donations, with Harvard receiving $614 million.[118] The top 10 percent of private institutions possess more than ten times the assets of the wealthiest public institutions. Sandy Baum, an economist at Skidmore College, states: "Higher education has always been stratified but the disparities were never as large as today. In the early 1990s, endowment income represented a small part of revenues at most colleges and universities." Harvard's 1990 endowment of $4.4 billion has grown to $35 billion in 2008, which is larger than all but fourteen other universities' total endowments. Today, Yale's $22.5 billion endowment provides $1.2 billion in annual revenue, or 45 percent of its annual budget expenditures. Fewer than 400 of the approximate 4,500 American colleges and universities possess endowment funds of $100 million or more, with most being less than $10 million.[119]

Clearly, the majority of Americans exhibit little concern, other than envy, for any possible long-term negative cultural effects resulting from the expanding affluence of wealthy celebrities, sports figures, and major corporate executives or from the widening gap in the country's distribution of wealth. One may view this twenty-first-century American social environment of the accumulation and questionable use of unprecedented levels of wealth as an inevitable consequence of an inherent competitive human aggressiveness, a necessary cultural ingredient of successful capitalism. The central issue is whether such a social climate contributes to the long-term vitality and viability of the social order. To what degree can a capitalist society remain successful relying on incentives and rewards that lavishly compensate their entertainers, economic leadership, and successful investors for their creative and competitive performances while ignoring an equitable sharing of the culture's productivity with all members of society? Currently, the American culture clearly exhibits the same attitudes, values, and behavioral characteristics that afflicted the deteriorating empires of Greece, Rome, Imperial Spain, and Great Britain.

Guided by history, it appears that mounting cultural prosperity inherently produces a continuously expanding divergence of wealth and its related material rewards within the populations of great civilizations. Typically, the prevailing cultural cohesion and common purpose that was an initial strength of a civilization's growth stage erodes and evolves into increasing internal economic competition and social strife, as well as external animosity and warfare. Durant's descriptions of Greek and Roman life appear to be apt characterizations of the twenty-first-century American culture:

> **The wealth of the wealthy grew beyond any precedent in Greek history. Homes became palaces, furniture and carriages more sumptuous, servants more numerous, dinners became orgies, and women became show windows of their husbands' prosperity.**[120]

> **A great civilization is not conquered from without until it has destroyed itself within. The essential causes of Rome's decline lay in her people, her morals, her class struggle, her failing trade, her bureaucratic despotism, her stifling taxes, her consuming wars.**[121]

It has repeatedly been demonstrated that a civilization's early economic growth and productivity is related to its investment in knowledge, education, and innovation: the cultivating elements of a continuously advancing and socially stable environment. From the Industrial Revolution to America's twentieth-century industrial and economic successes, investments in pure and applied scientific research, engineering, and technological advancements have produced socioeconomic prosperity. This environment of economic affluence typically fosters an expansion of the fine and performing arts and literature, supported from the small segment of society that directly benefits from financial success. However, when mature societies, as illustrated by twenty-first-century America, experience inadequate revenues and heavy debt, government funding for basic research, infrastructure, innovation, public education, and social services constitutes a low priority. Former great civilizations, reaching their stagnant stage, found themselves greatly in debt, were conducting expensive warfare, and were unable or unwilling to invest resources in the fundamental activities initially responsible for achieving their socioeconomic prominence. Entering the twenty-first century, America finds itself slipping from its previous high global rankings in education, investment in basic research and development (R&D), and scholarship.

In 2006, according to a report by the reputable science and technology company Battelle, U.S. research and development spending increased by 2.9 percent, reaching the $330 billion level, or about half the average growth rate of the late 1990s. While the federal government increased spending on developing military weapons systems and spacecraft, the investment in basic and applied research declined. In the 1970s, government's research and development investment accounted for 67 percent of the nation's total R&D spending but steadily declined to 30 percent in 2006. Thus, corporate America is assuming a larger role in R&D but with the motivation of increasing profits, not necessarily for the primary objective of the acquisition and distribution of more advanced basic knowledge. The American Association for the Advancement of Science's director of the R&D budget and policy program stated, "The government is spending a smaller and smaller amount of our economy on R&D and that comes at a time when other countries are dramatically increasing their investments.... It raises big concerns for the future of

U.S. innovation." China and South Korea are increasing their spending by 10 percent or more annually, and China and India have increased their combined share of the world's research and development personnel from 19 percent in 2004 to 31 percent entering 2008.

A significant parameter in the determination of a sound, futuristic economy is a nation's investment in non-defense R&D as a share of its economy, for which America currently ranks twenty-second. The federal government continues to spend more on developing weapons systems and less on basic science research. Increasingly, American corporate R&D spending is taking place overseas. Battelle and *R & D Magazine* report: "40 percent of the American high-tech industry already had a research and development presence in Asia, and, of that share, about one third plan to increase their presence." Such data highlight the question of the anticipated future source of innovation and global economic standing.[122]

A microcosm of America's misplaced investment priorities for its national wealth is provided by a 2007 snapshot of government support for cancer research at the time of a $500 billion Defense Department budget and two wars projected to cost in excess of $2 trillion. An interview with the Deputy Physician-in-Chief for Breast Cancer Programs at Memorial Sloan-Kettering Cancer Center provides insight into the priorities of the American culture. While cancer researchers make impressive advances in understanding this disease, the National Cancer Institute (N.C.I.) has reduced by 10 percent patient clinical trials as a result of inadequate funding. Less than $150 million is spent annually on all types of cancer clinical trials. While the total National Cancer Institute budget is $5 billion, the pharmaceutical companies spend about $4.5 billion for cancer research, and an additional $1.5 billion supports private noncommercial research efforts. This total of about $11 billion is only two thirds of the amount spent by the tobacco industry on advertising (i.e., $15.5 billion), and Americans spend six times as much, $68 billion, on soft drinks. Finally, government research funding for cancer research is so limited that only 18 percent of research proposals that N.C.I. wishes to support are funded.[123]

In 2005, the National Science Board found that 38 percent of doctorate holders in America's science and engineering workforce were foreign born and that foreigners constitute more than half of the students enrolled in doctoral science and engineering programs. In 1974, the United States ranked third as a nation in the proportion of college students majoring in science and engineering but slipped to seventeenth by 2005. According to the National Science Foundation, the U.S. share of knowledge generated by scientific and engineering research as measured by published manuscripts has steadily declined from about 40 percent in 1988 to 30 percent in 2001. This trend is influenced by the increasing scientific input from China, South Korea, Singapore, and Taiwan. In recent years, Asian countries are graduating more science and engineering Ph.D.s than the United States, while Europe graduates 50 percent more. In 1995, the United States produced about 200,000 research papers, but by 2001, Western Europe had increased its output to 230,000.

The increases in research among the Asian countries could be considered as a catch-up phase to the historically well-developed American and European programs as a component of the global migration of knowledge and commercial applications. In recent times, a number of specific political and ideological issues appear to be stifling American research efforts, which have reduced and restricted research productivity. Increasingly, the values and objectives of special interest groups have been politically inflicted on government's administration and funding of research. In addition, symptoms of political interference are evident by the selection of federal scientific advisory appointees, which appears to be frequently based on political and moral ideology rather than on appropriate research credentials. Topics related to climate change, stem cell research, and biological evolution are examples. Since the beginning of the Industrial Revolution, scientific, technological, and engineering creativity and exploration has fueled successful national economies and should not be universally muted by political and religious ideologies of a few nations.[124]

Western civilization's relative decline in worldwide science, technology, engineering, education, and commercial innovation has increased in recent decades. For example, America has only six of the most highly successful information technology companies in the world, while Asia now has fourteen. Half of IBM's engineers and technical experts are now located in foreign countries. Chinese automobiles are about to enter the American market. Meanwhile, the nature and extent of America's corporate priorities for investment in future productivity is illustrated by the expenditure of over $200 billion on tort litigation in 2005, which exceeded its investment in research and development.

Two of the essential elements of a progressive society, beyond the population's basic security, nutrition, and shelter, are health care and education. The quality and availability of these five supportive elements of a society not only influence potential economic success, but also affect human thought, attitudes, and behavior applied to a broad range of social, political, and cultural issues. One would assume that America, being the wealthiest society in the history of the world, could afford to establish quality health care and public education systems as primary priorities. However, an increasing population of the lower segment of America's socioeconomic hierarchy is unable to afford adequate health care and lacks access to quality public education.

The Commonwealth Fund, a nonpartisan organization, found that 41 percent of non-elderly American adults with annual incomes between $20,000 and $40,000 were without health insurance during 2005. This is an increase from 28 percent in 2001. *The Wall Street Journal* reports that rising premiums and medical costs have produced a curtailment of employer health benefits. The Institute of Medicine, a component of the National Academy of Sciences, estimates that the lack of health insurance leads to 18,000 unnecessary deaths each year and undoubtedly detracts significantly from the quality of life for many more. Meanwhile, it is estimated that as of 2006, over 46 million Americans were uninsured. *The New York Times* reports that people may select from over 300 different private health insurance plans, the complexity of which requires doctor's offices to employ staff members just to administer the maze of complex rules and options. These private plans utilize about 80 cents from each dollar of insurance premiums for medical care, with the remaining 20 percent being profit, marketing, and administrative costs. Interestingly, despite its reputation for bureaucratic confusion and patient complaints, Medicare is reported to use 98 percent of its funding for medical care. The article suggests, "You do the math; it becomes clear that covering everyone under Medicare would actually be significantly cheaper than our current system."[125]

The universal American view is that health care costs are not affordable for many citizens and that government and private health care programs are extremely complex and confusing, resulting in people being unable to rely on available options to adequately serve their health care needs. Meanwhile, sociopolitical forces block implementation of a government plan for universal health care, labeling the concept as unaffordable while government funnels billions of dollars annually to the military and debt payment.

America has the most highly privatized system of health care as well as the highest health care costs in the world, which is understandable given the American profit incentive and its competitive economic culture. Thus, it is difficult to envision a for-profit health care insurer aggressively striving to reduce unnecessary services, providing full coverage to the lowest income population, and enrolling those with risky medical preconditions. Creating loopholes with carefully crafted legalistic, self-serving contracts that create mammoth, cumbersome paperwork exercises, permitting administrative bureaucracies to maximize deniability of claims, has been the operational standard. This system is inherently incompatible with the concept of a private, high-quality, cost-effective health and medical insurance system that would cover all Americans.[126]

The country's refusal to provide universal health care for all Americans parallels the failure of America's public education system over the past half century. In 1983, a national education commission concluded that the nation's elementary and secondary schools faced a "rising tide of mediocrity." A quarter century later, about a third of ninth graders fail to graduate from high school, another third graduate but are not prepared for college, and only a third are considered qualified for college. Internationally, America ranks sixteenth in high school graduation rates and fourteenth in college graduation rates among the twenty developed nations. This study did not include the less-developed nations of China and India, both of which are rapidly becoming America's most significant economic competitors.[127]

The Asia Society report "Math and Science Education in a Global Age: What the U.S. Can Learn from China" notes that Chinese students spend twice as many hours studying in school and out of school as Americans do. Typically, Chinese students are assigned several hours of homework each day during their summer vacation of two months. Nearly all high school students study advanced biology and calculus, compared to 13 percent of American students taking calculus and less than 18 percent taking advanced biology. American schoolchildren average 900 hours a year in a schoolroom, less time than the average 1,023 hours sitting in front of a television screen.[128] The achievement level and established standards of American high schools are also an issue.

The "national report card" from the National Assessment of Educational Progress based on the most recent test results finds American twelfth-grade reading skills worse than in 1992. The 2005 data revealed that only 35 percent of the twelfth-graders were proficient in reading (i.e., the majority of twelfth-graders have difficulty comprehending reading material sufficiently to establish meaning, draw inferences, and make conclusions). Additionally, only 23 percent met or exceeded the established mathematics proficiency level.[129]

America's half century of a "rising tide of mediocrity" in K-12 education is a result of the continuous decline in the quality of the nation's system of public education. This "mediocrity" is due to the unwillingness of the American public to insist that the nation's highest priority be placed on the creation of a world-class public education system appropriately funded by tax revenue. Of greatest consequence is that the quality of the America's teaching profession has been fatally eroded during the last half of the twentieth century.

The continuing inability of America's sociopolitical institutions to create the world's premier public education system illustrates the extreme to which the American culture has shifted toward self-serving materialistic values. The prevailing cultural theme, as with previously deteriorating civilizations, is that providing for the less-fortunate socioeconomic segment of society is of lower priority than enhancing the wealth and influence of the more economically fortunate. It is significant that in twenty-first-century America, upper- and middle-class high school and college students view the profession of public school teaching as an undesirable career option. College graduates entering the career of public school teaching are overwhelmingly the less academically talented students, unable or unwilling to compete in the more rigorous and financially rewarding academic majors such as math, science, engineering, technology, finance, and business. Thus, not only has the quality of those entering the teaching profession declined in the last half decade but also the shortage of minimally qualified teachers has been a chronic national problem for decades.

In desperation, the New York City public school system attempted to hire 800 math teachers, 450 science teachers, and 1,300 special education teachers for fall 2006 to fill routine vacancies by offering housing subsidies of up to $14,600. One of its officials stated, "There are no math and science teachers coming out of universities. This is not anything that's new, it's across the nation. It's been talked about for the last twenty years. We are all faced with the alternative structures." These "alternative structures" include the assignment of English and history teachers, which are in greater supply, to teach math and science courses. According to the National Academy of Sciences, nearly 60 percent of eighth-graders in American schools are taught math by teachers who neither majored in math nor studied it sufficiently to pass a certification exam.[130]

Textbook publishers, realizing the plight of school systems and their ill-prepared teachers in all disciplines, have profitably produced helpful unified instructional systems consisting of textbooks, CDs, workbooks, and testing materials. Such modalities establish and manage the instructional process, requiring less teacher involvement in formal teaching and in the preparation of classroom materials. Additionally, such materials with accompanying teacher manuals minimize the need for teachers to fully comprehend their subject matter. This instructional methodology compensates for the lack of the teacher's knowledge of academic content, minimizes the teacher's need to directly cover the intricacies of academic concepts, and redefines the traditional teacher's role as a *coordinator for commercially produced instructional media systems*. As the corporate world has long recognized, quality and productive organizations depend on hiring and retaining bright, educated, and creative individuals possessing a positive work ethic, capable of thinking and adjusting to changing circumstances. The American culture insists on such rigorous hiring criteria for the corporate competitive for-profit environment, but not for its public school system.

The unattractiveness of the teaching profession that evolved during the latter portion of the twentieth century is reflected by the longevity of the careers of newly hired teachers. Approximately half of new teachers leave the profession during the first five years after college graduation.

While the main reason for leaving the teaching profession is often cited as inadequate compensation to raise a family, not attaining the expected emotional satisfaction from helping young people learn and the lack of a civilized working environment are also major factors. This dissatisfaction with the learning environment is related to a general lack of self-motivation, self-discipline, ambition, civility, and a positive work ethic among a growing population of today's public school students. It particularly reflects the urban American environment of the twenty-first century.

A 2005 study by University of Pennsylvania researchers reports that U.S. students are "falling short of their intellectual potential" primarily due to "their failure to exercise self-discipline," not "inadequate teachers, boring textbooks, and large class sizes." A University of Michigan study comparing attitudes of Asian and American students inquired as to student perceptions of the most important factors in achieving success in mathematics. Japanese and Taiwanese students responded with the answer "Studying hard" twice as often as did the American students. A clear majority of American students placed the responsibility for learning primarily on their teachers while citing native intelligence and home environment as important factors. Interestingly, in American high schools and colleges, non-Asian students acknowledge the self-imposed discipline and ambition found among a large portion of their Asian-American peers toward academic achievement as well as their peers' parental expectations of achievement and of respect for the teacher's role.[131] Compared to the general American population, the first and second generations of Asian-American students more often adopt values that reflect self-reliance, a Puritan work ethic, determination, and initiative while eagerly and creatively responding to challenges and making their own future opportunities.

The federal government's No Child Left Behind Act and various state efforts to establish improved learning requirements in the public schools have resulted in significant time and money expended teaching minimum competencies, objectives, and standards appropriate for the lowest academic achievers. Meanwhile, the more motivated and able students suffer in a classroom environment of mediocrity. A 2008 study by the Institute of Education Sciences, a unit of the Department of Education, found that the $6 billion Reading First program of the

2002 No Child Left Behind law failed to make any difference in the reading comprehension scores between students who participated in the program and those who did not.[132] The program was intended to increase student performance in low-income elementary schools by utilizing a common, highly structured, and restrictive instructional methodology for reading skills.

The last half century of decline in the American public education system is a microcosm of the erosion of American cultural values and the increasing adoption by the general population of misplaced and self-oriented priorities. In twenty-first-century America, *the education of a given child may be of utmost importance to his or her parents, but Americans have demonstrated an unwillingness to financially support a high-quality system of public education for all students.* In large part, this erosion of the K-12 public school system originated in the 1950s and 1960s with the post-WWII rapid expansion of public higher education. To that point in time, public higher education had generally consisted of teacher-training institutions that evolved from "normal schools," which were based on the French *Ecole Normale Superieure* model founded in 1794. Such schools were first established in the United States in 1839 solely to train schoolteachers. The two-year curriculum was expanded to four years when twentieth-century teacher training requirements were extended. Consequently, beginning in the 1930s, normal schools gradually evolved into single-purpose teacher colleges and eventually, by the 1950s, were being assimilated into multidisciplinary state colleges and universities. Some would argue that this assimilation stymied the elevation of teaching to the level of a true profession. Up to this era of the 1950s through the 1960s, a large segment of the brightest high school graduates, as a result of their family financial circumstances, had few college alternatives other than attending a public college. This, de facto, meant attending a low-cost public normal school to become a public school teacher.

However, the post-WWII mission of public higher education has offered broader academic opportunities to the lower and middle socioeconomic segment of American society. Access and affordability was achieved to pursue the same broad spectrum of academic disciplines and professions that had previously been restricted predominantly to affluent white males attending private colleges and universities. Males of above-average high school academic achievement with career aspirations, regardless of family financial circumstances, quickly took advantage of new opportunities for more lucrative occupations of higher prestige. This was the beginning of a mass migration of the nation's best and brightest young people from entering the teaching profession.

Meanwhile, with the gradual cultural acceptance of women as professional engineers, lawyers, doctors, scientists, and accountants, the brightest and most ambitious female high school graduates quickly fulfilled their aspirations toward better salaries, greater independence, and enhanced professional status and working conditions. By the 1970s, the continuing expansion of public higher education programming at the undergraduate and graduate levels offered both men and women new opportunities for college programs that reflected the prosperity of the national economy. This new economy of the late twentieth century emphasized the applications of science, engineering technologies, evolving computing hardware and software, and the new technologies and practices of financial and business management. The resulting socioeconomic environment provided the financial motivation and support for the rapid development of the well-compensated professions in law, engineering, health care, and financial services.

Accordingly, during the latter portion of the twentieth century, employment opportunities in the competitive and expanding American economic system became available to all high school graduates via a college experience, regardless of family income. Large numbers of parents with little formal education now viewed, with great hope and pride, the prospect of unlimited academic and economic opportunities for their children. This new culture, not unexpectedly, resulted in fewer high academic achievers selecting a career in public school teaching. Fortunately for many talented and ambitious young people, by the 1970s the goal of access and affordability to public higher education with an unlimited range of academic majors became fully available to all segments of American society, regardless of financial status. Unfortunately, the teaching profession became increasingly unable to compete for the more academically talented young people in the new American culture of abundant opportunity and the new social order of materialism and private sector employment rewards. In the twenty-first-century culture of thriving materialism, public school teaching is associated with relatively inferior incentives and rewards compared to the many alternatives.

Consequently, many teaching positions go unfilled or are filled by uncertified and unqualified persons. This has prompted school districts to import international teachers, and Congress has approved the necessary special immigration provisions. In 2002, a large Virginia school district initiated a multiyear program of contracting with Philippine math teachers who were able to significantly improve their incomes by teaching in America. Philippine immigrants, utilizing the same economic and political mechanisms, are also providing relief for the nursing shortage. This is the same familiar rationale used to justify Mexican immigration; that is, public school teaching and nursing are also becoming "jobs that American workers don't want." The nation avoids the obvious question of "Why?"

Society has subtly conveyed to its young people that it does not value rewarding public school teachers as it does the professions, corporation employment, professional athletes and entertainers, and the skilled labor force, all associated with the for-profit economic sector. As the Oklahoma University athletic director stated regarding the football coach's 2007 $6 million annual salary: "The people who are generating the revenue, why on earth wouldn't they share in it? ... Because without them, we wouldn't have any of it." This extreme cultural attitude toward the primacy of directly generating revenues and profits as the basis for one's compensation while important public service careers, as for example the teaching profession, become a less desirable, second-class profession is a sad commentary on the modern American culture.

The gradual, half-century-long demise of America's public school system is just another symptom of cultural degradation, a cultural defect that is unfortunately consistent with the gradual erosion of the communal, altruistic values and principles of the Colonial era. The evolving and escalating culturally ingrained materialism of American society has become the basis for an unwillingness to shoulder the necessary level of taxation to support such important initiatives as world-class public education and universal health care that the United States could easily afford. Typically, middle- and upper-income families are increasingly willing to pay for their children to receive a private elementary and secondary education but unwilling to incur a lifetime's higher "tax burden" to support a high-quality public education system that *benefits all children* and the long-term future of American society. Meanwhile, politicians take this cue from the public and recognize the theme of "no new taxes" as the pathway to successful elections while, after the election, blaming inadequate funding for education on the lack of tax revenues. An affluent society's attitude of indifference, if not open opposition, toward adequately funding public education, health care, and other public services that could benefit all citizens regardless of socioeconomic status is consistent with the perverted values of a deteriorating civilization.

American public higher education has followed a much different developmental pathway than elementary and secondary education but, in the beginning of the twenty-first century, is also suffering from the same issue of indifference of the masses toward appropriate support for the public good with adequate tax support. In recent decades, financial support for public higher education has been significantly reduced, restricting student access and affordability established in the 1960s and 1970s. After many decades during which state governments have reduced tax support per full-time equivalent student, corrected for inflation, four-year public colleges have found it necessary to dramatically increase tuition and fees to compensate for this reduction of tax support. Consequently, the affordability of a public college education has been significantly reduced for lower- and middle-income families. As a result, these families must increasingly rely on inexpensive local community colleges rather than have access to four-year residential colleges. These circumstances are the political consequences of elected officials unwilling to raise taxes sufficiently to support public higher education at a level that approximates state-approved enrollments and program expansions. Consequently, over the last few decades, significant tuition and fee increases were the only alternatives for college administrators and their governing boards. In the two decades prior to 2006, college tuitions increased by almost 300 percent. Since 1988, 14 percent of undergraduates have been from the lowest socioeconomic quartile, while 64 percent are from the top quartile. This translates into 8 percent from the lower quartile and 46 percent from the upper quartile receiving their degrees. At the most selective universities only 3 percent are from lowest quartile and 74 percent from the top quartile.[133]

This profile demonstrates that an increasing proportion of young people have little choice but to attend local community colleges and thereby are denied the broader learning opportunities afforded by a four-year residential college experience. It has been estimated that 73 percent of students entering community colleges aspire to a four-year degree, but only 22 percent accomplish this goal.[134] Unfortunately, many academically capable high school graduates decide that their personal situations will not permit them to pursue a college education, thus accepting a lifetime of fewer personal and financial opportunities.

America's financial support for elementary, secondary, and higher public education has been sadly neglected so as to maintain lower taxes, maximize disposable personal income, and accommodate higher-priority budget items that provide a more conducive political environment for politicians to be re-elected. However, the creation of new jobs to stimulate a continuously renewed and revitalized economy is inherently dependent on education and the acquisition of new knowledge and skills for a continuing evolution of new economies.

It is well recognized that as the nature of a national economy changes, research and educational programming must appropriately evolve in order to provide for new careers and well-trained employees to support advancing production objectives. Over the last few decades, the U.S. economy has become increasingly dominated by the provision of specialized services, as opposed to the production of goods, thus creating the "services science sector." In 2006, services represented more than 75 percent of the American work force. Since 1980, the number of jobs producing goods has declined slightly from 20 million, while by 2005, the number of service jobs increased from 60 million to about 110 million. The major contributors to growth have been health care, consulting, retail, financial services, and the food industry. These economic shifts from 1980 through 2005 correspond to a 10 percent increase in the services category of the country's gross domestic product, with a corresponding decline of 10 percent in the goods category.[135]

These expanding employment areas have increasingly utilized technological advances in computer software, high-speed Internet capabilities, wireless networks, electronic sensors, and electronic equipment. A major issue that affects the rate of economic growth is the extent to which appreciable growth in the "service workers" category is capable of contributing to the net creation of new national wealth. Consequently, America's service workers must be functioning at the cutting edge of professional services, equipped with the latest knowledge, innovation, and technology, if the nation's productivity is to sustain a high level in the competitive global economy. The impact of America's decline in producing goods, historically the more profitable industry, in favor of an increase in the service work industry suggests to some a net loss of income per capita and a lesser role in the major industries of the new century. Given the continuing economic crisis of late 2008, including the failure of major financial institutions and industrial components of the economy, America faces unprecedented challenges in reducing and possibly reversing its current rate of societal degradation so as to more successfully compete with the resurgent Asian cultures.

Based on historical precedent, it is doubtful that the American socioeconomic system will be able to meet the necessary criteria for cultural "rejuvenation," as established by the authors cited in previous chapters. First, such a reversal would require the consumption of wealth-energy resources according to priorities that maximize economic productivity so as to continuously generate a *prescribed level of new wealth* while minimizing the creation of national debt. Second, this *prescribed level of regenerated wealth* is defined as the amount needed to maintain a continuum of sufficient wealth regeneration to support public services at a level that maintains a socially and economically stable social order (i.e., Galbraith's theories of social balance and investment balance). Third, cultural survival and longevity depend on prioritizing and controlling the variables of wealth and energy consumption, while employing ideational human values, attitudes, and behavior toward all segments of society. Fourth, the current cultural flaw of chronic greed is an addiction to unrealistic, materialistic rewards controlling the world of work that is dedicated to short-term revenues, profits, and individual wealth regardless of the ethics and values utilized.

The concept that twenty-first-century Americans would be agreeable, in large numbers, to *redistribute* a painful level of personal wealth in the form of taxes or to nonprofit private entities for the general well-being of other Americans is an anathema to the materialistic self-interest of the twenty-first-century American culture. Unfortunately, for the last

Wealth, Energy, and Human Values

4,000 years, this prevailing human attitude of self-interest and greed has been a major characteristic of severe cultural degradation. The America culture has also fallen victim to this human frailty. "Around twenty-five years ago, American business—and the American political system—bought into the idea that greed is good."

> **For almost a generation, America has cheated our future and lived only in the here and now. Economic growth depends on the level of investment in both physical capital---machines, infrastructure, technology---and human capital, which consists of the combined skills and health of our work force.**[136]
>
> **When it shall be said in any country in the world, "My poor are happy, neither ignorance nor distress is to be found among them; my jails are empty of prisoners, my streets of beggars; the aged are not in want; the taxes are not oppressive; the rational world is my friend, because I am a friend of happiness": when these things can be said, then may that country boast of its constitution and its government.**[137]
>
> **Harvey J. Kaye, <u>Thomas Paine and the Promise of America</u>**

Chapter 12
References and Notes

1. Brooks, David. "The Great Seduction." *The New York Times*. 10 June 2008.
2. Morgenson, Gretchen. "How the Thundering Herd Faltered and Fell." *The New York Times*. 9 November 2008.
3. Toynbee, Arnold J. *A Study of History*, ab. Sommervell, D. C. New York: Oxford University. 1957. I-VI: 273.
4. Ibid., 365.
5. Berman, Morris. *Dark Ages America: The Final Phase of Empire*. New York: W. W. Norton. 2006.
6. Berman, Morris. *Twilight of American Culture*. New York: W. W. Norton. 2006.
7. Ibid., xi.
8. Ibid., xiv.
9. Berman, Morris. *Dark Ages America*. 15. Citing Bauman, Zygmunt. *Liquid Modernity*.
10. Berman, Morris. *Twilight of American Culture*. 2.
11. Ibid., 14.
12. Berman, Morris. *Dark Ages America*. 17.
13. Virgin, Bill. *Seattle Post-Intelligence*. "That Latest Whiz-bang Version May be More Than You Need." *The Virginian-Pilot*. 2 October 2005. J2.
14. Berman, Morris. *Dark Ages America*. 17-18.
15. Ibid., 42. Citing Carter, Stephen. *Civility*.
16. Ibid., 27. Citing Putnam, Robert. *Bowling Alone*.
17. Ibid., 26. Citing Gitlin, Todd. *Media Unlimited*.
18. Ibid., 25. Citing de Zengotita, Thomas. *The Numbing of the American Mind*.
19. Ibid., 31-32.
20. Ibid., x. Citing Postman, Neil. *Amusing Ourselves to Death*.
21. Kristof, Nicholas. "With a Few More Brains." *The New York Times*. 20 March 2008.
22. Ibid. citing Jacoby, Susan. *The Age of American Unreason*.
23. Berman, Morris. *Dark Ages America*. 47.
24. Roach, Stephen S. "Dying of Consumption." *The New York Times*. 28 November 2008.

25. Dash, Eric. "The Last Temptation of Plastic." *The New York Times*. 7 December 2008.
26. "Oklahoma's Stoops Set to Earn $6 Million in '08." *USA Today*. 22 February 2008.
27. Dean, Cornelia. "Determined to Reinspire a Culture of Innovation." *The New York Times*. 10 July 2007.
28. Berman, Morris. *Dark Ages America*. 45-46.
29. Ibid., 51.
30. Krugman, Paul. "Banks Gone Wild." *The New York Times*. 23 October 2007.
31. Anderson, Jenny. "Atop Hedge Funds, Richest of the Rich Get Even More So." *The New York Times*. 26 May 2006. Also Anderson, Jenny. "Wall Street Winners Hit a New Jackpot: Billion-dollar Paydays." *The New York Times*. 16 April 2008. Story, Louise. "Smiling Through a Down Year." *The New York Times*. 25 March 2008.
32. Herbert, Bob. "Working Harder for the Man." *The New York Times*. 8 January 2006.
33. Dash, Eric. "C.E.O. Pay Keeps Rising, and Bigger Rises Faster." *The New York Times*. 9 April 2006.
34. Dash, Eric. "Off to the Races: Many Left Behind." *The New York Times*. 9 April 2006.
35. Berman, Morris. *Twilight of American Culture*. 21.
36. Ibid., 23.
37. Greenhouse, Steven and Leonhardt, David. "Real Wages Fail to Match a Rise in Productivity." *The New York Times*. 28 August 2006.
38. Zuckerman, Mortimer B. "Rich Man, Poor Man." *U.S. News & World Report*. 12 June 2006.
39. Hall, Kevin G. "Study Shows Larger Gap Between Rich and Not." McClatchy-Tribune News Service. 5 November 2006.
40. Porter, Eduardo. "After Years of Growth, What about Workers' Share?" *The New York Times*. 15 October 2006.
41. Berman, Morris. *Twilight of American Culture*. 24.
42. Ibid., 63.
43. Huntington, Samuel P. *The Clash of Civilizations and the Remaking of World Order*. New York: Simon and Schuster.

1996. 81.
44. Gergen, David. "Great to Good." *U.S. News & World Report.* 26 June 2006.
45. Dairymple, Mary. "Tax Help Needed as Tangle." *USA Today.* 16 April 2006.
46. Layton, Lyndsey. "IRS Losing Millions Using Private Debt Collectors." *The Washington Post.* 15 April 2008.
47. Zuckerman, Mortimer B. "Playing Fair on Taxes." *U.S. News & World Report.* 1 May 2006.
48. "The Alternative Minimum Tax." PBS Report. 14 April 2006.
49. Johnston, David Cay. "With Tax Break Expired, Middle Class Faces a Greater Burden for 2006." *The New York Times.* 16 April 2006.
50. "Taxing Thoughts." *USA Today.* 14 April 2008.
51. Shapiro, Carolyn. "Surcharges, Taxes, Fees." *The Virginian-Pilot.* 5 March 2006.
52. Durant, Will. *The Story of Civilization: Part II: The Life of Greece.* New York: Simon and Schuster. 1954. 659.
53. Contained in a January 3, 1919, letter by Theodore Roosevelt written, three days before he died, to the president of the American Defense Society.
54. Center for Immigration Studies. "Immigrants in the United States: A Profile of America's Foreign-Born Population." November 2007. 12.
55. Ibid., 1-3.
56. Center for Immigration Studies. "Immigrants at Mid-Decade: Snapshot of America's Foreign-Born Population in 2005." December 2005. 28.
57. Center for Immigration Studies. "Economy Slowed, But Immigration Didn't: The Foreign-Born Population, 2000-2004. November 2004. 3.
58. Ibid., 6-7.
59. Camarota, Steven. "Immigration From Mexico." 49, 53.
60. Ibid., 54.
61. Center for Immigration Studies. "Economy Slowed, But Immigration Didn't: The Foreign-Born Population, 2000-2004.

November 2004. 13-14.
62. Ibid., 19.
63. Center for Immigration Studies. "Immigrants in the United States." 9.
64. Center for Immigration Studies. "Immigrants at Mid-Decade." 28.
65. Huntington, Samuel P. *The Clash of Civilizations.* 306.
66. Toynbee, Arnold J. *A Study of History.* I-VI: 245.
67. Mouawad, Jad. "Rising Global Demand for Oil Provoking New Energy Crisis." *The New York Times.* 9 November 2007.
68. Zuckerman, Mortimer B. "The Energy Crisis." *U.S. News & World Report.* 9 October 2007.
69. Zuckerman, Mortimer B. "Getting Serious About Oil." *U.S. News & World Report.* 7 August 20006.
70. Krauss, Clifford. "Oil-Rich Nations Use More Energy, Cutting Exports." *The New York Times.* 9 February 2007.
71. Streitfeld, David. "Global Need for Grain That Farms Can't Fill." *The New York Times.* 9 March 2008.
72. Krauss, Clifford. "Exports Are Rising and So Is the Price." *The New York Times.* 19 March 2008.
73. Foroohar, Rana. "Goodbye, Manhattan." *Newsweek.* 20 March 2006.
74. "U.S. International Trade in Goods and Services Highlights." U.S. Census Bureau. 11 March 2008.
75. Wolf, Richard. "Growth in Federal Spending Unchecked." *USA Today.* 4 April 2006.
76. Stolberg, Sheryl Gay. "Bush Presents a Budget That Would Raise the Deficit." *The New York Times.* 5 February 2008.
77. Cauchon, Dennis. "Growth Historic in Federal Aid Rolls." *USA Today.* 14 March 2006.
78. "Pulling Pork Can Be Unappetizing." CNNPolitics.com. 28 January 2008.
79. "Study: Most in Congress Pile on the Pork." CNNPolitics.com. 14 February 2008.
80. Lende, Heather and Utt, Ronald D. "The Bridge to Nowhere: A National Embarrassment." *Anchorage Daily News.*

The Heritage Foundation. 20 October 2005.
81. www.publicdebt.treas.gov/opd/opdint.htm. Bureau of Public Debt.
82. Hanley, Charles J. "Iraq War Will Cost $12 Billion per Month, Researchers Claim." Associated Press. 10 February 2008.
83. Herbert, Bob. "The $2 Trillion Nightmare." *The New York Times*. 4 March 2008.
84. Armour, Stephanie. "More Firms Phase Out Pension Plans." *USA Today*. 6 December 2005.
85. Martin, Patrick. "Corporate Bankruptcies Exhaust U.S. Pension Guaranty Fund." January 29, 2003.
86. Zuckerman, Mortimer B. "Saving the Fairies." *U.S. News & World Report*. 30 May 2005.
87. Porter, Eduardo. "Japanese Cars, American Retirees." *The New York Times*. 19 May 2006.
88. Will, George. "Good (non)Work, If You Can Get It." Washington Post Writers Group. 13 April 2006.
89. Maynard, Micheline. "Toyota Considers 4 Southern States for Site of New Assembly Plant." *The New York Times*. 15 April 2006.
90. Galbraith, John Kenneth. *The Affluent Society*. New York. Houghton Mifflin. 1998. 109.
91. Toynbee, Arnold J. *A Study of History*. I-VI: 190.
92. "Failure of Initiative." U.S. House of Representatives Select Committee Executive Summary of Findings. February 2006.
93. Borenstein, Seth. "Report Faults Chertoff, Brown for Lack of Leadership During Katrina." Knight Ridder Newspapers. 31 January 2006.
94. Quigley, William P. "Six Months After Katrina." Institute for Southern Studies. 21 February 2006. 30-31.
95. "The Mardi Gras Index: The State of New Orleans by Numbers Six Months After Hurricane Katrina." The Institute for Southern Studies. February 28, 2006.
96. "One Year After Katrina: The State of New Orleans and the Gulf Coast." *Southern Exposure* 34, 2. 2006. A special report

by Southern Exposure and the Institute for Southern Studies. August 2006.
97. Krugman, Paul. "Broken Promises." *The New York Times*. 28 August 2006.
98. Whoriskey, Peter. "New Orleans Inundated by Incarcerated After Katrina." *The Washington Post*. 15 April 2006.
99. Parker, Laura. "New Orleans Plans First Criminal Trials Since Katrina." *USA Today*. 23 May 2006.
100. Jervis, Rick. "New Orleans' Homeless Rate Swells to One in 25 Residents." *USA Today*. 17 March 2008.
101. Heath, Brad. "Most of $4.5 B in Gulf Aid Unspent." *USA Today*. 11 January 2008.
102. Felix Rohatyn, as quoted by Herbert, Bob. "Our Crumbling Foundation." *The New York Times*. 5 April 2007.
103. "A Bridge Collapses." Editorial, *The New York Times*. 5 August 2007 and www.asce.org/reportcard/2005.
104. Lyman, Rick. "Census Reports Slight Increase in '05 Incomes." *The New York Times*. 30 August 2006.
105. Copeland, Larry. "Blacks Losing Ground Financially, Study Finds." *USA Today*. 28 March 2006.
106. Roberts, Sam. "Most Children Still Live in Two-parent Homes, Census Bureau Reports." *The New York Times*. 21 February 2008.
107. Pear, Robert. "Gap in Life Expectancy Widens for the Nation." *The New York Times*. 12 March 2008.
108. Harrison, Paige. "Nation's Inmate Population Increased 2.3 Percent Last Year.î The Associated Press. 25 April 2005.
109. Liptak, Adam. "More than 1 in 100 Adults Are Now in Prison in U.S." *The New York Times*. 29 February 2008.
110. Ortiz, Jorge L. "An Expensive Lesson." *USA Today*. 30 November 2005.
111. Nightengale, Bob. "Sudden Impatience Sees Slumping Vets Cut Quickly." *USA Today*. 13 May 2008.
112. Bell, Jarrett. "Technology Drives NFL New Plan." *USA Today*. 29 March 2006.
113. Elliott, Stuart. "30 Seconds of Fame At Super Bowl XLI Will Cost $2.6 Million." *The New York Times*. 5 January 2007.

114. Weeden, Curt. "What Would Carnegie Do about Corporate Pay, Charity?" *USA Today.* 29 March 2006.
115. della Cava, Marco R. "Hailing the King of Car Auctions." *USA Today.* 29 March 2006.
116. Sheets, Hillarie M. "Parting with the Family van Gogh." *The New York Times.* 22 April 2006.
117. Vogel, Carol. "A Picasso Sells for $95 Million as Spring Art Auctions Begin." *The New York Times.* 4 May 2006.
118. Arenson, Karen W. "Wealth Gap Growing Among American Universities." *The New York Times.* 20 February 2008.
119. Arenson, Karen W. "Soaring Endowments Widen a Higher Education Gap." *The New York Times.* 4 February 2008.
120. Durant, Will. *The Story of Civilization: Part II: The Life of Greece.* 563.
121. Durant, Will. *The Story of Civilization: Part III: Caesar and Christ.* New York: Simon and Schuster. 1954. 665.
122. Bernasek, Anna. "The State of Research Isn't All That Grand." *The New York Times.* 3 September 2006.
123. Marchetto, Marisa Cocella. "Op-Art." *The New York Times.* 1 April 2007.
124. Bagley, Sharon. "Losing its Edge." *The Wall Street Journal.* 11 August 2005.
125. Krugman, Paul. "Death by Insurance." *The New York Times.* 1 May 2006.
126. Krugman, Paul. "Voodoo Health Economics." *The New York Times.* 4 April 2008.
127. Gergen, David. "Will America Slip from No. 1?" *U.S. News & World Report.* 4 April 2005.
128. Kristof, Nicholas. "Chinese Medicine for American Schools." *The New York Times.* 3 July 2006.
129. "A Bad Report Card." *The New York Times.* Editorial. 27 February 2007.
130. Hersezenhorn, David M. "New York to Lure Teachers with Housing Subsidies. " *The New York Times.* 19 April 2006.
131. Welsh, Patrick. "For Once Blame the Student." *USA Today.* 7 March 2006.
132. Zuckerbrod, Nancy. "Study: Bush Administration's Reading

Program Hasn't Helped". Associated Press. 1 May 2008.
133. Zuckerman, Mortimer B. "Fairness and the Future." *U.S. News & World Report*. 24 April 2006.
134. Schemo, Diana Jean. "At 2-Year Colleges, Students Eager but Unready." *The New York Times*. 2 September 2006.
135. Lohr, Steve. "Academic Dissects the Service Sector, but Is It a Science?" *The New York Times*. 18 April 2006.
136. Bradley, Bill. *The New American Story*. New York: Random House. 2008.
137. Kaye, Harvey J. *Thomas Paine and the Promise of America*. Toronto: Douglas & McIntyre. 2005. 75-76; direct quote from Paine's *Rights of Man*, Part 2, p. 446.

Chapter 13
Shifting Twenty-First-Century Civilization Patterns: The Asian and Islamic Resurgence

> "A civilization, then, is neither a given economy nor a given society, but something which can persist through a series of economies or societies, barely susceptible to gradual change. A civilization can be approached, therefore, only in the long term, taking hold of a constantly unwinding thread—something that a group of people have conserved and passed on as their most precious heritage from generation to generation, throughout and despite the storms and tumults of history."
>
> Fernand Braudel[1]

The beginning of the twenty-first century reveals a deteriorating Western civilization, an impressive Asian cultural resurgence fueled by an expanding economy, and a global, spiritually driven Islamic revival distinguished by poverty, religious and ethnic intolerance, and an uncivilized militant minority. This time in history may be portrayed as a mature civilization continuing to self-destruct while formerly dormant cultures strive to become revitalized and to displace the old, worn-out Western civilization as the dominant global economic and sociopolitical force.

The World is witnessing an extension of Sorokin's two cycles of the "triple rhythm" of cultural progression that began for the Graeco-Roman and Western civilizations' with the Minoan period of 1600 B.C.[2] This expansion of human history into a new age represents impressive socioeconomic advances of Asia and rapid growth of an Islamic population via its spiritual appeal and by providing a sense of hope and common cause for historically demoralized, oppressed, and geographically dispersed groups of people.

The West's historically successful *instruments of expansion*—military power and political, economic, and religious philosophies and influences—have led to historic economic success and a long-standing, dominant role in international affairs. However, its pathway follows the familiar script of past declining civilizations, characterized by enormous achievements followed by the gradual loss of self-discipline, altruistic values, and integrity on its journey of cultural degradation and declining global economic stature. Typical of mature civilizations, Western socioeconomic systems have overemphasized the production and consumption of nonessential, luxury goods and services and have maintained costly, global military and political influence and interventions. In previous chapters, it has been established that such a cultural mentality ultimately leads to a slowly faltering economy and a demoralized population as wealth-energy resources are squandered and human greed, misplaced priorities, and unrealistic aspirations gradually erode past success.

The first decade of the twenty-first century reveals Western civilization in the transition from Toynbee's Stage III, Breakdown and cultural stagnation, to Stage IV, Deterioration. Meanwhile, Asia has entered a cultural "rejuvenation" period, corresponding to a new genesis-to-growth phase and the beginning of a new cycle of Sorokin's triple rhythm of sociocultural transitions. Additionally, the formerly "ossified" Islamic civilization attempts a narrowly defined cultural rebirth via a spiritually catalyzed mechanism as defined by Toynbee, Sorokin, and others. Consequently, the world is witnessing a pivotal sociocultural transition due mostly to Asia's unanticipated, explosive socioeconomic revitalization. Conversely, the sociopolitical militancy and worldwide chaos from Islamic terrorism has produced social and economic instability worldwide. In 1996, Huntington described these sociocultural transitions, which have intensified in the last decade:

> **The West is increasingly concerned with its internal problems and needs, as it confronts slow economic growth, stagnating populations, unemployment, huge government deficits, a declining work ethic, low savings rates, and in many countries, including the United States, social disintegration, drugs, and crime. Economic power is rapidly shifting to East Asia, and military power and political influence are starting to follow. India is on the verge of economic takeoff, and the**

Islamic world is increasingly hostile toward the West. The willingness of other societies to accept the West's dictates or abide its sermons is rapidly evaporating, and so is the West's self-confidence and will to dominate.[3]

The global mood and agenda of the twenty-first century is created primarily by six major powers: the United States, Europe, China, Japan, Russia, and India, and a number of less-potent nations and multinational sociopolitical movements, reflecting distinctive views and stages of modernity. These primary global players constitute inherently divergent ethnic, cultural, and religious histories. One subset of perspectives represents nations that dominated the twentieth century via economic and military power but which have faltered as they enter the twenty-first century. Another subset symbolizes the twenty-first-century phenomenon of highly inspired, resurgent Asian and Islamic cultures, which have selectively adopted and modified elements of twentieth-century Western civilization. A vastly different, more evenly balanced world order has emerged as a consequence of the global diffusion of industrialization and capitalism, whereby the West's traditional military and imperialistic socioeconomic domination has been diminished. A global technological and economic migration has occurred in the context of a rising worldwide focus on ethnicity; philosophical and religious ideology; and hostility toward Western philosophies, policies, and practices.

The world's escalating dependence on energy, and more specifically on the availability and cost of such resources as oil, coal, and food, has reshaped the parameters characterizing and affecting global economic, political, and cultural issues. It is noted that currently the world's major players represent five different cultures, which, in itself, inherently increases the probability of misunderstanding, mistrust, and conflict. During the late twentieth century, the developed Western countries, by virtue of their self-serving need for cheap immigrant labor and profits, became more ethnically and religiously diverse, but with little evidence of the nineteenth- and early twentieth-century constructive cultural assimilation. Consequently, both legal and illegal immigration have created major social problems for Western nations.

Huntington contends that American multiculturalists reject the country's cultural heritage and desire to create "a country of many civilizations ... not belonging to any civilization and lacking a cultural core." He concludes: "History shows that no country so constituted can long endure as a coherent society. A multicivilizational United States will not be the United States; it will be the United Nations."[4] This issue also currently applies to European nations. Additionally, the non-Western societies of Japanese, Chinese, Hindu, Muslim, and African cultures constitute a diversity of spiritual and ethnic values and traits among and within themselves, thus creating multiple sources of global, non-Western-focused cultural tensions, issues, and political and economic competitiveness. Thus, the basis for the global mood of clashing cultures is the vast network of multicentered values, tensions, and hostilities within and between the Western and non-Western civilizations.

Cultural, ethnic, and religious diversity and related historic hostilities are a catalyst for global conflict and disorder, often providing irrational moral and spiritual justification for military action, societal chaos, and the most despicably inhumane acts of violence. The modern history of Eastern Europe, the Middle East, Ireland, and Africa provides numerous well-documented examples. Unfortunately, throughout the world, the same respect, entitlements, and prerogatives that groups violently demand for their inherited or adopted culture (e.g., assumed freedoms to select their own ideologies) are not respected and extended to others of differing philosophies. Organized religions, cultures, and nation-states have increasingly abandoned civility, intellectual and moral integrity, and a respect for societal diversity and basic human freedoms. Previous chapters have addressed the relationships among human behavior, ethnicity, and religious and cultural values that have proven important in establishing vital socioeconomic processes and stable societal organizations and functions.

The twenty-first-century concept of modernity is defined in terms of progressive societal ramifications of scientific, technological, and production innovations; capitalism's evolving economic principles and practices; and expanded urban life styles and associated socioeconomic consequences. The twenty-first century is increasingly being affected by the resurgent civilizations and their philosophies and practices associated with religion, ethnicity, cultural traits, and historic values. These evolving cultural modifications have promoted a diversity of industrialized and capitalistic economics (i.e., a spiritual and cultural hybridization that includes selective Western socioeconomic principles). During the nineteenth century, the initial development of industrialization and capitalism was centered in Europe and the United States, with prosperity relying on entrepreneurship and free market economic concepts. Germany and Japan created their own versions of modernism involving significant government and political involvement in the planning, financing, and control of their new economies. Even later, Russia, Asia, and Latin America created even greater authoritative and restrictive governmental models for their modern economies, but under strict political management. Also, new economic models evolved from the independence movements related to colonial conquests as found in India and China. Thus, various twenty-first-century models of modernity have emerged from differing political environments, reflecting historic cultural traits of past civilizations including, most importantly, church-state relationships. It is of great significance that the modern politics of some non-Western cultures reflect historic social and economic disadvantages, religious and ethnic intolerance, and selective, hybridized democratic principles, some of which reflect few principles of true democracy. These have been adopted by, or inflicted upon, populations by both internal and external forces.

Meanwhile, the major dominant economic, military, and political powers of the twentieth century have attempted to maintain their historic prosperity and global leadership in an increasingly culturally homogenized and rapidly changing world. The more modern, advanced, and broadly based concept of modernity emphasizes science, technology, and advanced financial and business strategies and has attracted new, major economic players from the historically less-developed countries. Given modern communication and transportation capabilities and the inherent nature of global economic expansion, societal complexities and disorder of modern life have increased exponentially in recent decades, necessitating a wild escalation in the world's wealth and energy consumption. As a consequence, the global societal entropy production is increasing at a breathtaking pace. Not surprisingly, the acquisition, distribution, and cost of fuels, food, and other basic resources has become, and will continue to be, a critical issue for the world's continuing economic development and for providing a quality of life that has come to be expected. The world consumes greater amounts of food and fuels, increasing energy consumption per capita as the population increases. Strong worldwide economic growth since 2002, particularly in China, India, and other developing countries, has escalated raw material prices for corn by 70 percent, oil 177 percent, and steel 117 percent, while basic metal price increases have ranged from 95 percent (aluminum) to 452 percent (nickel). It is estimated that China now accounts for over 60 percent of the annual increases in the world's demand for many metals.

Energy, food, financial and economic stability, international tensions and related armed conflicts, internal security, and wealth distribution are among the primary twenty-first-century global issues and concerns. Energy is the essential element that enables a society to function, whether to support the lower-level rate of consumption of a less-developed society or that of a more highly complex mature social order. The global diffusion of the nuclear energy knowledge base and its technological applications for warfare has reached the historic have-not nations and militaristic sociopolitical, religiously affiliated groups. The threat of readily available nuclear and high-powered conventional weapons, combined with sophisticated military tactics capable of providing large-scale destructive capabilities, has significantly challenged traditional Western-style military power and tactics and the nature of international diplomacy. Thus, not unexpectedly, human behavior and values that drive global socioeconomic and political perspectives are being forced to adjust to an environment of unique and more highly complex twenty-first-century threats, fears, and conflicts, particularly associated with economic globalization and terrorism. The greater availability of more destructive weaponry and greater exposure to sophisticated terrorist tactics and advanced technologies have resulted in national security becoming more extensive and a costly budget item for the nations of the world.

A variety of cultural contexts exist among economic players seeking international leverage and access to material resources and commercial markets in order to achieve their own versions of success and prosperity. Islam, without benefit of a unified focus or a major nation-state sponsor, and China, a recognized economic power, represent disparate ancient civilizations in differing stages of attempted cultural renewal. Thus, these cultures would be expected to exhibit differing characteristics and potential in approaches to achieving socioeconomic success. China's twenty-first-century progress reflects more ideational cultural values of self-discipline, authoritative social control, self-reliance, societal-oriented priorities, and a relative socioeconomic simplicity. The United States and European nations represent a more mature and broader cultural growth phase, exhibiting a more sensate mentality while approaching cultural stagnation. Representative Western characteristics, contrasting those of China's, include greater materialism and incivility; more highly structured, bureaucratic policies and practices; self-oriented priorities and incentives; and a declining rate of economic growth. Japan may be described as closer to the midpoint of Sorokin's continuum of cultural mentalities (i.e., *idealistic*, positioned in the continuum between China and the United States).

These differing and coexisting stages of cultural development continue to evolve as open-ended complex systems, absorbing, refining, and utilizing newly discovered and created rudiments of intellectual, social, economic, political, technological, and artistic life. Such societal processes depend on basic organizational and functional variables that define unique cultures, which establish fluctuating *rates of cultural growth*. In the first decade of the twenty-first century, such relative growth rates range in order from the highest, China, followed by India, Japan, and America.

The unexpected emergence during the last two decades of China's global economic and political ascension to prominence has had profound global consequences. Historically, in addition to favorable social and financial conditions, economic effectiveness has been related to scientific discoveries and inventions, new application technologies, creative financing, and the development of successful strategies for the marketing and distribution of products. Examples of major stimuli for socioeconomic transitions are computer-assisted design and manufacturing, and telecommunications. For the lesser-developed nations, the global transfer of such knowledge and technology during the late twentieth century created new industrial and business opportunities that provided the world with quality consumer products at lower cost than the Western nations could provide.

However, for any society to sustain an economic advantage in the competitive global marketplace, the search for the next level of innovation becomes a continuous, challenge to producing the next generation of successful commercial products and services before the competition. Innovative products are actively and universally sought but quickly recognized, rapidly diffused, and easily replicated. Thus, a society requires a cultural ecology of continuous innovation, "strategies of permanent innovation," in order to remain competitive. In the first decade of the twenty-first century, this challenge pits the West versus the East, and thus far Asian nations have surged ahead of America and other Western nations with escalating investments in education, basic research, and technological and engineering innovation.

The history of cost-effective, quality manufacturing technology during the last half century reflects the migration of industrialization from the more technically advanced, originating Western nations to the lower cost, less-developed nations. In modern competitive commerce, the significant differences of labor and material costs between the East and West divide has produced negative economic consequences for the West, as should be expected from the inevitable global diffusion of technology to less-developed countries. This is illustrated by the migration of automobile manufacturing production from the United States and Western Europe in the late twentieth century to the twenty-first-century supremacy of Japan and recent developments in China and Eastern Europe.

The current, more profitable economics of Japanese and Korean automobile production is in stark contrast with that of the declining twenty-first-century American automobile industry. Additionally, Chinese cars are expected to invade the U.S. market in the near future. Other global economic shifts in the automobile industry will continue to negatively impact the developed Western nations as European automobile production expands from Western to Eastern Europe. Slovakia, Poland, Hungary, Romania, and the Czech Republic are projected to produce 3.4 million cars annually by 2010, a 33 percent increase since 2005. Russian production is expected to increase to 1.6 million cars a year from the 2006 level of 1.2 million. This compares to the U.S. production of 12.9 million; Japan's 10.3 million; a British decline from 1.8 to 1.5 million; and France's stable production of 3.6 million vehicles.

In Slovakia, Peugeot Citroen recently inaugurated an $890 million automobile factory that will employ up to 3,500 workers, producing 300,000 cars a year. According to Peugeot, engineers in Slovakia earn about half the comparable wage of Western European engineers, and assembly-line workers earn about one third to one fourth of Western workers. By 2010, Slovakia is projecting automobile production that matches France's annual output and twice that of the British, which is expected to decline to 1.49 million vehicles. Suppliers are beginning to establish themselves in the region, including U.S. Steel, which is bringing a $160 million plant online to manufacture automobile-grade steel sheet. General Motors, while losing money elsewhere, is expected to double its sales in Central and Eastern Europe, including Russia, within the next three years by expanding its dealerships to almost 500. Meanwhile, in Western Europe, Volkswagen has eliminated 4,000 jobs and Peugeot has cut 11,000 employees as it plans to open the new Slovakia plant. The shifting of production sites from Western to Eastern Europe is consistent with Central Europe, Russia, India, and China becoming the world's greatest untapped automobile consumer markets as well as possessing a highly profitable, low-cost labor force.[5]

Nissan Motors has announced plans to build its first passenger vehicle factory in India, a $1.1 billion complex, jointly with Renault to produce 400,000 units a year by 2010. Indian wages are about one tenth those paid in Japan, consequently production will rely more on human labor and less on the world-class Japanese automated processes and technology. Based on such low cost production factors, Japan's big three automakers, Toyota, Honda, and Nissan, are expanding their production capabilities from Saharan Africa to the former Soviet Union to southern India. India's population of 1.1 billion is just entering the modern automotive world, after China, "the world's next megamarket." CSM Worldwide, an auto market research firm, estimates that the developing regions of the world will consume about 10 million vehicles over the next six years, 76 percent of the global increase. The potential market is illustrated by Nissan's vehicle sales to the developing nations, which have tripled to 1 million units during the last seven years. Japanese automakers have announced plans for factories in the Middle East, South America, and Latin America. While Ford and General Motors have modest efforts in such markets, American automotive companies, possessing little investment capital and significant debt, are clearly not in the forefront of this global effort to provide affordable vehicles to expanding consumer markets.[6]

This half century of evolving automotive production and consumer markets, being systematically displaced from one geographical region of the world to another, is the result of the natural diffusion of developing technology and capitalist principles and practices into less-developed regions capable of lower production costs. The global migration of the know-how to profitably manufacture automotive products is a spontaneous socioeconomic phenomenon comparable to the probability, as the driving force, of perfume molecules spontaneously diffusing from an opened bottle to uniformly occupy the totality of an enclosed room. Nature abhors a vacuum, and people will not miss an opportunity for economic profit. Competitive societies see opportunities when others appear no longer able to maintain their economic dominance. Cultural forces representing variables of wealth, energy, and human values and behavior are continuously disrupting, modifying, and reorienting the processes of economic productivity. This is the basis for the high probability of India's and China's continued impressive socioeconomic success and potential economic domination during the early part of the twenty-first century. This global scenario of automotive history is a microcosm of the cultural development of human history, providing great insight into the continual rise and fall of socioeconomic systems and why for millennia civilizations have come and gone (and occasionally have been reborn).

This history of the migratory worldwide automotive production brings to mind Karl Marx's comment: "The country that is more developed industrially only shows, to the less developed, the image of its own future." Thus, as less economically developed regions of the world displace established Western product lines and markets, the West must respond creatively and quickly with more advanced, affordable, and desirable products, becoming more commercially diverse and maintaining global niches. An expanding array of more advanced goods and services should be continually replacing older product lines that are about to be hijacked by lower-cost producers in less-developed countries. The spirit of continuous innovation and quality improvement are necessary cultural attitudes as the new drives out the old. However, in the beginning of the twenty-first century, as the migration of global technology and economic practices continues at a rapid pace, the gap in production capabilities and the rates of growth between the West and East have rapidly diminished from twentieth-century levels. The West has been found lacking in cultural values, traits, and behavior, resulting in inadequate investments of wealth-energy capital in productive outcomes relative to excessive nonessential and unwise consumption of available resources. China's massive capital reserves are available for internal and external investment, while America now attempts to deal with spiraling historic debt, a failing economy, and a lack of available investment capital.

The West's material success of the twentieth century became the envy and model for the less-developed, non-Western populations. Ultimately, it became clear to the non-Western world that potential economic prosperity would require adoption of modern technologies and economic concepts which, in some societies, would clash with historic cultural and religious philosophies. For such cultures, adapting to a new global economy and sense of modernity would inherently require modifications or even rejection of some elements of their historic cultural thinking. Japan, during its post-WWII period of prosperity, successfully modified ancient cultural traits and created its own version of modernity by successfully combining the basic components of the Western economic model with its historic ethnic, religious, and cultural principles and practices. Japan, India, and China have each recognized the realities of seeking cultural modifications and adopting societal change by defining their own culture's concept of modernity. However, it is clear that the Islamic resurgence is floundering, as it fails to respect, based on religious grounds, socioeconomic principles and practices as well as cultural values foreign to their history but endemic to success in the new world order.

Cultural moderation, driven by the reality of global economic competition, will be essential to any future Muslim socioeconomic success. However, the fundamental Islamic principles regarding the roles of church and state, appropriately described as "God is Caesar," is in stark contrast to civilizations that have successfully incorporated and adopted more secular versions of modernity. The significance of religion, or perhaps more accurately a society that lives its spirituality, is universally accepted by scholars as a vital constructive internal ingredient for a stable and successful social order. As stated by Christopher Dawson, religion is "the foundations on which the great civilizations rest."[7] The converse is also true: Religion may become, or may be selectively used as, a negative element, destabilizing the social order of a nation or region. The modern world of increasing, unavoidable interdependence affects not only global commerce, but also solutions to critical, more intricate twenty-first-century international issues of hunger, environment, health, and mutual security. Ideally, this requires, more than ever, that the nations of the world extend the golden rule of all religions to all people in order to cooperate in establishing greater social, political, and economic stability on a global scale.

Max Weber and others argue that the sixteenth-century Protestant Puritan work ethic and values represent a desirable cultural mentality and traits that maximize the probability of socioeconomic success.[8] Recall that this view is one rationale offered as to why the Industrial Revolution occurred in England, Germany, the United States, and the Netherlands at the time it did. The manner and conditions under which Puritan communities conducted their socioeconomic affairs represent the most favorable social conditions for the effective consumption of wealth-energy capital and for the longevity of the social order. This cultural environment, or more specifically the values guiding community attitudes, behavior, and practices, produces effective investments (i.e., a high return on their labor and wealth and energy investments) that minimize the creation of waste, cultural complexity, and disorder within the community. In general, these Puritan characteristics maximize the potential for a society to achieve a high marginal productivity or return on investments in societal progress. This includes minimizing expenditures for nonessential commodities and investing wisely and prudently in the production of goods and services capable of generating new forms of wealth-energy capital. This societal attitude and behavior constitutes a force of moderation for inherent, conflicting human instincts and promotes constructive spiritual and moral values (i.e., ideal economic influences). Thus, such characteristics maximize the probability of a positive rate of cultural growth, less waste or frivolous use of resources, and general social satisfaction, all conducive to long-term social and economic stability.

These cultural traits, practiced by sixteenth- and seventeenth-century Puritan Protestant communities, are consistent with and similar to those of modern Chinese and Japanese cultures. However, these cultural traits are in sharp contrast with those practiced by twenty-first-century Islamic populations. Additionally, there exists a significant disparity in the important roles of secularism among the three cultures.

The prevailing global dynamics of the twenty-first century continue to conflict with, modify, and displace more traditional systems of truths, values, and ethics of populations representing less-developed nations. Such populations have increasingly been forced to contend with external value systems and philosophies underlying the socioeconomics of the more competitive, historically aggressive and successful developed nations. However, to survive when confronting a more culturally diverse worldwide economic system, a population may pragmatically compartmentalize traditional values and religious beliefs, separating them from their nonspiritual world. The realities of socioeconomic survival and prosperity may overcome inherent fears of differing cultures, religions, and races. Such emotion may range from hostility to tolerance to reluctant respect.

The current Islamic resurgence enjoys a rapidly increasing religious membership, largely from those suffering from generations of poverty, neglect, and abuse. However, the revival faces significant challenges to socioeconomic success for its various geographical populations. These include the lack of a primary nation-state dedicated to their cause, insignificant available wealth, extreme cultural disunity and dysfunction, and unchallengeable spiritually based political control by militant religious leaders with their own personal agendas. Such major negative factors preclude achievement of the constructive cultural aspects necessary to revitalize the Islamic movement. Conversely, China and India do not possess these limiting attributes and have developed an achievable vision of success, positive rates of cultural growth, and relative social stability, all factors capable of contributing to prosperity as measured by twenty-first-century standards.

In this era, the less-developed nations of Asia must have access to modern knowledge and information, wealth-energy capital, new discoveries, and innovative technologies, while understanding and utilizing the forces and opportunities of a realistic global view of modernity. Additional requirements include an organized and effectively governed nation-state, continuous access to natural resources, competitive and innovative economic strategies, a skilled workforce, and an educated nonsectarian business and political leadership. Accordingly, the emerging Asian cultures, exemplified by China and India, have made rapid and impressive progress in developing such a cultural base, while the Western cultures have continued to stagnate.

Japan's modern cultural evolution, specifically the technological and industrial advancements driving much of its late twentieth-century socioeconomic progress, is testimony to the effectiveness of fundamental and historic ideational cultural values. The linkage of Japanese cultural traits with modern science, engineering, and technology has been evident in the precision and efficiency of the country's extremely profitability automobile manufacturing industry. Less well known is Japan's significant achievement in becoming the world's most energy-frugal developed nation since the global oil crisis of the 1970s. According to the International Energy Agency, in 2005 Japan consumed one half as much energy per dollar of economic activity as the European Union countries or the United States and one eighth that of China and India. Japan's annual energy consumption of 200 million tons of oil has remained unchanged since the early 1970s, while its economy has doubled to reach an inflation-adjusted figure of $5 trillion. Credit for this most impressive feat is due to the nation's consensus of embracing traditional altruistic values including a conservative cultural ethic and willingness to follow leadership's direction. Japan imposed government-mandated energy efficiency targets and high taxes on oil products while significantly investing in energy-efficiency technology.

In the past thirty-six years, Japan's steel industry has also invested $45 billion in energy-saving research and technology.[9] One major accomplishment has been to harness waste heat and gases to generate electric power for manufacturing. JFE Steel at Keihin utilizes gases and heat previously released into the air or burned off to generate its electricity, producing 90 percent of its electricity needs. This innovative technology extracts a larger percentage of heat content from a given amount of the consumed fuel from industrial production, corresponding to additional wealth-energy capital. Additionally, this constructive use of wealth, energy, and human resources increases the probability that, not only will such a culture prosper in the short-term, but its long-term viability as a prosperous and stable culture is also enhanced.

Contrast this progressive Japanese national energy policy with America's inability to achieve a major nuclear power capability since the 1970s, its increasing consumption of energy for nonproductive priorities, and its lack of national consensus on reducing its dependence of foreign oil. The Japanese culture clearly represents a more favorable environment in which to achieve a more rational solution to increasing global energy issues and to socioeconomic prosperity.

With reference to Toynbee's four stages of civilization development, illustrated by Figure 8-1, it would appear that Japan is at the mid-range of Stage II, Growth (i.e., approaching an advanced adolescent maturity phase). However, as Japan enters the beginning of the twenty-first century, benefiting from a successful socioeconomic history, it also exhibits typical characteristics of cultural degradation. These are similar properties, but less severe, to those found in the more mature cultures of America and the European nations.

Japan has historically cherished its egalitarianism, a society that views everyone as being middle class, a "100 million, all middle-class society." However, economic prosperity has increasingly produced a widening disparity in wealth, referred to by the Japanese as "Divided Japan" and "Light and Darkness," a familiar consequence of a maturing economic system and more specifically of mature capitalism. Government policies of deregulation, privatization, spending cuts, and tax breaks for the wealthy have been able to produce impressive corporate profits, additional jobs, and a high stock index. Beginning in 2006, commercial property values in the three largest urban areas began to appreciate for the first time in over a decade, and urban high-rise building construction increased. Meanwhile, the Japanese household savings rate has significantly declined, and the number of people receiving welfare has increased since 2000 by 37 percent to over 1 million. From 1970 to 1990, the number of Japanese families without savings was stable at about 5 percent, but by 2006 it increased to over 30 percent. From 2000 to 2004, the number of schoolchildren receiving aid rose by 36 percent to about 13 percent of all elementary and secondary students. Additionally, Japan is witnessing a large gap in the quality of public and private education, which is just another aspect of a widening social disparity.[10]

According to Masahiro Yamada, a Japanese sociologist and author of *Society of Disparities in Expectations*, Japan is unable to maintain its historic program of paternalistic capitalism. Companies have abandoned traditional employment practices such as lifetime employment and now lay off employees and relate promotion to performance. Additionally, the government has lifted many restrictions on hiring temporary workers and on reducing health and pension benefits. Consequently, workers are increasingly facing lower wages, fewer benefits, and less chance to acquire full-time employment. Also, significant changes in the tax code have signaled the government's abandonment of its long-held philosophy to aggressively pursue policies of wealth redistribution. The highest personal income tax rate has been reduced from 75 percent to 37 percent, the capital gains tax on stock sales has been lowered from 20 percent to 10 percent, and inheritance tax laws have been changed to ease the transfer of large assets.[9] Such evolving societal characteristics, familiar to Americans, are thought necessary to salvage an economic system and its profitability that is under stress, an inherent inevitability for a maturing capitalist society.

This 2006 snapshot of Japanese cultural trends is consistent with past experiences of industrialized, capitalistic countries, as for example, the United States and Britain. The characteristics of both cultural growth and decay exhibited in the past by previously successful civilizations, described in previous chapters, are also emerging during the modern evolution of Japanese society.

Twenty-first-century India also provides a graphic illustration of a culture undergoing rapid, successful advancement while at the same time experiencing the consequences of progress that reduce the favorable impact of socioeconomic success. India's rapid and flourishing economic development has evolved from an emphasis on providing global services to a focus on its domestic economy, particularly manufacturing. India's annual increase in manufacturing output is 9 percent and closing the gap with its services industry growth rate of 10 percent. The nation's exports of manufactured goods to the United States are increasing at a faster rate than China's and generally represent higher-quality products using more advanced technologies. Over two thirds of India's increase in foreign investment last year was in manufacturing. India's future development is expected to be significantly influenced by its expected population growth of young workers aged twenty to twenty-four, which will number 116 million by 2020, compared to China's projected 94 million. Also, India's general population will surpass China's by 2030 as a result of China's "one child" policy of population containment.

Foreign investors are attracted by India's well-educated and trained labor force of assembly-line factory workers, who earn an entry-level wage of $90 a month. Typically, such workers have a high school education plus technical college training. This projects an adequate, better-educated, lower-wage labor pool than found in China. As a result, India's rate of economic growth has gradually increased to 8 percent a year, compared to China's annual growth of about 10 percent. However, India's infrastructure is not advancing sufficiently to adequately serve a socioeconomic system that must adapt to modern industrialization and a large increase in population. Inadequate roads, port facilities, and power plants are increasingly complicating India's modernization, but its current agriculture problems, the consequence of its success, have become a major element of cultural deterioration.[11]

India's arable land acreage is the second largest in the world, and its economy is one of the most rapidly advancing, successful, and industrially innovative in the highly competitive global economy of the first decade of the twenty-first century. However, India's agricultural system has become inadequate for its current population base due to lagging infrastructure support and inadequate available resources, both inevitabilities of rapid cultural maturity and accumulating cultural complexity and bureaucracy. Hence, cultural progress is visibly taking a toll on Indian society as a result of its rapid rate of growth. This is an example of accumulating societal entropy production gradually becoming apparent during a relatively early Stage II cultural Growth phase, while cultural success generally overshadows elements of accumulating cultural degradation.

During the 1960s and 1970s, India made great progress in agriculture by implementing the foreign-assisted Green Revolution program that essentially drove hunger and famine from being a dominant, critical threat to the general population. This program introduced high-yield varieties of rice and wheat, modernized and expanded irrigation systems, increased the use of pesticides and fertilizers, and provided farmers with technical assistance. However, in the last decade, the resources of the nation have not been sufficient to maintain the constant growth rate of crop yields achieved since the 1960s. Particularly since the 1980s, governmental irrigation programs, financial assistance to farmers, and agriculture research have substantially declined. Major groundwater depletion has also been an important factor in declining crop production, as groundwater tables have shrunk by as much as 100 feet during the last thirty years. In Punjab, more than three fourth of the districts draw more groundwater than nature restocks. Only 40 percent of Indian farms are irrigated, and cultivatable land is declining. Consequently, farming increasingly relies of small, rain-fed plots of land, and with inflation at 11 percent, family farms are declining as farmland is increasingly sold to developers.[12]

China's rejuvenated culture exhibits a similar growth pattern to that of India and Japan but more closely corresponds to a less-mature Stage II Growth compared to Japan's more advanced phase. However, the combined accomplishments of these three nations illustrate the rebirth of Asian socioeconomic systems. In the past two decades, China's economy has been growing at an annual rate of about 9 percent, based in part on labor-intensive, low-tech manufacturing products that are attractive to an extensive worldwide consumer market. China's labor force exhibits constructive cultural values and provides a low-wage, hard-working, and motivated workforce, appropriately skilled for the country's level of industrialized maturity. The educational system provides the necessary engineering, science, and management graduates, enabling the economy to mature technologically and to effectively compete in world markets. As a result, the expanding Chinese socioeconomic system continues to shift toward a more mature, market-driven economy. However, China's escalating productivity has required major increases in the consumption of energy. In 2006, China consumed 90 percent of the increase in the world's coal consumption, which has risen as much in the last three years as in the previous twenty-three years. Additionally, global energy consumption is expected to increase by over 53 percent by 2030, with oil increasing from a 2006 level of 85 million barrels a day to 116 million barrels, mostly due to consumption by developing countries. Meanwhile, by 2009 China is projected to surpass the United States as the world's biggest emitter of the main global warming gas, carbon dioxide.[13]

China's role and prestige in international affairs has risen in parallel with its commercial prominence. In 2001, China became a member of the World Trade Organization and, by 2006, significantly lowered its tariff barriers from 41 percent to less than 6 percent, a lower tariff rate than any developed country. In addition to bringing a desirable work ethic and, in general, a favorable set of cultural values to the workplace, the Chinese take seriously sound financial investments and the avoidance of frivolous personal spending. In 2006, approximately 50 percent of the Chinese gross domestic product was being reinvested, which compares to about 30 percent in Japan and South Korea and 0 in the United States.[14] Thus, the Chinese cultural characteristics are favorable to produce a healthy rate of economic growth.

China is following an important principle of economic success; that is, minimizing the frivolous consumption of national wealth and energy so as to achieve an effective economy, generate a profit to be utilized for wise reinvestments, and maintain ideational cultural values. These characteristics should permit China to sustain a projected economic growth rate of at least 7 to 8 percent for the foreseeable future and thus become the largest world economy by 2050.

As experienced by all civilizations in early Stage II Growth, China's rapid rate of economic development and societal advancement has increasingly bred and accumulated societal complexity, disorder, and irresolvable issues constituting significant symptoms of cultural degradation that require new efforts and resources. For example, during the next two decades, about two thirds of the Chinese population will be over sixty-five years of age, with 100 million over the age of eighty. China is aging faster than any other country in history. Thus, the costs of supporting the retired population and their health costs will dramatically escalate during the next fifty years.[14] More generally, the inherent cost increases associated with a maturing adolescent civilization to support a growing bureaucratic government and public entitlements of health, education, social services, and transportation can be expected to parallel those of Western civilization during its twentieth-century period of rapid economic growth. In particular, human expectations for public services inherently proliferate during times of economic prosperity.

While China currently has a personal savings rate of about 40 percent during this early rejuvenation growth phase, recent data and the history of past civilizations suggest a decrease over time, as economic prosperity continues. This has been the record of modern America and Japan, as the attraction of urbanization and materialistic consumerism intensifies. Over the last few decades, China's urbanized industrial economy has been accompanied by the familiar migration of employable young adults from the rural countryside to better economic opportunities and a more desirable life style. It has been estimated that in 2006, this floating population of internal migration to east coast cities numbered about 150 million, overwhelmingly young adults. It is noteworthy that the migration has resulted in 80 percent of China's rural population being over the age of sixty; consequently, the 2,500-year-old Confucian doctrine to venerate and to care for the elderly is being tested.[15] This established Chinese tradition of children providing for elderly parents has been significantly eroded as adult children migrate to the cities, seeking unprecedented economic opportunity. China's economic development has produced many such cultural conflicts and pragmatic accommodations for an increasingly materialistic culture (e.g., enticing young people with the potential rewards of socioeconomic affluence and individual fulfillment).

Confucian doctrine is giving way to sensate values, as did the similar Protestant Puritan values of Western civilization centuries before, early cultural traits following the same behavioral patterns of ancient Greek and Roman. China's shift to a more modern technological, capitalist economy and its inherent participation in today's global economy will undoubtedly result in the continuing infiltration, diffusion, and absorption of Western values and practices within the Chinese population. This would suggest that China would increasingly utilize its national wealth for more nonessential, luxury goods and services with a commensurate decline in personal savings and sound long-term societal investments. This constitutes the potential seeds of deterioration that can wastefully consume national wealth and reduce economic effectiveness and financial returns on societal investments.

The same pathway led to Western civilization's gradual and insidious socioeconomic deterioration.

Meanwhile, in recent decades, China has achieved impressive economic growth and has emerged from its past isolation and cultural "ossification" to increasingly become a dominant twenty-first-century international player. The characteristics of this period of cultural growth are the same as those noted by Toynbee and other authors to describe the "rejuvenation" or "rebirth" of formerly successful civilizations.

Meanwhile, during past decades, the worldwide membership of the Islamic faith has dramatically increased and has focused attention on a potential resurgence of the Islamic civilization. However, the socioeconomic fortunes of the Islamic Middle East and other Muslim populations have demonstrated little socioeconomic progress over the last fifty years, in sharp contrast to the Japanese and Chinese populations. Therefore, it is instructive to review the root cause for this lack of socioeconomic advancement and to assess and contrast the future potential of the Islamic resurgence with that of Asia.

In the Middle East, diverse and hostile spiritual, ethnic, and tribal issues, in addition to historic poverty and sociopolitical disadvantages, have stymied efforts to stabilize Muslim social, economic, and political priorities, policies, and aspirations. Unfortunately, for many generations, conflicting religious, cultural, and political ideologies and antagonisms have dominated Middle East life and constituted a major negative socioeconomic force that precludes modern-day advancement being experienced by China and India.

In contrast, China's leadership, absent such extreme emotional, religious, and tribal-based divisiveness, has been singularly committed to a more unified sociopolitical ideology and economic agenda. China is achieving economic progress by creating its own opportunities through an intensely dedicated, energized, and technologically adaptive population. This realization of societal common purpose and economic success is due, in a large degree, to the culture's self-discipline, self-determination, and creativity that has been historically imbedded in, and reinforced by, its moral and ethical values.

In particular, Islamic factions within the Middle East have not exhibited the necessary self-determination, creativity, work ethic, and resourcefulness to create sufficient cooperative opportunities among

Arab communities and nations to achieve regional socioeconomic success in support of a global Islamic resurgence. This is due, in large part, to Islam's unwillingness to constructively, cooperatively, and effectively devise its own particular model of *Muslim modernity* capable of political and economic success while contributing to a stable twenty-first-century regional and global environment. Reality necessitates that significant modification of Muslim attitudes and behavior, including greater civility, respect, and accommodation for cultural and spiritual diversity, will be necessary to successfully function in an increasingly competitive worldwide political and economic environment. Mutual cooperation among nations in a rapidly evolving global economy to coexist and prosper and to effectively combat mutual health, environmental, and security threats is an imperative.

The need among nations of the world for greater mutual respect for diverse cultural values, religions, and basic human welfare will continue to intensify as normal human advancement inherently accelerates complexity, competition, dysfunction, and conflict among nations. Mutual assistance and cooperation will no longer be an optional or polite gesture of international diplomacy but will become a survival strategy for regional and global stability during increasingly more challenging times.

The Islamic transition from colonial domination to twentieth-century independence to the current twenty-first-century spiritual awakening is a very short cultural time frame compared to that required for cultural rejuvenation. In 1987, Braudel described the necessity for the Muslim world to struggle with, and address, solidarity, self-discipline, and nationalism:

> **Nationalism has a role to play in the near future: All Muslim countries will have to adopt and apply strict austerity plans. They will need, in fact, plans for solidarity and social discipline; and nationalism will help all these young countries to deal with the serious economic difficulties they face. It will make it easier to accept essential innovations which clash with very ancient social, religious, and family structures—age-old ancestral habits, perpetuated in Islam's traditionalism, and only likely to spark off violent reactions.**[16]

Since the time of Braudel's analysis, the Muslim movement has not achieved greater "solidarity and social discipline" but rather

Shifting Twenty-First-Century Civilization Patterns: The Asian and Islamic Resurgence

created more numerous and horrific "violent reactions." Other than the potential wealth of its oil reserves, the twenty-first-century Middle East is essentially devoid of the historic, broad-based properties and characteristics that enabled the Greeks, Romans, British, and Americans to create successful civilizations. The lack of a sense of nationalism is a major deterrent to Islam's potential rebirth. Thus, despite its impressive eighth- to twelfth-century era of greatness, Islam's ability to achieve significant socioeconomic progress and cultural rebirth in the twenty-first-century environment is severely restricted and highly improbable during the next few generations. Fundamentally, the Middle East is centuries behind the modern world in science, technology, education, and economic innovation, while remaining in the grips of tribalism, cultism, religious leaders, and a violent mob psychology. Islam is, or should be, struggling with the fundamental and practical aspects of defining a version of modernity that will permit social and economic stability for future generations. It is often stated, and accurately so, that twenty-first-century high-profile Muslim conflicts are fundamentally clashes of Eastern and Western civilizations and among internal religious and cultural heritages. However, at the root of these clashes are divergent human values that pit the Middle Ages versus the twenty-first century, religion versus reason, and barbarism versus rationality.

The current status of Islamic "rejuvenation," based on the characteristics and properties of each of the four stages of civilization development described in Chapter 8, corresponds to a primitive Stage I Genesis. Fundamentally, the inability of the current Islamic movement to meet Stage II cultural Growth criteria is a direct consequence of the Muslim world's insistence that no distinction may be made between spiritual and temporal affairs and that socioeconomic and political processes should be managed according to Muslim theology. This "God is Caesar" philosophy of cultural, religious, and governance leadership has stifled Islam's potential rebirth and socioeconomic progress. The continuing long history of negative consequences should, at some point in time, provoke Muslims to question their sense of reality relative to the successes of the Western nations and the more recent prosperity of Asian cultures. Specifically, what socioeconomic achievements have been realized since the seventeenth-century Ottoman Empire collapsed, the subsequent deterioration of the Islamic civilization, and the nineteenth-century humiliation of foreign domination? The modern Islamic strategies and tactics are inherently unable to create the cultural characteristics and properties recognized as favorable for cultural growth and prosperity, particularly given twenty-first-century global socioeconomic realities.

Clearly, the economic prosperity of Western civilization has not been fairly shared, to any appreciable extent, with Muslim populations. The former Muslim colonies did not receive adequate education, training, or economic assistance necessary to prepare them for establishing stable and successful socioeconomic systems independent of colonial powers. Historic Muslim organizations and functions were eliminated without the establishment of appropriate transitions to be implemented at independence from colonial nations. Importantly, the native populations have not minimally benefited from the region's chief natural resource of oil. Essentially, the experience of the Middle East populations has followed the same imperialist patterns established by the English and Spanish during their conquest of the Americas. The Spanish confiscated precious metals and enslaved the people for the benefit of Europeans, abused indigenous populations, and destroyed civilizations.

Thus, while Muslims celebrate their twentieth-century achievements of liberation from colonialism and a greatly expanded spiritual membership, they face enormous, if not impossible, challenges. Fundamentally, an archaic, traditional Muslim civilization has been in an "ossified" phase for centuries and finds itself without the tools and resources required to compete in the socioeconomic and political systems of the twenty-first century. Addressing these deficiencies will require finding some form of congruency and acceptable accommodation among Muslim subcultures and with the cultures of the world.

The basis for East-West conflict of religion versus reason and barbarism versus rationality has become most evident and more intense as a consequence of the West's 2003 invasion of Iraq that has been justified, depending on U.S. political circumstances, as a defensive strategy against Iraqi weapons of mass destruction, an anti-terrorism strategy, or an opportunity to spread democracy and protect economic interests. Most Iraqis view the nature and manner of America's invasion and occupation as being aggressive and in conflict with their spiritual and cultural principles. The consequences of this intervention once again illustrate that cultural unity and the sense of obligation to clans, tribes, and religious ideologies and leaders are more potent in less-developed cultures than the modern concepts of multicultural unity, law and order, and nationalism. In such a culture, it is permitted, even expected, to violate laws so that tribal customs are observed and spiritual leaders are obeyed. Today, the concept of nationalism among Muslim populations is generally secondary and perhaps has become an irrelevant concept in the Muslim world. In contrast, allegiance to a nation-state and to law and order is considered an essential component of survival and socioeconomic success in the more developed countries.

The parallel of cultural contrasts between primitive and modern civilizations and between Muslim theological precepts and its twenty-first-century practices is exemplified by the 2005 Iraqi celebration of Ramadan, which is similar to the Jewish Yom Kippur (i.e., a period of atonement). On the first day of Ramadan, Islam's holiest day, a Sunni Muslim suicide bomber blew up a Shiite mosque in the middle of a memorial service, killing twenty-five worshipers. The following year, on the first day of Ramadan, a Sunni suicide bomber killed thirty-five people in a Shiite neighborhood.[17] The same day, heads from nine murdered Iraqi policemen and solders were found. The context for this violence is that Muslims are commanded by God to abstain from evil thoughts and deeds during Ramadan, as well as from the pleasures of food and drink from dawn until dusk throughout the month. Instead, Iraqi Muslims of one clan slaughtered Iraqi Muslims of another clan on Islam's most holy day. However, such human slaughter, Muslim against Muslim, was occurring in Iraq on a daily basis, not by random acts of misguided individuals, but by organized religious subcults seeking political control and domination of the country. Religious militant leaders, some of whom have a controlling influence in official government affairs, command these groups. This is a civilization that places primacy on spiritual guidance commanding every aspect of daily living. Meanwhile, the non-Muslim world is subjected to Muslim criticism, threats, and violence, resulting from newspaper cartoons and from comments by the pope considered as anti-Muslim. These actions by non-Muslims arouse the Muslim world to rage and violence, but Iraqi Muslims continuously slaughtering each other in large numbers and in the most violent and cruel manner is ignored, supported, or at least tolerated through silence by a large number of non-Iraqi Muslims worldwide. Assassins have struck down at least thirty-five Lebanese politicians, clerics, and journalists in Lebanon's sixty-three years of independence. Bombs killed two presidents, one just before and one after their inaugurations. One prime minister was blown up in a helicopter; another was killed four months after leaving office in a suicide truck bombing.[18] These incidents were not actions of deranged individuals but of organized political, religious organizations.

Constructive but misguided efforts of Western occupation forces in Iraq to promote democracy, centralized military and governmental control, and a unified sense of nationalism applied to Muslim tribes, religious sects, and cults contradict the lessons and sensibilities of history and have intensified the East versus West conflicts. About 2 million Iraqis have fled their homeland to escape the warfare and find shelter in neighboring countries, usually living in great poverty and at the mercy of less-than-friendly foreign governments. As of mid-2008, America has allowed only 1,600 Iraqis into the United States, while European countries have welcomed many more. Meanwhile, the U.S. tactics employed in Iraq create much unhappiness and disorder among Muslim populations throughout the world and in America. In addition to U.S. causalities, America's participation in the Iraqi invasion and the prolonged occupation is expected to eventually cost America between $1 trillion and $2 trillion. No debate or consideration of competing priorities of financial and social costs occurred within the American government prior to invasion. Additionally, the funding of this war is acquired through debt financed by China and other economically prosperous countries, adding to U.S. annual budget deficits and the soaring, accumulating national debt.

In contrast, Asian nations, illustrated by China, India, and Japan, are utilizing their national wealth-energy resources in a more constructive manner that promotes effective economic outcomes and generates additional national wealth for reinvestment. Thus, the leading Asian nations have significantly greater rates of cultural growth than Western nations.

It is of interest to contrast the twenty-first-century cultural characteristics of the Asian and Islamic populations with those that have historically appeared to stimulate periods of rapid socioeconomic progress. A parallel appears to exist between the spiritually based cultural environment that existed just prior to and during the early industrial phase of Western civilization's socioeconomic affluence and the current Asian economic prosperity. However, these similarities are at variance with the cultural environment of the Islamic world.

Periods of economic strength, growth, and satisfaction become more probable when a population's attitude and behavior reflect a strong humanism and morality and when a strong individual work ethic and deep sense of common purpose exist. Such characteristics appear to coincide with a strong desire to conserve wealth-energy resources and to use profits to regenerate national wealth (i.e., reinvest a significant portion of the gross domestic product). These characteristics, prevalent in Colonial America, have gradually and substantially diminished in America during subsequent centuries.

Such cultural properties associated with the Puritan Protestant culture, found among sixteenth- through nineteenth-century developing Western nations, are consistent with the dynamics contributing to recent Asian economic resurgence. This socioeconomic rejuvenation may be traced to a spiritually oriented philosophy that provides a constructive cultural environment conducive to promoting societal advancement. The philosophical and spiritual tenants of Confucianism constitute altruistic individual and societal values and traits, as also found within the Protestant Puritan philosophy, that have provided Asian populations with an underlying philosophy of life for thousands of years.

Spengler, Sorokin, and other sociologists place great significance on the need for spiritual and community-oriented values necessary to promote the successful development of primitive cultures and the rebirth of deteriorating or stagnating civilizations. Such a cultural mentality emphasizes self-respect, self-discipline, civility, low levels of cultural disorder, conservative economics, and societal-oriented priorities. These characteristics should be placed in the context of the Confucian philosophy as a way of living, as opposed to being defined as a formalized authoritative religious theology. This philosophy of life embraces unique spiritually based ethics, political ideologies, and cultural traits. This Asian cultural mentality is in stark contrast to the twenty-first-century Islamic culture.

That Confucianism, a more philosophical, spiritually based way of life, rather than a religion, could provide a highly favorable cultural environment for the rejuvenation of Asian civilizations is consistent with accepted sociological thought. Sorokin describes "the formula" for cultural resurrection as following sequentially crisis, ordeal, cleansing, grace, and ultimately resurrection. In modern times, Asia has achieved the transition from cleansing to grace, and has entered resurrection. Brander, noting Sorokin's formula, cites historic examples of other cultural resurrections and their spiritual nature:

> **This chain of stages leading to cultural renewal unfolded several times in ancient Egypt; in Babylon around 1200 B.C.; in India when major crises came to an end with the revival of Hinduism or the emergence of Buddhism; in China of the sixth century B.C. when Taoism and Confucianism arose out of social disintegration; in the Hebrew nation when ordeal brought great religion prophets like Elijah, Isaiah, and Jeremiah; and in the days Rome when social decline cleared the way for the Christian faith and culture.**[19]

To Sorokin, cultural resurrection or rebirth becomes more probable in a cultural environment that emphasizes spiritual regeneration. He views "altruistic ethics" as the pathway to cultural rebirth, "moral adjustment I regard as paramount above any technological change.... Political, economic, and technical changes should be carried out; but morality is basic to all secure social life. Without morality, there can be only struggle and hatred."[20] Toynbee, Sorokin, and other respected social thinkers clearly view cultural rebirth as requiring major social movements based on moral and spiritual altruistic values.

Since the fifth century B.C., Confucianism has brought philosophical guidance and influence to China and has been widely absorbed by Asian populations, significantly influencing many sociocultural transitions over the millennia. An inherent self-cultivating sense of humanity has provided a nourishing cultural environment independent of formal religious affiliations and has become a source of core values and a social code. Confucian ethics, characteristic of traditional Chinese moral character, existed even in periods when Taoism and Buddhism were the predominantly observed religions. In modern times, Confucianism has remained an integral part of the Chinese culture as it undergoes the latest transformation and adaptation to global modernism, providing a pathway of cultural modification.

Since the time of Confucius (551-479 B.C.), Confucian philosophy has given emphasis and been committed to social and political concepts that would improve the quality of life for the masses. With the overthrow of the Ch'in dynasty, the newly established Han imperial dynasty (206 B.C.-A.D. 220) formally adopted the Confucian ideology that emphasized moderation and virtue in its governance system. The highly successful Han dynasty ruled longer than any other Chinese empire, and its official ideology continued for almost 2,000 years.[21] Today, Confucian concepts, values, and practices are still fundamental to Asian populations.

Max Weber (1864-1920), who cites the Protestant Puritan work ethic and related values as being a major contributing factor to the European- and American-inspired Industrial Revolution, labeled Confucianism as the least able of the world's religions to contribute to capitalist development. During the early twentieth century as Asia debated its adaptation to modernity and sought economic advancement, Confucian philosophy was considered a negative influence, with Marxism and capitalism conceivably being the only realistic alternatives. However, the rapid, unpredicted Asian economic successes during the latter decades of the twentieth century prompted social scientists to rethink the Confucian influence. Bell and Hahm point out that "the need for a new theoretical framework became all the more acute primarily because the social scientists, both liberal and Marxist, failed to predict or explain the economic success of these 'Confucian' states while the Weberian thesis regarding the alleged incompatibility between Confucianism and capitalism rapidly lost credibility." Also, they note: "Unlike Islam, Hinduism, and Buddhism, there has never been an organized Confucian resistance to modernization"; the goal has been to achieve some realistic form of modernity that promotes progress. Eventually, the Chinese Communist government altered its policies and reversed its anti-Confucian position in an attempt to improve the ethics of the governance system.[22] It would appear that Confucianism and Puritan Protestantism provide similar cultural values and characteristics conducive to stimulating and maintaining an environment of socioeconomic stability and advancement.

Given the importance of human values and cultural traits in decision-making responsible for vital societal functions and history, it is instructive to examine the elements of Confucianism and Puritan Protestantism for commonality and mutual significance in contributing to cultural growth and socioeconomic prosperity. That is, are the major Confucian ideals and concepts that now affect twenty-first-century Asian attitudes and behavior fundamentally different from the Puritan-inspired philosophy of the seventeenth- to nineteenth-century era of the West's successful cultural development? Perhaps Confucian philosophy, which emphasizes moral leadership and "Cardinal Virtues," provides an effective and efficient cultural foundation for twenty-first-century Asian socioeconomic success comparable to that of the West's early Puritan Protestant values, attitudes, and practices.

The shifting of Western values during its period of cultural prosperity from a predominantly ideational to sensate mentality has been discussed in previous chapters. It may be argued that the Confucian philosophy, or way of life, is of broader scope and has been more successfully ingrained for a longer period of time into a larger Asian population than the comparable Protestant Puritanism history. Additionally, Confucian values and philosophical principles include attitudes, behavior, and practices related to the investment of resources that are of greater socioeconomic benefit to a larger segment of the population.

The Confucian principles that appear to provide this unique cultural environment beneficial to socioeconomic effectiveness include (a) *a ritual propriety,* "*li,*" which establishes a political norm and provides a restraint on political rulers (i.e., a system of horizontal public accountability); (b) "*junzi,*" or *exemplary persons*, behavior expected of rulers intended to check abuse of power; (c) "*three-min*" *principles* of nationalism, citizen rights, and the welfare of human beings; (d) *the ethic of mutual help or the community compact*; and (e) "*rang,*" *or self-criticism*, politely yielding, giving concessions to others, or compromising in lieu of assertion of one's rights.[23]

The net result of these cultural principles, intended to guide human behavior, is a strong social force than promotes ideational values and frowns upon sensate values to an extreme not observed in Western civilization, particularly in more mature cultural stages. Given the region's growing prosperity, the twenty-first-century issue is whether, as in previous great civilizations, the Asian cultural rebirth will systematically breed a high level of materialistic, sensate attitudes and behavior, resulting in a proliferation and domination of individual self-interest and an emphasis on nonessential priorities and nonproductive investments of national wealth.

It is instructive to explore the Confucian ideals significant to Asian socioeconomic success, specifically the determination and control of priorities affecting national wealth consumption and of the distribution of the wealth emanating from society's productivity. The question is whether these cultural priorities for the consumption of national wealth and for wealth distribution differ substantively from those initially utilized by Western civilization. Gilbert Rozman is quoted by Bell and Hahm: "Confucianism provides the ideological underpinnings for further dynamism in Asia."[24] He views Confucian principles as effectively maintaining a desirable balance between centralized and decentralized economic controls that encourage local economies to become major participants in overall marketing mechanisms, including major complex urban networks. In addition, rapid economic development has been accompanied by a more egalitarian distribution of income than found in Western nations, with the most developed countries also being the most egalitarian.[25]

> **The ongoing, living Confucian tradition helps to explain the distinctive characteristics of East Asian economic systems and welfare states, ... economic reform that strikes the right balance between productive economic activity and meeting the needs of the "worst-off" in an overall capitalist framework.**[26]

As previously noted in Chapter 3, John Kenneth Galbraith argues that a "balance between productive economic activity and meeting the needs of the 'worst-off'" must protect the production capability of the economy's private sector. This is based on the premise that public services are ultimately funded by private-sector productivity, thus excessive public services, including necessary welfare, may threaten the stability of private enterprise and a society's economic future. In the Asian culture, such a *balance* is derived from over 2,000 years of Confucian philosophy reflecting the amalgamation of the ideals of ritual propriety, exemplary persons, nationalism, citizen rights, welfare assistance, mutual help, and self-criticism. It is not clear that Western civilization has ever embraced and demonstrated such a broad and enduring, reinforcing, and all-inclusive set of ideational values. However, as evident in Weber's work, a number of seemingly conflicting Confucian ideals have historically clouded viewpoints of the contributions of Confucianism to the current Asian concept of modernity. It is not clear as to the extent these values and ideals have become hybridized with capitalistic socioeconomic principles that are being implemented in the Asian cultural rebirth in the twenty-first century.

Debates have raged regarding what some interpret as conflicts between Confucian values supporting ruler authoritarianism and the checks and balances of the people regarding the centralized powers of their rulers. Rozman makes "the case that Confucian ideals justify limitations of central power, encouraging social solidarity below to balance controls from above." For example, he supports the thesis that Confucian concepts provide for the maintenance of a socioeconomically healthy equilibrium between urban and regional stakeholders. It is noted that historically, eight dynasties in China, Korea, and Japan, guided by Confucian ideals, achieved over two centuries of stability attributable to this balance of power among the elite, mid-level, and low-level participants (i.e., economic and political decentralization with a central state in a support role).[27]

Historically, extensive and creeping centralization of power and control have been a common element of civilization extremes associated with urbanization; increased bureaucratic management; and cultural complexity, stress, and disorder, all of which tend to stifle socioeconomic progress. Also, such extremes promote significant increases in consumption of national wealth. However, it appears that Confucian ideals may have created an inherent cultural barrier to the extremes of centralized and expensive bureaucratic management, characteristic of Western civilization during its initial Stage II Growth phase. In such Western systems, a continuously increasing influence of centralized power and control has resulted in a concomitant withering of decentralized participation, influence, and benefits for the people. Rozman describes an Asian attitude toward commerce whereby urban life and the countryside form a productive economic decentralized network, "free to take place periodically in thousands of transportation junctures and in urban locations more convenient for exchange with the countryside or for daily consumption." He suggests: "Unlike European countries, where state building largely followed commercial development, Asian states faced the challenge of drawing on the vitality of local economies and societies to revitalize already powerful states and to unleash the potential for more dynamism."[28] Consequently, it would appear that during the latter decades of the twentieth century, China and Japan developed an appropriate and efficient decentralized distribution of economic power founded on an ingrained Confucian moral foundation, conducive to socioeconomic productivity beneficial to a large segment of the population. The question remains as to the extent that economic success will erode this historic cultural foundation during future generations and to what extent Asian nations will be able to form and maintain these economic and political regional bonds of mutual support and influence in the face of a faltering Western civilization.

Rozman highlights a very significant point: "It is important to view Confucian ideas as more than an equivalent to religion and ethical principles. While comparisons of Confucianism and Christianity cover many themes, including attitudes toward society and the individual, they usually overlook the ways value systems and religions were used in state building."[29] He also describes an "elite Confucianism" as a general moral compass for society and a "mass Confucianism" for families, with both types separated from the "cult of state power in Asia."

Thus, Confucian ideals appear to support a desirable and constructive economic decentralization in Asia that is conducive to socioeconomic prosperity, emphasizing a more benevolent distribution of wealth than has historically occurred in Western civilization. This is portrayed as unique in the twenty-first century compared to the competitive systems, which Rozman describes and contrasts as:

> **Anglo-Saxon liberalism, dominant in the West, glorified individualism.... Soviet-launched socialism insisted on the supremacy of the state, rejecting decentralization in favor of vertical hierarchies under firm control.... East Asian societies clearly contrast with these two ends of the spectrum.... Confucian societies gravitated to a community-centered approach.**[30]

One is left with the view that the Asian culture is based on a historically more intense and broader scope, altruistic mentality than Western cultures. Interestingly, Asian cultural values and traits also appear to be associated with twenty-first-century educational and economic achievements observed among the Asian-American population. The 2005 American Community Survey reports that Asian Americans, on average, have higher incomes and education levels than American whites, blacks, or Hispanics. In 2005, 49 percent of Asian-American adults possessed at least a bachelor's degree, compared to 30 percent of whites, 17 percent of blacks, and 12 percent of Hispanics. Asian-American households had an average income of $60,367, white households averaged $50,622, black households averaged $30,939, and Hispanic households averaged $36,278.[31]

The ancient Asian cultures have re-emerged from their long-standing cultural atrophy and have initiated their own versions of a modern industrial and technological revolution, competing satisfactorily in global commercial markets. Their rapid rebirth can be expected to dominate the twenty-first century as the weakening economies and influence of Western nations reduce their former competitive advantage. The twenty-first century should see enormous relative productivity in Asian nations that should spawn greater international recognition and respect as dominant international economic and political powers.

> **The dominant system prepares its own downfall and paves the way for the ascendance and domination of one of the rival systems of truth and reality, which is, under the circumstances, more true and valid than the outworn and degenerated dominant system.**[32] **Pitirim Sorokin**

Chapter 13
References and Notes

1. Braudel, Fernand. *A History of Civilizations.* London: Penguin Press. 1992. 35.
2. Sorokin, Pitirim A. *Social and Cultural Dynamics.* Boston: Porter Sargent. 1970. 676.
3. Huntington, Samuel P. *The Clash of Civilizations and the Remaking of World Order.* New York: Simon and Schuster, 1996. 82.
4. Ibid., 306.
5. Tagliabue, John. "There's Detroit and There Trnava." *The New York Times.* November 25, 2006. Also Tagliabue, John. "Would Stalin Drive a Peugeot?" *The New York Times.* 5 January 2006.
6. Fackler, Martin. "In India, a New Detroit." *The New York Times.* 26 January 2008.
7. Dawson, Christopher. *Dynamics of World History.* New York: Sheed and Ward. 1956. 128.
8. Rabinbach, Anson. *The Human Motor: Energy, Fatigue, and the Origins of Modernity.* Berkeley: University of California Press. 1990. 85.
9. Fackler, Martin. "Japan Sees a Chance to Promote Its Energy-Frugal Ways." *The New York Times.* 4 July 2008.
10. Onishi, Norimisu. "Revival in Japan Brings Widening of Economic Gap." *The New York Times.* 16 April 2006.
11. Bradsher, Keith. "A Younger India Is Flexing Its Industrial Brawn." *The New York Times.* 1 September 2006.
12. Sengupta, Somini. "India's Growth Outstrips Crops." *The New York Times.* 22 June 2008.
13. Bradsher, Keith. "China to Pass U.S. in 2009 in Emissions." *The New York Times.* 6 November 2006.
14. Zuckerman, Mortimer B. "A Giant's Growing Pains." *U.S. News & World Report.* 23 January 2006.
15. French, Howard W. "Rush for Wealth in China's Cities Shatters the Ancient Assurances of Care in Old Age." *The New York Times.* 3 November 2006.
16. Braudel, Fernand. *A History of Civilizations.* 99.

17. Friedman, Thomas L. "Islam and the Pope." *The New York Times*. 29 September 2006.
18. Karam, Zeina. "Nation Has Long History of Killing Politicians." The Associated Press. November 2006.
19. Brander, B. G. *Staring into Chaos: Explorations in the Decline of Western Civilization*. Dallas: Spence. 1998. 358.
20. Ibid., 361-362.
21. Bell, Daniel A. and Hahm, Chaibong (Eds.). *Confucianism for the Modern World*. New York: Cambridge University Press. 2003. 1.
22. Ibid., 2-3.
23. Ibid., 7-9, 18.
24. Ibid., 13.
25. Ibid., 15.
26. Ibid., 17.
27. Ibid., 182.
28. Ibid., 185-186.
29. Ibid., 197.
30. Ibid., 199.
31. Ohlemacher, Stephen. "Economic Disparities Lingering Among Races, Census Data Show." The Associated Press. 14 November 2006.
32. Brander, B.G. *Staring into Chaos*. 265.

Chapter 14
The Mechanistic-thermodynamic Paradigm: A Unifying Perspective of Civilization Prosperity and Failure

> "Human societies and political organizations, like all living systems, are maintained by a continuous flow of energy. From the simplest familial unit to the most complex regional hierarchy, the institutions and patterned interactions that comprise a human society are dependent on energy. At the same time, the mechanisms by which human groups acquire and distribute basic resources are conditioned by, and integrated within, sociopolitical institutions. Neither can exist, in a human group, without the other, nor can either undergo substantial change without altering both the opposite member and the balance of the equation. Energy flow and sociopolitical organizations must evolve in harmony."
>
> Joseph Tainter[1]

The objective of this work has been to utilize fundamental science-based principles, relationships, and variables that permit the formulation of a unifying perspective and rationale for the periodicity of cultural growth and deterioration of the Graeco-Roman Western civilizations from B.C. 1600 to the beginning of the twenty-first century. Mother Nature's design of the universe established controlling and restrictive principles for the necessary consumption of energy by all physical, chemical, and biological processes that may exist in nature or be created by people in the pursuit of cultural survival and prosperity. Thus, scientific laws mandate specific energy requirements for a society's processes of existence and each of their step-by-step mechanisms. This *mechanistic-thermodynamic paradigm* provides conceptual glue for the coalescence of accepted sociological, economic, and political thought into a unified perspective of two cycles of the "triple rhythm" of prosperity to failure of the Graeco-Roman Western civilizations. These scientific principles (i.e., Mother Nature's management rules) are the enablers for all of society's efforts and processes designed to advance humanity. Thus, the purpose of this final chapter is to more fully develop the concept of the mechanistic-thermodynamic paradigm utilized in previous chapters.

Joseph Tainter, in *The Collapse of Complex Societies,* recognizes the underlying role of sociopolitical institutions and the fundamental nature of mechanistic, energy-driven economic activity in the functioning of a social order, specifically cultural collapse. "Human societies, ... like all living systems, are maintained by a continuous flow of energy.... The mechanisms by which human groups acquire and distribute basic resources are conditioned by, and integrated within, sociopolitical institutions. Neither can exist ... without the other, nor can either undergo substantial change without altering both the opposite member and the balance of the equation."[2] Frederick Soddy also addresses a society's need for a "continuous flow of energy" as he references John Ruskin's nineteenth-century definition of wealth as a "dynamic flowing stream" of a finite physical form.[3]

The interrelated mechanistic and thermodynamic aspects of this paradigm jointly establish ironclad, science-based rules and protocols that regulate a civilization's socioeconomic growth as well as its degradation. Consequently, permissible and prohibited mechanisms or pathways exist as a society attempts to complete the difficult journey toward the goals of survival and economic prosperity. Obviously, not all processes that people aspires to advance are permitted, due to Mother Nature's energy restrictions and mechanistic limitations. Based on this paradigm, the evolution of a civilization's progression may be viewed, in a mathematical sense, as a continuous multivariable function of numerous, complex socioeconomic and political mechanistic processes that conform to scientific principles, including the laws of thermodynamics. Importantly, these mechanisms, while requiring wealth-energy resources, are also subject to management and thus the behavioral elements of human wisdom, values, limitations, and eccentricities. Such behavioral aspects of sociocultural transitions are addressed, for example, by Sorokin's modeling of human values and sociopolitical behavior.

Conceptually, cultural development may be envisioned as involving an extremely large, complex system continuously consuming new materials and energy while creating additional subsystems, each possessing specific mechanisms and properties, all subject to human behavior and to nature's rules. Consequently, based on human priorities, decisions, and actions; probability; and Mother Nature's restrictions, cultures may grow, decay, stagnate, ossify, or collapse. These processes, constituting an evolving culture, may exhibit a net positive or negative rate of growth (i.e., the summation of numerous, interrelated, but yet mostly independent social, cultural, political, and economic variables or factors). Matthew Melko refers to such variables as "fluxes"; thus, the resultant summation of all cultural factors or "fluxes" will vacillate over time, giving rise to periods of cultural progress and deterioration.

In Melko's book *The Nature of Civilizations,* Chapter 1, "The Need for a Model," he notes, "Spengler, Toynbee, Sorokin, and Kroeber have each offered impressive systems … that have come to be called 'civilizations.'" But, in contrast to these authors' historical and sociological approaches, Melko's "integration" model also includes broad-based mechanistic interrelationships between politics and economics while dismissing the narrow "cause-and-effect political history" methodology of other authors. Melko's stated purpose is the creation of a "model of civilization":[4]

> **Civilizations, like symphonies, retain characteristic patterns notwithstanding fluxes of formation, disintegration, and reconstitution.... There are periods of high integration and periods of low integration. In the latter there are both disintegrating and formative forces in action, and only in retrospect can you be certain which proved dominant.**[5]

It is suggested that Melko's depiction of "characteristic patterns" is essentially that of the "mechanistic" portion of the mechanistic-thermodynamic paradigm. Melko's model represents civilization development and human history as a sequence of potential transitional patterns, dependent on a vast array of variables including elements of human behavior. However, while his model cites "formative forces," it does not address the necessary energy counterpart that constitutes the driving force for the mechanisms of the "integration" processes he defines and utilizes. Instinctively, if not scientifically, Melko's mechanistic integration model of civilization development must rely on inherent energy requirements, availability, and processing techniques, which must conform to the laws of thermodynamics.

Melko cautions that "determinism" or human control of destiny "frequently encourages attitudes of pessimism and hopelessness" and that concern for "historical inevitability," as found in the comparative history literature, is "possibly excessive." However, he notes that "it would be surprising, after all, if writers comparing conditions and forces were not inclined to emphasize developments beyond man's control." While Melko apparently has in mind the behavioral limitations of man and chance occurrences of man and nature, the restrictions of thermodynamics could have appropriately been included as also being "beyond man's control." In addition, he considers that his concept of societal "reconstitution" eliminates the possibility of "a civilization hurtling inevitably to its doom."[6] On the other hand, Melko's civilization model ultimately and *inevitably* relegates a civilization to "ossification" or to deterioration into a feudal state, yet always with the possibility, if not a high probability, of "reconstitution" or rebirth. What is *inevitable* is the thermodynamic directionality and spontaneity of the second law of thermodynamics, which inherently controls the mechanisms of any such model.

Thus, the word "inevitability" is used here in the thermodynamic sense to signify that while society, for better or worse, provides a management function for the utilization of its wealth-energy capital, the principles of thermodynamics will ultimately, albeit perhaps in the very long run, control socioeconomic outcomes. *Inevitably,* some form of a "failed civilization" will occur. Specifically, *failed* is defined, in accord with Melko's model, as either becoming "ossified" (a dead-end status beyond survival) or deteriorating to a less integrated state that may someday be "reconstituted," but nevertheless has deteriorated and failed as a more advanced socioeconomic system.[7]

The concept of "reconstitution," or the rebirth of a dying civilization, is found in the work of many authors, and this quotation of Will Durant is one example:

> **Civilization does not die, it migrates; it changes its habitat and its dress, but it lives on. The decay of one civilization, as of one individual, makes room for the growth of another;**

> **life sheds the old skin, and surprises death with fresh youth. Greek civilization is alive; it moves in every breath of mind that we breathe, so much of it remains that none of us in one lifetime could absorb it all. We know its defects—its insane and pitiless wars, its stagnant slavery, its subjection of woman, its lack of moral restraint, its corrupt individualism, its tragic failure to unite liberty with order and peace. But those who cherish freedom, reason, and beauty will not linger over these blemishes. They will think of Greece as ... our nourishment and our life.[8]**

While Durant recognizes the characteristic "defects" of cultural decay, he also expresses the view that a civilization "does not die" but "changes," it "lives on" to "make room for the growth of another, ... surprises death with fresh youth." Thus, the foundations of today's cultures have been established by the experiences of past civilizations. Also, he is correct that the "Greek civilization is alive" in the sense that the *cultural legacy provides nourishment* to twenty-first-century life by way of its philosophy, architecture, literature, and art. However, when a civilization begins to decay and undergoes disintegration, it loses much of its basic vitality, inspiration, and creativity, the stimulus for cultural growth as well as for the pursuit of dreams and aspirations. Thus, a civilization "lives on" after collapse in the sense that its legacies of thoughts, words, deeds, and visual images are utilized by future generations to enjoy and build upon. However, as Latin is referred to as a dead language, the cultures of ancient Rome and Greece ceased to grow in the socioeconomic and thermodynamic sense. Depending on one's point of view, these cultures either ceased to exist or, by virtue of their cultural legacies, contributed to the pillars of future generations. Either way, history demonstrates that the ultimate phase of a maturing civilization is one of cultural deterioration and, at best, ossification based on human limitations, capabilities, and Mother Nature's resources and scientific laws.

Numerous authors, cited in previous chapters, have made the observation that all known past civilizations have been unable to sustain continuous growth and ultimately have succumbed to cultural decay.

> **No social trend is eternal. Just as civilization never has enjoyed unlimited ascent, neither can its decline go on forever. Modern humanity might be witnessing the decline**

Wealth, Energy, and Human Values

of a world civilization, but too we might be seeing a dual transformation: the clearing away of the old social order so that fresh creative life can blossom in its place.[9]

Deteriorating civilizations have three potential fates: complete collapse, ossification, or rejuvenation, which will also ultimately lead to another phase of cultural degradation. Typically, authors speculate on the reasons for observed cultural deterioration being based only on flawed human behavior. Some admit to being clueless in explaining why all past civilizations observed the same sequence from genesis to growth, breakdown, stagnation, and deterioration while contributing to a multi-rhythm pattern of human history. Some authors suggest there must be some nonbehavioral, related missing link to such a universal phenomenon occurring in very different eras under similar circumstances.

This work has attempted to demonstrate that this missing link is the commanding influence of thermodynamics over the mechanistic steps of the almost infinite number of societal processes (i.e., the "fluxes of formation, disintegration, and reconstitution"), all of which require wealth-energy capital. Thermodynamics necessitates that the consumption of energy will produce societal entropy in many forms including burned fuel residue, an economically disadvantaged segment of society, societal complexity, bureaucracies and disorder, and political abuses by persons intending to gain a wealth and power advantage over others.

Thus, this chapter will reformulate Matthew Melko's model of civilization development by interjecting thermodynamics into his mechanistic model or "patterns of a civilization, ... an inevitable progression from feudalism to empire." Inherent in this reformulation is that the second law of thermodynamics, or the entropy concept, specifies and mandates explicit process *directionality*, "an inevitable progression" from an orderly state to a more disorderly state, which Melko's model follows.

Fundamentally, Melko's model is based on Sorokin's concept of "transitional periods" that simultaneously involve both cultural growth, which increases societal integration, and cultural decay, which lowers the degree of integration. Such transitions occur when a society becomes energized and invigorated by bursts of progress via creative leadership, overcoming major challenges, new technologies, and/or the unexpected acquisition of new wealth-energy capital. Thus, new leaders, new opportunities, new creativity, and new vitality often emerge to produce a blip in the rate of cultural growth and social advancement. The history of mankind demonstrates the human ability to overcome hardship and to survive, adapt, and prosper in a new version of a former social order. Failure to become energized and creative and to stimulate and accomplish such transitions is equivalent to an inability to evolve, to create opportunities, and to overcome unresolved challenges, all of which lead to cultural stagnation. Thus, periods of transition are considered necessary for achieving cultural progress, leading to a higher degree of societal integration or crystallization; otherwise, disintegration of the system will occur. That is, each of Melko's crystallized phases must lead to a subsequent transition, or else the culture becomes "institutionalized."[10]

"Institutionalized" is Carroll Quigley's term indicating that growth is not occurring; an acute crisis is in progress and, if not resolved, is capable of contributing to disintegration of the socioeconomic system.[11] Melko considers, for example, that a developing society undergoing crystallization will eventually reach a stage that he describes as the "limits of its possibilities" and therefore "must lead to another transition unless patterns become institutionalized and their forms ossify and become meaningless."[10] That is, a fully crystallized social order has achieved a status of full potential, is completely integrated or institutionalized, and has achieved all that is possible, given existing resources and conditions. Subsequently, available options include a reversion to a more primitive, less integrated society; advancement via a burst of significant cultural advancement; or becoming a failed state (i.e., ossified).

All civilizations eventually outlive their developmental potential, and this will mean inevitable loss of control of some areas and the possibility for the development of new civilizations with a minimum of pseudomorphosis.[12]

It is reasonable to assume that at some point in time, certain functions of a civilization may have reached the limit of what is possible to achieve while others continue contributing to cultural progress. The specific inadequate functions may become the basis for serious socioeconomic and political issues and may seriously diminish the net rate of culture growth. Furthermore, the same positive and negative cultural growth factors influence a civilization's struggle to make the transition to a more advanced phase and avoid becoming a stagnant society. The requirement of cultural change is eternal.

It is instructive to examine the characteristics of Melko's model at the critical balancing point between being able to achieve a transition to a more advanced social order and becoming ossified via inactivity. First, failure of a socioeconomic system to attain expected positive outcomes and to resolve nagging, long-term issues eventually deflates a society's faith, loyalty, and support of leadership and the potential for achieving an acceptable level of cultural advancement and satisfaction. If existing institutions fail to satisfy society's needs and ambitions, they are eventually rejected by the population and replaced with newly created institutions, perceived as more appropriate to accomplishing desired outcomes. Inadequate leadership, failed policies, cultural instability, and economic failure may sufficiently energize a population to achieve a quantum leap of progress, seeking a transformation to a superior social order with a greater degree of cultural integration. Otherwise, *cultural degradation will gradually take its toll as societal deterioration expands, the reverse of cultural integration, crystallization, and stratification.*

Melko illustrates a fully integrated or crystallized feudal system failing to make the transition to a state system and returning "to the previous situation" or being "resolved" when it suffers a "reversion" due to a failed political system. The insecurity of the political system is replaced by the promise of new direction and security from radically new leadership or from a new spirituality or religion. Alternatively, a transition from a feudal to a state phase is successful when subsistence farming disintegrates as the town concept emerges or when a challenge to existing authority produces a new, more appropriate form of governance.[13] Toynbee views existing institutional structures of a society reacting to new social forces or stresses as producing one of three possible transitions or outcomes: "Whenever the existing institutional structure of a society is challenged by a new social force, three alternative outcomes are possible: either a harmonious adjustment of structure to force, or a revolution, or an enormity. It is also evident that each and all of these three alternatives may be realized in different sections of the same society."[14] He cites industrialism and democracy as examples.

At the personal level, Toynbee views four "ways of life" for individuals to emotionally escape the disorder and uncertainty of transitional periods. Archaism, futurism, detachment, and transfiguration represent retreating to a past happier sense of reality, getting lost in a wishful future, ignoring the present circumstances, and becoming totally consumed in the serenity of spirituality.[15] A new religion or sense of spirituality may provide a more secure pathway, as illustrated by Christianity arising from the decaying Roman civilization. All of these options are detachments from reality, endeavoring to avoid the stress and insecurity of the realities of individual and cultural transitions.

In more advanced civilizations, the complexities of life and the effort, patience, and ability required to absorb and comprehend the details of significant societal issues result in a large portion of the population being unable or unwilling to become intellectually engaged in sociopolitical issues and processes. The ability to cope with large-scale complexity has its human limitations. It is often easier and more comfortable for the individual to support a superficial but emotionally appealing ideology or personality or to adopt a new system of values that safeguards one's self-interests. For Melko's model, Sorokin's continuum of shifting human behavioral traits and cultural mentalities is an important element in the determination of whether a fully integrated or crystallized society is able to make a constructive transition (e.g., from a feudal to a state phase), or suffers "reversion."

Sorokin points out the dilemma for a society experiencing a deteriorating cultural condition to either continue its "dangerous drift, dry up and perish, or to make a great effort and restore a fuller and more genuine truth and system of values. Such a restoration means a reintroduction of the other systems of truth, reality, and value."[16] That is, cultural survival and prosperity necessitates modification of society's system of values, renouncing its current sensate behavior, and adopting a more ideational way of life. Such a shifting of cultural values, attitudes, and behavior is capable of recapturing long-lost characteristics and principles of more primitive societies, which place greater significance on a more altruistic and spiritual philosophy. Sorokin describes the "rebirth and regeneration" process "to restore a fuller and more genuine truth and system of values" as "the experience through which humanity passes: crisis, ordeal, cleansing, grace, resurrection." A new or renewed spirituality and value system that provides emotional, economic, and social stability and a living philosophy is the key for personal "cleansing" and "grace" and thus a pathway to "resurrection."[17]

Sorokin describes the "inadequacy of each system of truth" from ideational to idealistic to sensate and the need to avoid "a growing domination of one of the systems ... carried away from the true reality and from the real knowledge of it.... Adaptation becomes less and less possible."[18] Thus, change is a necessity for cultural growth and survival, as well as rejuvenation.

Sorokin notes:

> **"The exclusively theologico-supersensory mentality of medieval culture that emerged and developed as a remedy to the hollow Sensate culture of the late Graeco-Roman period, after several centuries of domination, also dried up, failed, and buried itself in the catastrophes of the end of the Middle Ages. So did the one-sided rationalistic mentality of the culture of the sixteenth to the eighteenth centuries [the mentality of the Renaissance and the Enlightenment. Mankind appears to consistently obliterate the] boundary line between truth and falsehood, reality and fiction, validity and utilitarian convention, ... mere expediency."[18]**

Self-serving socioeconomic objectives dictate and alter human behavior, which in turn establish political and governmental attitudes, values and policies.

Despite the difficulties of achieving such transitions, some deteriorating cultures, as for example, found within the Graeco-Roman Western civilizations, successfully rebounded from cultural disintegration to become "rejuvenated." They "continue to live and to pass through the recurrent rhythm studied; the others either perish and disappeared, or were doomed as a stagnant, half-mummified existence, with their hollow and narrowed truth, reality, and value becoming a mere 'survival' or 'object of history' instead of being its creative subject. Such cultures and societies turn into mere material for other—more creative and alive—cultures and societies."[19] Sorokin's view of the alternatives for a highly stressed culture is consistent with Melko's mechanistic options of crystallizing, transitioning, disintegrating, ossifying, or collapsing.

Melko considers a civilization developmental sequence as evolving through *phases,* from Primitive to Feudal, to State, to Imperial. Each *phase* may potentially undergo *stages* of crystallization, transition, disintegration, or ossification, each representing a different mechanistic process.[20] Consequently, the Melko model of civilization development, an expansion of Sorokin's hypothesis of societal transitions and crystallized phases, may be interpreted as a thermodynamic process.

The following summary represents a modified and expanded version of Melko's model with the addition of wealth-energy resource considerations:

1. Each of the Feudal, State, and Imperial *phases*, regardless of its particular *stage*, will experience competing forces of cultural growth and decay, simultaneously creating elements of both order and disorder in a continuous disruption of society's attempts to achieve dynamic equilibrium. At any given time, these forces promote both societal integration and disintegration related to socioeconomic and political functions that consume wealth-energy resources and generate complexity and dysfunction.

2. Normal societal dynamics create higher levels of social, economic, and political integration, which is also referred to as an increase in the degree of crystallization representing greater cultural complexity and orderliness, as is found in increasing large bureaucracy. This is accomplished by the continuous acquisition and consumption of resources including human labor to support the various socioeconomic and political systems in accordance with the principle that a society constitutes an open thermodynamic system. Note that *the creation of greater organizational and functional orderliness within a society* also produces a by-product of even greater disorder: a net increase of societal entropy.

3. A *transitional stage* is an unstable condition of a specific *phase* (e.g., *the Feudal phase*), attempting to overcome its "time of troubles" to migrate to the more highly integrated and stable *State phase*. Conversely, the given transitional stage may revert to the less integrated phase (e.g., return to a less mature Feudal phase).

4. Each transitional stage evolves into a more advanced, highly integrated phase, or else the social order is assumed to revert, ossify, or crease to exist. Ruston Coulborn is credited with establishing the

Wealth, Energy, and Human Values

concept of such a "reversion" and also that a primitive society has never developed into a State phase without first experiencing a Feudal phase, which is incapable of returning to a Primitive phase.[21]

Importantly, Melko's model illustrates the *scientific concepts of both irreversibility and randomness* or probability associated with the processes of nature and society. To the chemist, Melko's "political and economic pattern" is a mechanistic pathway analogous to the fundamental molecular reactivity concept of the physical, biological, and chemical world (i.e., a thermodynamic process involving sequential steps of a mechanistic nature). His civilization development model is represented by the following equation:[22]

$$P \rightarrow F \Leftrightarrow S \Leftrightarrow I \rightarrow F \qquad \text{Equation 14-1}$$

The transition of a social order in a Primitive phase (**P**) proceeds unidirectionally to a Feudal phase (**F**), which becomes in dynamic equilibrium ($\Leftrightarrow$) with a State phase (**S**), which may, in turn, become in dynamic equilibrium with an Imperial phase (**I**) that ultimately deteriorates into a Feudal phase. Each phase may also result in dead-end ossification (**O**).

Melko's model defines a condition or *stage* where a particular *phase* is fully developed (i.e., fully integrated), relative to its potential, and can either make a transition to the next higher phase of development or undergo reversion to a lesser state of sophistication. This condition is referred to as a "highly integrated" or "fully crystallized" stage of a particular phase; for example, **SC** represents the fully crystallized State phase that must successfully make the transition to the Imperial phase, revert to as lesser phase (i.e., Feudal), or become ossified. Melko presents a mechanistic analogy for these sequential phases shown in Equation 14-1 in their crystallized forms, which are denoted by **C**:

> It [*the mechanism*] is rather like a pin-ball machine. The ball, when shot, has a universe of possibilities ... but as soon as it passes through the first slot on its downward curve, the possibilities are narrowed. Thus, having reached FC, it can no longer return to PC. But the machine also has rubber bumpers, so that the ball can reverse its course. Having reached slot SC, it can no longer return to PC, the initial ejecting slot, which has been closed off by a one-way valve.[23]

Melko's pinball machine "mechanism," **PC→ FC ⇔ SC,** containing elements of directionality, probability, and random and restricted motion associated with his "periods of transition," is analogous to the transition state theory of chemical kinetics. Consider a chemical gas phase reaction of a reactant **A** forming product **B**. In molecular reactivity parlance, an analogous molecular reactant species **$\underline{A}$** becomes a highly energized or activated "transition state" molecule **$\underline{A}^{\ddagger}$** that either transitions into the molecular product **$\underline{B}$** or becomes deactivated and returns to the original **$\underline{A}$** status. This concept is fundamental to the kinetic theory of gases and the representation of the distribution of gas molecule velocities and kinetic energies within a confined volume. The mechanistic representation is as follows:

$$\underline{A} \Leftrightarrow \underline{A}^{\ddagger} \Leftrightarrow \underline{B}$$

The transition state, **$\underline{A}^{\ddagger}$**, represents a high-energy, unstable molecular species that must react in a short time frame due to its instability, a condition analogous to a culture's higher degree of "crystallization" or "integration," existing in an unstable "time of troubles." Both the agitated cultural state and the high-energy molecular species represent unstable conditions with finite probabilities for various transitions, including reversion to its initial status. These mechanistic and associated thermodynamic principles that are routinely applied to physical and chemical events apply equally to sociocultural processes, all of which require wealth-energy resources.

The details of Melko's pattern of civilization development[24] begin with a primitive society **P** undergoing a process of cultural growth and integration. When a highly "crystallized" Primitive phase is achieved, denoted by **PC**, the realities of inevitable cultural change include ossification, **O**, collapse, or achieving the transition state **(PT)‡** on the pathway to the Feudal phase, **F**. This sequence is schematically represented as follows:

$$\text{Collapse} \leftarrow \text{PC} \Leftrightarrow (\text{PT})^{\ddagger} \rightarrow \text{F} \rightarrow \text{FC} \qquad \text{Equation 14-2}$$
$$\downarrow$$
$$\text{O}$$

Melko's model or vision of civilization development is consistent with the mechanistic-thermodynamic paradigm. That is, if a civilization does not maintain a net positive rate of cultural growth over the rate of decay ($R_g > R_d$), as defined in Chapter 9, socioeconomic stagnation and cultural arrest will occur. This unavoidable outcome of gradual cultural disintegration ultimately leads to ossification or collapse if improved socioeconomic conditions and a prolonged net positive rate of cultural growth do not emerge from a rejuvenated cultural environment. Thus, a civilization that is not continuously and progressively evolving to maintain socioeconomic growth will ultimately cease to exist in a viable form, if for no other reason than reaching the limitations of its inherent and acquirable wealth-energy resources.

It is noted that **PC** and **(PT)**‡ are represented as being in a dynamic equilibrium between cultural growth and decay (i.e., integration and disintegration, denoted by the double arrow ⇔). Thermodynamics establishes the spontaneous ultimate mechanistic direction and indicates the path to a more highly crystallized stage of a given phase (e.g., **FC**). Also, once a society reaches the Feudal phase, the model does not permit a return to the Primitive phase, which is justified by Melko and others based on such a transition having never been detected.

Under favorable socioeconomic conditions, the Feudal phase, **F**, experiences a net positive rate of growth, increasing cultural advancement and societal "crystallization," thereby reaching a limiting degree of integration, represented by the unstable transition state **(FT)**‡. The possible outcomes from this fully crystallized form **FC** are the transition to the State phase, **S**, return to a less integrated Feudal state, and ossification, **O**.

$$\mathbf{F} \Leftrightarrow \mathbf{FC} \Leftrightarrow \mathbf{(FT)}^{\ddagger} \Leftrightarrow \mathbf{S}$$
$$\downarrow$$
$$\mathbf{O}$$

Equation 14-3

Continuing cultural advancement and integration will propel the culture to the next sequence in Melko's model, the fully integrated or crystallized State phase, **SC**. This stage evolves through the transitional form **(ST)**‡ to the Imperial phase, **I**, becomes ossified, or reverts to the Feudal phase, **F**.

$$\mathbf{F}$$
$$\uparrow$$
$$\mathbf{S} \rightarrow \mathbf{SC} \Leftrightarrow \mathbf{(ST)}^{\ddagger} \Leftrightarrow \mathbf{I}$$
$$\downarrow$$
$$\mathbf{O}$$

Equation 14-4

Finally, the Imperial phase, **I**, upon achieving full integration, **IC**, becomes the unstable form **(IT)‡** and either disintegrates, reverting to **I**, ossifies, or collapses to a Feudal stage, **F**.

$$IC \Leftrightarrow (IT)^{\ddagger} \rightarrow F$$
$$\downarrow$$
$$O$$

Equation 14-5

All possible transitions in the sequence of civilization development from the Primitive phase, **P**, through the Imperial phase, **I**, may be represented as the following:[25]

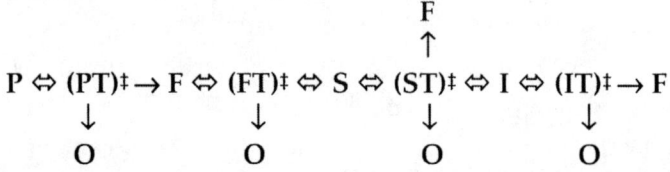

Equation 14-6

Note that both the State and Imperial phases may disintegrate into the Feudal phase and that all phases may ossify.

From the mechanistic-thermodynamic perspective, Melko's mechanistic model would ultimately produce only a feudal society, but recall that thermodynamic outcomes are time independent, thus the time frame for reaching the ultimate feudal society is irrelevant and may theoretically approach infinity. Thermodynamically, one could argue that at some point in time, the Earth's increasing population will reach the point where insufficient total energy is available or inadequately distributed, thus precipitating large-scale cultural deterioration, resulting in a proliferation of feudal environments. No civilization has ever experienced continuous growth; that is, all known civilizations have ultimately suffered cultural disintegration after achieving their peak of socioeconomic prosperity.

Various assumptions are made in the application of mechanistic-thermodynamics principles to civilization maturation that primarily affect socioeconomic systems but which, in turn, also dictate political and governance structures, functions, and behavior. These include the following:
- A society's social, economic, and political processes require wealth-energy resources and provide numerous opportunities for a broad range of cultural successes and failures.

- Each process will have a specific, complex pathway or mechanism as well as its own set of conditions and material requirements, with outcomes greatly dependent on human behavior and random events.
- A large number of societal variables or factors will influence these outcomes, such as the nature and quality of leadership, the intensity of common purpose within the population, the availability of wealth-energy resources, human values and intelligence, personal priorities, behavior and goals, and internal and external competition.
- The nature, intensity, and significance of these societal factors will be determined by the maturity of the society; that is, Feudal, State, and Imperial phases possess increasingly higher order socioeconomic and political capabilities, knowledge, and functional complexities.
- Over time, an infinite number of societal processes and events produce fluctuating rates of cultural growth, with peaks and valleys reflecting variations in socioeconomic advancement and regression.
- Ultimately, the rate of cultural decay R_d will exceed the rate of cultural growth R_g, resulting in a civilization entering a period of cultural degradation, leading ultimately to ossification, decline to a feudal state, or rejuvenation through a "rebirth" process that achieves and maintains the condition of $R_g > R_d$.

Having incorporated Melko's mechanistic model within the context of the mechanistic-thermodynamic perspective, it is now instructive to address the wealth-energy capital component of civilization development with emphasis of the "inevitability of failed civilizations."

Joseph Tainter points out that societal collapse has long been of great scholarly interest, but also a "mystery for historians and social scientists.... Explanations of collapse have tended to be ad hoc ... so that a general understanding remains elusive."[26] Tainter discusses minor differences among the conflict, integration, and complexity models of collapsing societies. He notes that *societal complexity is a response, a solution to the conflicts of class competition as well as to increasing cultural integration designed to meet societal needs and expectations.* In order to become more responsive to the forces of change, it is necessary to create a more fully integrated and complex society that inherently requires a continually increasing energy investment.

> **Human history as a whole has been characterized by a seemingly inexorable trend toward higher levels of complexity, specialization, and sociopolitical control, processing of greater quantities of energy and information, formation of even larger settlements, and development of more complex and capable technologies.**[26]

Thus, it is apparent that theories of cultural prosperity and failure must inherently assume that a society will attempt to continuously increase its complexity and degree of crystallization and integration in order to satisfy its aspirations toward "progress". Such models descriptively identify societies maximizing organizational and functional disorder and dysfunction (e.g., Toynbee's "Breakdown" stage and Melko's "limitation of man"). Such cultural breaking points have been described by Sorokin: "Sensate culture in its overripe and rotting condition has debased mankind and now is destroying him and his environment as well"[27] and by Spengler as a "period of contending states." Thus, there appears to be agreement that some limiting condition of extreme cultural deterioration results from the creation of an ultimate, unsustainable level of societal complexity, dysfunction, and instability.

Tainter emphasizes and incorporates the wealth-energy perspective into his civilization development model and, combined with Melko's model of mechanistic patterns, represents the essential components of the mechanistic-thermodynamic perspective presented in this work. Additionally, Sorokin and others provide the necessary sociological components that relate human values, motivations, and behavior to the creation, implementation, and management of cultural processes. Importantly, Galbraith's analysis of economic theory and practice, as discussed in previous chapters, provides a valuable understanding of the social and economic instability that has resulted during cultural maturation.

Thus, a successful, unified model of civilization development must provide for social, economic, and political mechanisms, policies, and practices with their appropriate wealth-energy resource ramifications, in addition to the behavioral characteristics of managing a civilization's processes of existences. Such a unified model of civilization development supports Tainter's view that "energy flow and sociopolitical organizations must evolve in harmony."[28] Accordingly, Tainter's perspective will be incorporated with Melko's transitional mechanism model of civilization development and with Sorokin's model of the continuum of cultural mentalities.

Tainter lists eleven "explanatory themes" that represent "various approaches" found in the literature to explain civilization collapse. He considers one of these "economic explanations" as "structurally and logically superior to the others," but notes: "While economic explanations are not universally accepted in the social and historical sciences, such scenarios remedy the logical deficiencies of the other approaches.... The major drawback to economic explanations ... is failure to develop an explanatory framework that is globally applicable." He considers the other ten "approaches" as "not necessarily wrong or misguided, ... simply inadequate as presently formulated."[29] It is suggested that the "framework" he seeks is the mechanistic-thermodynamic paradigm that provides a restrictive wealth-energy-based scientific context for selective linkages among the history, sociology, and economics of cultural development.

Tainter's approach of analyzing the relationships among economic productivity, sociopolitical systems, and energy consumption provides a more comprehensive view of cultural collapse. More specifically, he searches for "a more general explanation of collapse" that links the "energy flow required to maintain a sociopolitical system" to the "energy harvested by a human population" in its process of "cultural evolution."[30] The "general explanation" sought is the science of thermodynamics. Importantly, at any given time, the amount of energy flow required to maintain a stable sociopolitical system is proportional to the degree of new complexity currently being created plus the amount of accumulated, residual cultural complexity previously generated and being utilized.

> **More complex societies are more costly to maintain than simpler ones, requiring greater support levels per capita. As societies increase in complexity, more networks are created among individuals, more hierarchical controls are created to regulate these networks, more information is processed, there is more centralization of information flow, there is**

> increasing need to support specialists not directly involved in resource production, and the like. All of this complexity is dependent upon energy flow at a scale vastly greater than that characterizing small groups of self-sufficient foragers or agriculturalists.[31]

Tainter develops a foundation for his model of civilization collapse based on inherent increases in cultural complexity arising from cumulative organizational and functional issues, costs, conflicts, and stresses of normally developing societies. The creation of solutions to these cultural maladies requires societal expansion, which in turn produces new complexities that contribute to spiraling increases in wealth and energy consumption. Thus, a developing society is continuously and increasingly "investing in complexity," but each response to a challenge or opportunity creates additional issues, setting the stage for even more costly investments in complexity, ... and the vicious cycle continues ad infinitum.

Consequently, as a civilization's processes of cultural integration mature and expand, its related wealth-energy requirements increase exponentially, usually devouring an increasingly larger percentage of available national wealth just to support its continuously increasing, self-induced cultural complexity. Accordingly, fewer resources will be proportionally available over time for investment in the generation of new wealth-energy capital, thus creating the necessity for a continuous influx of larger amounts of resources just to maintain the status quo social and economic stability.

This concept is defined by Tainter as the *marginal return on investment in cultural complexity*, the ratio of productivity relative to cost of the investment. He notes, "At some point in the evolution of a society, continued investment in complexity as a problem-solving strategy yields a declining marginal return."[32] Achieving a lower return on investment while increasing the total cost of societal advancement is financially incompatible with long-term financial solvency and typically results in the creation of significant debt. This condition corresponds to an era when a civilization is having great difficulty maintaining a positive rate of socioeconomic growth due to inadequate wealth-energy capital and great debt. The rate of cultural deterioration increases to the point that it approximates the rate of cultural growth, a precursor to cultural decline. Thus, the net effect of a social order continually increasing its cultural complexity is to eventually deprive the socioeconomic system of wealth-energy resources sufficient for the continued advancement and maintenance of the society's most vital functions.

> **There is in complex societies a recurrent and seemingly inexorable trend toward declining marginal productivity in hierarchical specialization.**[33]

Accordingly, in order for a civilization to maintain a constant rate of development, an exponential growth in wealth and energy consumption is required to compensate for the spiraling costs of accumulating cultural complexity *and the declining marginal return on its investment*. Past great civilizations found the never-ending challenge of acquiring such nonlinear, incremental wealth-energy capital to be insurmountable. Even for the most prosperous, knowledgeable, and powerful civilizations of their era, the outcome was ultimately cultural dysfunction and decline. The American culture of 2009 is a prime example.

Tainter concludes: "Complex societies with large, well-developed economies have historically been able to sustain only rather inferior rates of economic growth."[34]

> **As the marginal return on complexity declines, complexity as a strategy yields comparatively lower benefits at higher and higher costs. A society that cannot counter this trend, such as through acquisition of an energy subsidy, becomes vulnerable to stress surges that it is too weak or impoverished to meet, and to waning support in its population. With continuation of this trend, collapse becomes a matter of mathematical probability, as over time an insurmountable stress surge becomes increasingly likely. Until such a challenge occurs, there may be a period of economic stagnation, political decline, and territorial shrinkage.**[35]

A civilization's economy that is lacking adequate wealth-energy capital and approaching stagnation will be further weakened by people abandoning the current leadership and prevailing cultural vision in order to seek greater security in a new pathway to survival and prosperity. While many important socioeconomic elements may assist cultural rejuvenation, new sources of wealth-energy capital, whether from newly discovered or conquered sources or modern innovations, must be acquired in order to stimulate the economy and provide people with inspiration and an upbeat vision of the future.

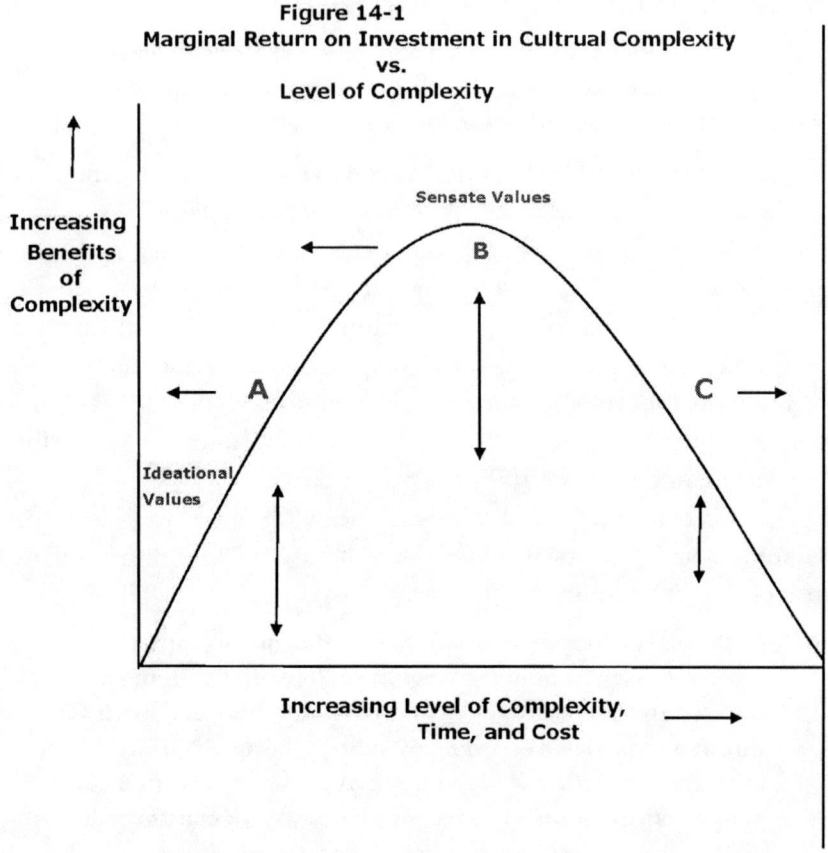

Tainter presents a "realistic expression of the economic history of some societies" that "emphasizes the recurring problem of marginal decline," including the Roman Empire and the United States since Colonial days. [36] Figure 14-1 is a generalized representation of such profiles for major civilizations. Each demonstrates an initial era of rapidly increasing marginal return or *benefits from increasing levels of cultural complexity*. However, at some point in time, represented by **B** in Figure 14-1, further investment in complexity, while still yielding returns, begins to suffer a *declining rate of marginal return*. That is, the slope of the plot begins to decline and approach a minimum at **B**. This era from **A** to **B** illustrates the beginning of a significant cultural accumulation of the seeds of ultimate degradation while still being economically productive. Point **B** represents Toynbee's cultural Breakdown stage, leading to socioeconomic stagnation. Tainter attributes the rapid increase in the initial rate of marginal productivity for each culture to an "energy subsidy" and/or a "technical innovation."[37] It is noted that at point **C** of Figure 14-1, the social order has a relatively much higher level of complexity and cost compared to an earlier time **A** but with a declining level of benefits.

Tainter emphasizes that the issue of an inherent, ultimately declining marginal return has been a "recurring problem" for the great cultures. The Roman Empire initially benefited from increasing complexity from 300 B.C. until peaking at A.D. 200, then suffered declining benefits at greater costs beginning about A.D. 400. Colonial America generated extremely rapid productivity from about 1700 until peaking about 1875, remaining constant until beginning its period of reduced benefits from increasing complexity about 1950.

It is noted that the profile of Figure 14-1 and Tainter's designation of the cultural characteristics and functions for each segment, **A**, **B**, and **C**, coincide with Toynbee's model of cultural development: Stages I through IV, cultural Growth through Disintegration (Figure 8-1). Additionally, connecting multiple profiles of Tainter's "realistic expression of the economic history of various societies" as illustrated by Figure 14-1, one obtains Sorokin's familiar pattern of recurring cultural mentality for Graeco-Roman Western civilizations (Figure 8-2). Thus, Figure 14-1 corresponds to one cycle of the triple rhythms of "shifting mentalities," from Ideational to Sensate to Ideational,[38] where peak **B** could represent the dominant sensate mentalities for the Roman Empire from A.D. 200 to 400 or for Western civilization from 1875 to 1950.

Human behavior is relevant to the design and management of a society's social, economic, and political processes, whose existence and success depend on wealth-energy resources, subject to the restrictive laws of science and probability. Combined, Tainter and Melko provide insights into the energy and mechanistic aspects of civilization development that broaden and supplement Toynbee's and Sorokin's historical and sociological perspectives, thereby identifying a more inclusive, appropriate array of variables. Specifically, Tainter's energy-based "complexity model" and concept of marginal return on investment in cultural complexity, Melko's mechanistic model of cultural development, Sorokin's model of shifting cultural values, and Toynbee's model of "challenge through response to further challenge" represent the observed sequence of societal growth, breakdown, and disintegration phases. These models and concepts link wealth-energy resources and mechanistic processes principles, sociological thought, and historical data to provide a more unified understanding of the prosperity-to-failure sequence of great civilizations.

An increasing intensity of Sorokin's designated Sensate mentality[39] corresponds to Tainter's period of the increasing rate and cost of complexity as well as to Toynbee's transition from Growth to Breakdown stages. All three models exhibit similar cultural characteristics and properties during these periods of cultural development, including a reliance on wealth, energy, and human values and behavior. Tainter identifies the necessary wealth-energy capital component to fuel this period of rapid growth as "technical innovation or an energy subsidy … to raise marginal productivity." However, most importantly, he makes the argument that declining marginal productivity is an inherent, recurring issue, and even if a deteriorating culture is rejuvenated, productivity "will ultimately begin to decline again."[40]

The significance of Tainter's work relating economics, societal complexity, and resource depletion to cultural development is illustrated by the application of his model to the collapse of specific civilizations. He investigates three mature societies that have collapsed so as to determine if his "explanatory framework" is helpful "to understand collapse in actual cases." The Western Roman Empire is the most famous and complex of the three. Tainter suggests that both Roman success and failure are "readily understandable by investigating its marginal return on investment in complexity during the periods of its rise and its decline."[41] This application of Tainter's thesis of marginal return to the collapse of the Western Roman Empire emphasizes his economic and energy perspective. Additionally, Melko's mechanistic perspective will be incorporated into the discussion so as to include an overview of the mechanistic-thermodynamic perspective of this aspect of Roman history.

While recognizing that a multiplicity of factors contributed to the collapse of the Western Roman Empire, Tainter primarily focuses his attention on the social, political, and economic consequences of Rome's policy of expansion and the resulting economic benefits and problems it created. Conquered territories provided many forms of immediate and long-term resources. As the Roman Empire expanded and its investment in complexity dramatically increased, the loot from conquests increasingly paid for the rapidly increasing cost of imperialism. Essentially, Rome's economic base relied on the continuous acquisition of wealth from the conquests of foreign territories. However, Augustus, who reigned from 27 B.C. to A.D. 14, ended the policy of expansion, which also eliminated the flow of incremental wealth that Rome had come to rely upon.[42] At this point in Roman history, the "marginal return" on the investment in its vast, expanding, complex empire declined, as former sources of incremental wealth became nonexistent. Essentially, the Roman Empire was now on the pathway from **A** to **B** in Figure 14-1, but to complete its journey would require a few hundred years.

For the long term, Rome's policy of territorial expansion, wars, and conquest was not profitable and contributed greatly to its ultimate demise. Melko's model, reflecting Sorokin's transitional periods, would describe the Roman civilization's situation as having become more highly integrated, undergoing "crystallization," and reaching the "limits of its possibilities." It was therefore forced "to revert" to a less integrated phase (i.e., a more primitive society), the genesis of Western civilization. Tainter's thesis, consistent with Melko's transitional mechanistic model, is that Rome had become so heavily invested in cultural complexity (i.e., overly integrated) that it became wealth-energy resource deficient beyond its capability to secure adequate resources. At this point in time, Tainter views Roman civilization as having reached the "point of diminishing returns" in its investment of cultural complexity and now proceeded on the pathway from **B** to **C** in Figure 14-1. Both authors' models agree that under such conditions, socioeconomic failure is imminent: The empire was rapidly deteriorating. Melko's mechanistic model leads to a Feudal phase, while Tainter's energy-oriented perspective is that "collapse becomes a matter of mathematical probability, ... political decline and territorial shrinkage." Likewise, Imperial Spain, the British Empire, and the United States have followed the same sequence and exhibited similar characteristics and properties.

Adopting more primitive cultural characteristics and practices could theoretically reduce cultural complexity and increase the rate of marginal return (i.e., increase cultural growth with less social complexity and cost). This could be accomplished by the transition from a State phase to a Feudal phase. Alternatively, new sources of wealth-energy resources, new creative leadership, technological innovation, or a new spiritual or political ideology could contribute to a cultural growth spurt and rejuvenation. However, consistent with Melko's transition model, Tainter notes that: "the marginal cost of evolution to a higher level of complexity, or of remaining at the present level, is high compared with the alternative of disintegration."[43]

The Roman civilization, by virtue of its extensive, continuous imperialistic expansion and related investments in socioeconomic and political "complexity," gradually created a cultural environment that ultimately precipitated a financial and economic crisis. Revenues sharply declined as a consequence of the discontinuation of imperialistic policies; consequently, inadequate resources were available to address normal expenditures and current problems, and invest in socioeconomic advancement. The impact was so structurally, functionally, and financially disruptive that the civilization rapidly self-destructed. By about A.D. 400, the Roman Empire collapsed. Melko's "transition" mechanism may be appropriately applied to this final stage of a highly integrated culture.

At this point, it is instructive to revisit Joseph Tainter's generalized view of civilization development that captures the essence of the mechanistic-thermodynamic paradigm:

> **Human societies and political organizations, like all living systems, are maintained by a continuous flow of energy. From the simplest familial unit to the most complex regional hierarchy, the institutions and patterned interactions that comprise a human society are dependent on energy. At the same time, the mechanisms by which human groups acquire and distribute basic resources are conditioned by, and integrated within, sociopolitical institutions. Neither can exist, in a human group, without the other, nor can either undergo substantial change without altering both the opposite member and the balance of the equation. Energy flow and sociopolitical organizations must evolve in harmony.**[44]

The *mechanistic and thermodynamic paradigm* applied to sociological and historical thought of civilization prosperity and failure provides a more unified perception of the long pathway of human history. This perspective supplements and complements the traditional viewpoints of cultural development. Based on this unified framework, civilization development may be expressed, in a mathematical sense, as a continuous multivariable function representing numerous, complex socioeconomic and political processes. These consist of mechanisms that require wealth-energy resources and are subject to human behavioral characteristics. The variables reflect thermodynamic principles, Tainter's energy concepts, behavioral aspects related to Sorokin's model of cultural mentalities, and Melko's sociopolitical and economic, mechanistic model of "transitional periods."

Conceptually, this paradigm is applied to an extremely large, complex cultural system for which new materials, energy, and subsystems, possessing their own mechanisms and properties, are managed and controlled, for better or for worse, by people. Based on human values, priorities, decisions, and actions; Mother Nature's restrictions; and pure chance, cultures may grow, decay, stagnate, ossify, or collapse and perhaps be reborn. Civilization changes, whether functioning positively or negatively, are dependent on numerous interrelated but yet independent social, cultural, political, and economic factors, referred to as "fluxes" by Melko. In addition, the resultant summation of all cultural variables will fluctuate over time, giving rise to people's observation of periods of cultural prosperity and failure.

Ancient Egypt is an appropriate example of a civilization that experienced a long history of such fluctuations, which may be viewed as a continuous, complex, multivariable function of civilization development. About a quarter of ancient Egypt's life span was spent vigorously advancing, after which cultural degradation dominated, which eventually evolved into a universal state. This was followed by a period of chaos that Toynbee labeled the "barbarian interregnum" that existed for some 2,000 years, after which, Egypt was suddenly reborn and existed for another 2,000 years, disappearing with the end of Roman dominion in the fifth century A.D. The latter half of ancient Egypt's history was a life-in-death existence with the culture paralyzed and inert, a cultural coma that took two millennia to overcome.[45]

These fluctuations of historic good times and bad have been noted and commented upon by many scholars, usually based solely on human behavioral grounds. However, the unified cultural mechanistic and thermodynamic paradigm achieves Tainter's desire for an "explanatory framework that is globally applicable" and constitutes "a more general explanation of collapse." It also supports Tainter's view of cultural collapse based on scientific principles: "The marginal return on investment in complexity is at present the best explanation of collapse."[46]

Tainter's argument and its reliance on marginal return on investment in cultural complexity as the best explanation of cultural collapse is another example of a nonscientific expression of the consequences of thermodynamics. Increasing cultural complexity, which seldom contributes to economic productivity, often diverts human and financial potential from useful applications capable of regenerating national wealth. As a result, over time a civilization suffers a gradual reduction in marginal return from its investment of national wealth and finds itself on a slippery slope toward stagnation and collapse. This is analogous to the degree of effectiveness of a given mechanical or electrical device due to the realities associated with its imperfect design, construction, and utilization by imperfect human beings. A culture that creates a continuous, high level of nonessential, highly integrated complexity will suffer from low economic effectiveness that will increasingly prevent it from accomplishing its mission of survival and prosperity. Collapse becomes inevitable as the burden of cultural complexity inescapably increases, thereby decreasing the effectiveness of the culture's socioeconomic engine. This continuously escalating wealth-energy consumption, generating declining benefits at higher costs, ultimately reaches the point where society is incapable of acquiring sufficient resources to maintain social and economic stability. Socioeconomic advancement inherently creates the complexity and chaos that ultimately produces cultural degradation and collapse.

Chapter 14
References and Notes

1. Tainter, Joseph A. *The Collapse of Complex Societies*. New York: Cambridge University Press. 2004. 91.
2. Ibid., 91-126.
3. Soddy, Frederic. "Economic 'Science' from the Standpoint of Science." *The Guildsman: A Journal of Social and Industrial Freedom*. July 1920. 3-4.
4. Melko, Matthew. *The Nature of Civilizations*. Boston: Porter Sargent. 1969. 1-3.
5. Ibid., 35.
6. Ibid., 41-43.
7. Ibid., 33.
8. Durant, Will. *The Story of Civilization: Part II: The Life of Greece*. New York: Simon and Schuster. 1954. 671.
9. Brander, B.G. *Staring into Chaos: Explorations in the Decline of Western Civilization*. Dallas: Spence. 1998. 19-20.
10. Melko, Matthew. *The Nature of Civilizations*. 54.
11. Quigley, Carroll. *The Evolution of Civilizations*. Indianapolis: Liberty Fund. 1979. 141.
12. Melko, Matthew. *The Nature of Civilizations*. 162.
13. Ibid., 117.
14. Toynbee, Arnold J. *A Study of History*, ab. Sommervell, D. C. New York: Oxford University. 1957. I-VI: 281.
15. Ibid., 439-440.
16. Sorokin, Pitirim A. *Social and Cultural Dynamics*. Boston: Porter Sargent. 1970. 692.
17. Brander, B.G. *Staring into Chaos*. 358-359.
18. Sorokin, Pitirim A. *Social and Cultural Dynamics*. 693.
19. Ibid. 692.
20. Melko, Matthew. *The Nature of Civilizations*. 47-57.
21. Ibid., 55-56.
22. Ibid., 57.
23. Ibid., 56.
24. Ibid., 54-57.
25. Ibid., 57.
26. Tainter, Joseph A. *The Collapse of Complex Societies*. 3.

27. Brander, B. G. *Staring into Chaos.* 367, 353.
28. Tainter, Joseph A. *The Collapse of Complex Societies.* 91.
29. Ibid., 90.
30. Ibid., 91.
31. Ibid., 91-92.
32. Ibid., 120.
33. Ibid., 106.
34. Ibid., 108.
35. Ibid., 127.
36. Ibid., 124-125.
37. Ibid., 124.
38. Sorokin, Pitirim A. *Social and Cultural Dynamics.* 676.
39. Ibid., 677.
40. Tainter, Joseph A. The Collapse of Complex Societies. 124.
41. Ibid., 127-129.
42. Ibid., 129.
43. Ibid., 121.
44. Ibid., 91.
45. Toynbee, Arnold J. A Study of History, ab. Sommervell. D. C. New York: Oxford University. 1957. I-VI: 360-361.
46. Tainter, Joseph A. The Collapse of Complex Societies. 203.

Appendix A
The Fundamentals of Thermodynamics Applied to Socioeconomics

In order to fully appreciate the ramifications of thermodynamic principles applicable to socioeconomic processes and cultural development, it is necessary to examine the physical nature of energy and its dynamic properties. Energy exists in many forms, having a broad range of specific applications. To comprehend the concept of energy as used in the sciences, it is important to visualize and conceptualize the many uses and meanings of the term *energy*.

The most widely recognized forms of energy are kinetic energy and potential energy. A car in motion is said to have kinetic energy by virtue of its mass and velocity; thus kinetic energy is said to be the energy of motion. Potential energy is associated with a body's position relative to the Earth's surface as it is subjected to the force of gravity. Thus, a ball released from a window twenty feet above the ground has potential energy as a result of the force of gravity acting on the ball at a twenty-foot height, causing it to fall. Thermal energy is the energy a body possesses by virtue of its heat content as measured by its temperature. The carbon dioxide gas molecules in a room possess thermal energy consisting of translational energy associated with random molecular motion and rotational and vibrational energies associated with the bonds between atoms.

These various degrees of translational, vibrational, and rotational freedom may be visualized as the molecules randomly moving in three dimensions within the room. While a molecule travels with a given velocity, each chemical bond between carbon and oxygen atoms is rotating and vibrating. The totality of these degrees of freedom represents molecular energy that will vary with the system's temperature and will theoretically approach zero at the absolute temperature of zero. Scientists associate *energy content of a molecule* with its particular chemical nature or atomic composition, and this energy is referred to as the heat of formation from its elements. The heat of formation of a molecule (e.g., carbon dioxide) is conceptualized as the sum of the bond energies of the molecule, which determines the maximum amount of energy to be potentially made available for a given chemical process.

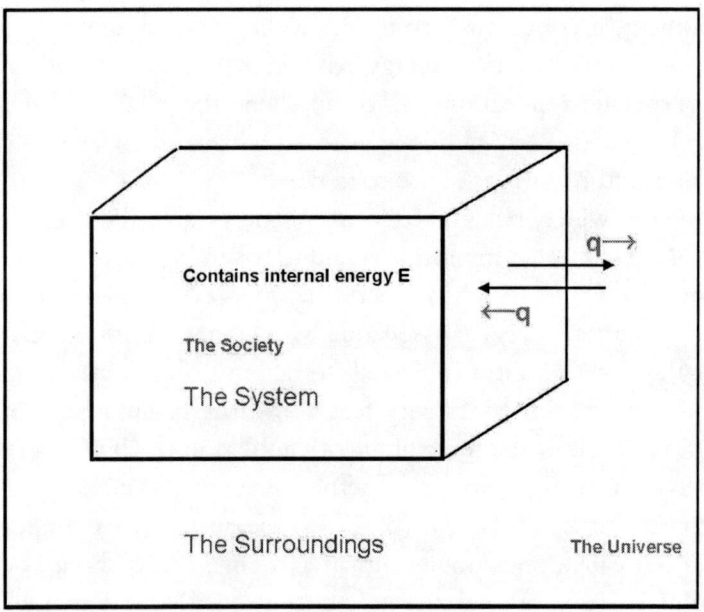

Figure A-1

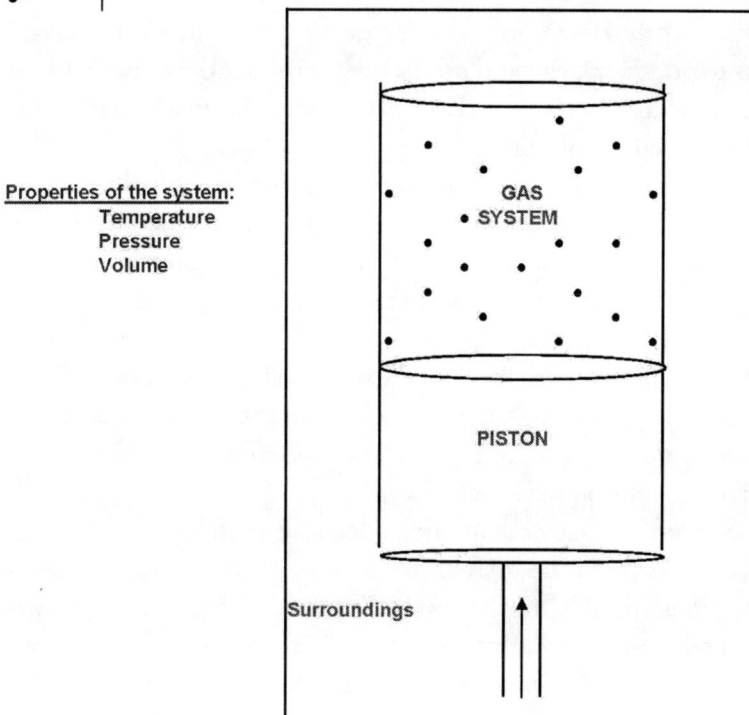

Figure A-2

There are various sources and forms of energy. Einstein's famous equation ($E = mc^2$) illustrates that energy, E, and mass, m, have an equivalency via the constant c, the speed of light. Accordingly, energy from a nuclear reaction such as the sun's process (i.e., nuclear energy) results from matter being converted into energy. Electrical energy may be produced by a generator associated with the mechanical energy of falling water at a dam or steam turning turbine blades. Electric motors produce mechanical energy, as in propelling a golf cart from chemical energy stored in the cart's battery being converted into electrical energy.

Energy has many forms, meanings, and applications. Also, the sciences have many conventions, rules, and restrictions that are, in part, a consequence of the need for mathematical precision in defining and expressing relationships involving the principles being utilized. As such, mathematics is essentially a precise language. More specifically, thermodynamics represents relationships between many forms and expressions of energy and their utility in the world of science. Consequently, it is important to express energy definitions and concepts so as to provide clarity and preciseness in thermodynamics relationships and their associated scientific principles and mathematics that are fundamental to societal processes.

First, a *system* and its *surroundings* must be formally defined and isolated by a distinct boundary, with the properties of the system formally described and specified with scientific and mathematical precision. Figure A-1 represents such a generic system and its surroundings; note that **q** calories could enter the system from the surroundings or be removed from the system. The piston-gas cylinder chamber, Figure A-2, illustrates a specific system of carbon dioxide gas molecules of a defined temperature, pressure, and volume (i.e., *the conditions*); the surroundings include the piston device and beyond.

Second, a thermodynamic process, a pathway from an initially defined energy state to a final state under specified and defined conditions, produces a *change of state* and corresponding changes in thermodynamic parameters of the system, including heat content. An example of an initial state is a one-liter volume of carbon dioxide gas at twenty-five degrees Celsius confined in a piston-cylinder arrangement at a pressure of one atmosphere while immersed in a large volume of water, also at constant temperature of twenty-five degrees (refer to Figure A-2). For thermodynamic purposes, the system, boundaries, surroundings, and properties of the initial state of the system are defined. The system (i.e., the gas) could be subjected to a process during which one or more variables of temperature, pressure, and volume are altered to transform the system along a particular pathway to a final state. The change in internal energy or heat content of the system (i.e., the gas), going from an initial state to a final state, may be determined and by definition is a thermodynamic quantity. This *change in heat content* is more formally referred as a *change in enthalpy,* for which the symbol **H** is utilized; thus the *change in enthalpy* for the specific process would be designated as ΔH.

(Changes in thermodynamic parameters, e.g., enthalpy, are expressed as ΔH, where the Greek letter delta, Δ, signifies "a change in or difference between" the final and initial states. Thus, ΔH is the change in enthalpy or heat content for the defined process, e.g., combustion of a fuel.)

Society's utility of thermodynamics emerges from attempts to perform useful work, which requires energy. Work is defined in terms of the quantity of energy that flows into or out of a system during a process that produces a change in its properties and impacts the surroundings. In the piston-gas cylinder arrangement, if energy in the form of **q** calories enters the system, work may be accomplished on the piston by the carbon dioxide gas as it expands, pushing the piston. That is, the added energy from the surroundings permits the gas to do work on the piston. An economic analogy could be described as a society investing national wealth-energy resources in a socioeconomic system with the intended purpose of accomplishing productive societal outcomes. On a large scale and over time, such investments of national wealth would generate both desirable productivity and undesirable outcomes within a nation's social, economic, and political systems.

Recall that thermodynamics mandates that consumption of energy, such as in the expansion of a gas to do work on a piston, is intrinsically a somewhat inefficient process and inescapably generates waste energy and material. This spontaneous direction unavoidably produces greater molecular disorder within the system. Likewise, a civilization's production processes follow an analogous thermodynamic pathway. As with all energy consumption, societal processes also generate consequences that include the wasting of energy and other natural resources and the creation of societal disorder. This discipline and management by Mother Nature applies to all of society's energy consumption tasks, with thermodynamics representing her guiding operational principles.

Mother Nature regulates energy transformations by virtue of specific restrictions represented by the law of the conservation of energy, also known as the first law of thermodynamics. This principle is Mother Nature's energy accounting system and is a very straightforward, simple concept. When a system undergoes a process that involves the acquisition of energy from the surroundings, the energy must perform work or be transformed into heat energy of the system (i.e., the temperature increases). A frequently used expression of the first law is that energy can neither be created nor destroyed; it is only transformed from a useable form to a less useable form and thus conserved.

The Fundamentals of Thermodynamics Applied to Socioeconomics

The first law can be formally stated as the change in internal energy, ΔE, of a system being equal to the energy absorbed by the system **q** minus the work done by the system **w**.

$$E = q - w$$

Equation A-1

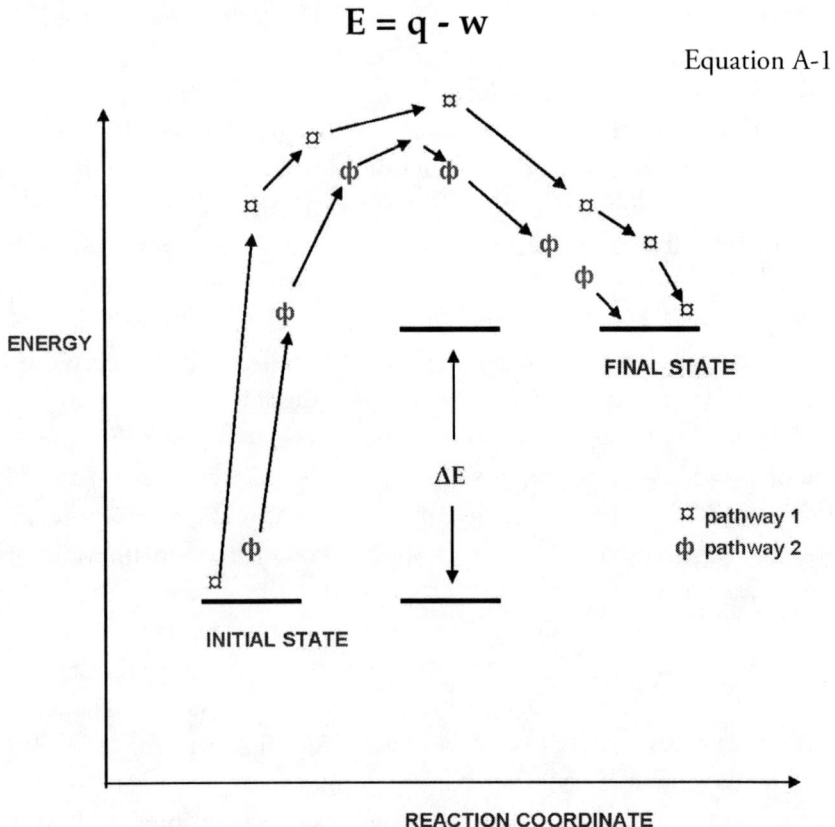

Figure A-3

Thermodynamic parameters (e.g., ΔE) are dependent only on the difference between the final energy state and the initial energy state and are independent of the particular pathway traveled in reaching the final state. Mathematically, such functions are referred to as state functions or exact differentials; changes in state functions are denoted by Δ, such as, for example, ΔE. This point has physical significance in that for a given change in a state function, there are many pathways that represent a given energy transformation, resulting in the same value for the change in the thermodynamic parameter. It follows from Equation A-1 that a large number of corresponding sets of values of **q** and **w** exist for a given value of ΔE.

Wealth, Energy, and Human Values

This may be illustrated by a gaseous system in the piston-cylinder arrangement where various possible combinations of the variables of temperature, pressure, and volume could produce different pathways from the same initial to the same final energy state, but all pathways will result in the same net thermodynamic change ΔE; see Figure A-3. Consequently, pathways 1 and 2 represent different mechanisms for progressing from the same initial energy state to the same final state, but with equivalent net changes in the thermodynamic parameter ΔE. This quantity, ΔE, represents the change in the total heat content of the gas molecules for the process (i.e., the sum of translational, rotational, and vibrational energies).

Based on Equation A-1, if q calories of energy enter a system, two examples illustrate the extremes in the degrees of usefulness of energy consumption by machines as well as by society's socioeconomic processes. First, if all of the energy absorbed by a system, q, goes to do work w, no change occurs in the internal energy of the system ΔE, that is, the heat content or temperature of the system remains unchanged. This corresponds to all of the energy q entering the gaseous system being completely utilized to do work on the piston while the temperature of the gas remains constant. This is referred to as an isothermal expansion of a gas and constitutes *optimal utilization of available energy* or maximum useful work, w_{max}.

In contrast, if the system uses none of the available energy, q, to perform useful work on the piston, all of the consumed energy is transformed into heat energy (i.e., a temperature increase of the gas). The piston doesn't move. This pathway constitutes *no useful work* being accomplished by the gas on the piston and is referred to as a process that constitutes minimum work, w_{min}. This continuum from maximum to minimum useful work represents the inherent theoretical limitation or range imposed by Mother Nature based on strict, practical limitations on process conditions. A pathway that is able to produce a quantity of useful work approaching w_{max} is more thermodynamically efficient than a pathway producing a value approaching w_{min}. The latter pathway inefficiently consumes energy, accomplishes little useful work, and wastes most of the supplied energy q by converting it to heat. This range of useful work that may potentially be accomplished illustrates that the manner and conditions by which a person chooses to conduct a simple piston-gas cylinder expansion process or a society's complex processes of existence possess inherent scientific limitations. Human knowledge and wisdom combined with thoughtful, efficient technology and techniques are capable of maximizing society's production of useful work and potentially benefiting all humankind.

It should be emphasized that technological capabilities, techniques, and advances, in addition to the availability of tools, materials, and the quality of labor, are factors inherently influencing the thermodynamic efficiency of any selected energy consumption pathway. Outcomes may range between the two extremes of minimum useful work (low thermodynamic efficiency) and maximum useful work (high thermodynamic efficiency). As a reference point, the steam engine is normally considered to have a theoretical thermodynamic efficiency limit of less than 30 percent; that is, a maximum conversion rate of 30 percent of consumed energy may be converted into useful work. It should also be noted that *thermodynamic efficiency* or the technical, engineering capability of converting a given quantity of energy into useful work is only one aspect of overall societal *economic effectiveness*. Economic effectiveness is also shaped by human behavior and societal priorities. Societal values, motivations, attitudes, and wise and unwise decisions are the basis for the prioritization of national wealth-energy resource allocations, which will affect overall economic productivity.

It is instructive to utilize the contrasting pathways of the maximum to minimum work continuum of the gaseous system in the piston-cylinder arrangement as an analogy of society's socioeconomic processes. National wealth, considered as **q** in Equation A-1, is consumed by a society utilizing a potentially wide range of human wisdom and capabilities, technological and production methodologies, and socioeconomic strategies and tactics (*process conditions*). The resulting social, economic, and political processes may hypothetically follow a large number of extremely divergent and complex pathways capable of outcomes that create, at one extreme, abundant new forms of useful work (e.g., significant capital regeneration), as opposed to ineffectively generating economically inconsequential products and services (e.g., luxury goods). The degree to which a culture's socioeconomic processes are *effective* (i.e., able to both *efficiently and wisely consume wealth-energy* *capital to regenerate national wealth*) will significantly impact cultural development and the prospects for continuous prosperity.

Thus, potential socioeconomic success will be restricted by the following: First, *economic limitations* resulting from a culture's priorities for the proportion of national wealth resources invested in generating new sources of wealth-energy capital, and second, *inherent technical and scientific limitations* (i.e., efficiencies) related to production processes, equipment, labor, and techniques. Society and Mother Nature impose their own unique limitations on the economic processes of a social order. However, in the final analysis, Mother Nature's discipline and management imposed by thermodynamics trumps society's wisdom, initiative, and technical capabilities.

Summarizing this analogy of a simple piston-gas cylinder expansion process with the interrelated complex processes of cultural development, the first law of thermodynamics allows the energy absorbed by a system, whether a gas confined by a piston-cylinder arrangement or a society's social and economic systems, to follow a number of comparable energy consumption pathways. The routes both entail similar issues of *technical efficiencies* and the *effectiveness of human management* designed to accomplish useful and productive work. However, the second law of thermodynamics requires that some portion of this absorbed energy *must and will* be dissipated as heat energy, that is, it is not possible for people or nature to conduct processes that attain 100 percent thermodynamic efficiency. As a result, a civilization has the highest probability for attaining a successful socioeconomic system when, first, it uses the most advanced technologically based production processes in order to maximize thermodynamic efficiency and useful work. Second, a civilization must prioritize its use of national wealth-energy resources so as to maximize wise investments and strategies capable of replenishing its consumed national wealth-energy capital. Essentially, this is the best argument for a society placing a high priority on basic research and education.

While the first law of thermodynamics is recognized as a valid statement of the conservation of energy (that is, energy can be neither created nor destroyed), it does not impose restrictions on the directional nature of energy transfers. However, experience convinces us that energy flows from hot objects to cold objects. If one end of a metal rod is very hot and the other end is very cold, intuition tells us that the rod will soon become uniformly warm. The reverse flow of energy, that a uniformly warm metal rod spontaneously becomes hot at one end and cold at the other end, would not be an expected occurrence. But the first law does not mathematically restrict the possibility of this event spontaneously occurring, even though it is universally considered an unlikely phenomenon, not permitted by the laws of nature.

However, the second law of thermodynamics, the entropy concept, does address nature's directionality (i.e., the spontaneity and irreversibility of physical processes such as the metal rod with hot and cold ends coming to a common uniform temperature). From lightning in the night sky to the blooming of flowers in spring, nature's processes are inherently irreversible and spontaneous. Thus, the entropy principle restricts the direction of spontaneous energy flow. Heat spontaneously and irreversibly flows from hot domains to cold domains as a system strives to achieve a uniform distribution of its thermal energy, which is the *thermal equilibrium condition*. Entropy's directionality, as indicated by the citation as *time's arrow*, is based on the driving force of a system spontaneously undergoing change in the direction necessary to achieve its equilibrium energy state. Work is accomplished by harnessing a system in a non-equilibrium condition as it advances toward the equilibrium state. This spontaneous journey corresponds to a transition to a higher level of disorder. *Such increasing freedom refers to more available degrees of motion* (i.e., greater translational freedom as well as increased rotational and vibrational freedom of motion). The journey ends at a new equilibrium condition that possesses no further usable energy content under the given conditions, and the system is unable to produce further useful work unless a new force from an additional energy source becomes available.

This journey toward equilibrium and greater disorder may be illustrated by the combustion of a fuel. Solid coal composed of carbon atoms is a more orderly arrangement of carbon atoms than exists in the combustion product state of gaseous carbon dioxide. Under appropriate conditions, solid carbon in the form of coal combines with oxygen from the air to be spontaneously and irreversibly transformed and liberated as carbon dioxide gas. The final energy state represents greater disorder and freedom for the system than the initial state of a solid piece of coal. The energy liberated from this transformation on its way to an equilibrium condition could be used to heat a house, cook food, or produce steel.

This concept of the transformation and degradation of heat content possessed by a fuel may be represented as the transition of a system from a given level of usable energy to a state of less usable energy. This may also be described as a transition from a defined state of orderliness to a state of greater disorder. An automobile burning gasoline as its energy source produces hot gases that exit the exhaust pipe. The fuel begins as a relatively orderly state of liquid gasoline molecules confined in the gas tank prior to combustion and is transformed to a post-combustion state of more randomly oriented or disorderly gaseous molecules free to roam the Earth's atmosphere. Thus, the chemical process of burning a fuel maybe described as an irreversible process producing greater disorder in the surroundings, while some amount of the original energy content is lost forever as a usable energy source. Hence, as Society consumes energy resources, the disorder of the universe will continually increase with the passage of time toward a state of maximum disorder. All energy flows unalterably from the orderly to the more disorderly state and from a utilizable form to a less usable form, resulting in an increase in entropy of the system. Thus, entropy is viewed as a driving force for processes spontaneously proceeding toward an equilibrium condition.

Consider two vessels of equal volume connected by a valve that can either block a gas from passing or permit a gas to pass from one vessel to the other. A gas is placed in one vessel and a vacuum is created in the other vessel. Upon opening the valve, the gas molecules will randomly and spontaneously diffuse back and forth between the two vessels and eventually establish an equilibrium condition. Probability indicates that at any one time, equal numbers of gas molecules will occupy each vessel (i.e., the gas pressures in each vessel are equal). This equilibrium condition for the system constitutes a state of greater disorder, and the process represents a transition for a fixed number of molecules to a larger available volume (i.e., fewer molecules per unit volume of space). Such a process is obviously spontaneous and irreversible and provides meaning to the reference that the entropy of the universe tends toward a maximum. The equilibrium condition possesses no net energy-based capability of acting on the system without an outside energy source becoming available and disrupting the existing equilibrium condition.

The probability that all molecules would ever reside in only one of the vessels at any given time approaches zero and represents a minimum entropy state. It is noteworthy that *this spontaneous molecular diffusion process is driven solely by probability* (i.e., entropy, as the heat content of the system, remains unchanged during the isothermal process).

However, the combustion of fuels (e.g. coal) is an example of a process for which both heat content and probability jointly constitute a net driving force. Coal in the solid form consists of carbon atoms in an orderly array or matrix mutually held together by a binding energy associated with carbon's atomic structure and forces. As coal is heated, its heat content continuously increases; the carbon atoms increasingly vibrate in their lattice until they possess sufficient energy to overcome the binding energy holding them together in the solid form. The process reaches the point where it becomes energetically more favorable (i.e., more probable) for carbon atoms to combine with oxygen to form gaseous carbon dioxide than to remain in its high-energy solid form. Carbon atoms become liberated from the more orderly solid form of coal to gain greater freedom in the gas phase as carbon dioxide; this process progresses toward greater disorder and represents an entropy increase. This combustion of solid carbon and atmospheric oxygen produces gaseous carbon dioxide and a quantity of energy, defined as the heat of combustion. The system undergoes an exothermic chemical reaction, resulting in an energy content decrease (i.e., energy is liberated) and a positive entropy increase, indicative of increased molecular disorder. These thermodynamic parameters of enthalpy **H** and entropy **S** will be shown to represent the variables of heat content and probability, respectively, for the physical, chemical, and biological processes of nature and society.

The entropy principle dictates that the directionality and irreversibility of the processes of human existence are impossible to reverse without the expenditure of more energy than is possible to obtain from nature's spontaneous process direction. However, cultures may and do create greater order within systems, including manufacturing energy sources, as for example manufacturing an automobile battery or refining crude oil into gasoline. Reversing nature's direction of energy flow requires the acquisition and consumption of an energy source from the surroundings; however, this outcome will generate even greater disorder in the surroundings. Such an example is an electric pump used to raise water to a higher level as stored potential energy for eventual use as a flow of water to turn a stone wheel to grind wheat. A greater amount of electrical work must be expended to get the water to the higher level than can be obtained from the process of the falling water performing work by turning the grinding wheel. The scientific principle involved is usually expressed as, *Whenever order is created in the universe, it is accomplished by an even larger disorder in the surrounding environment.*

Likewise, it is possible to force heat energy to flow from a lower temperature to a higher temperature. But that too requires work from an energy source external to the system. The home heating and cooling device known as the heat pump illustrates this situation. Electrical energy is used to cool the house indoors by pumping heat out of the house during warm weather. Left to nature, the inside temperature of the house (the system) would eventually come into thermal equilibrium with the temperature outside (the surroundings). In warm weather, as the house heats up, electricity provides the source of energy to transfer heat energy from a warm house environment to even warmer outside surroundings. This directionality opposes nature's spontaneous direction of energy flow from hot to cold, and the cost of such opposition to nature's direction is reflected in the monthly electric bill. In nature, there is no such thing as a free lunch; Mother Nature makes sure that the price is paid for accomplishing useful work and always at the time of the transaction; no credit is extended.

In the previous discussion of the combustion of coal, attention was focused on the energy content of the reactants and products that resulted in the release of energy (i.e., an exothermic, combustion process). In addition to this thermodynamic consideration, there is also a *mechanistic segment* that illustrates the involvement of mathematical probability. The reaction process of the reactants of solid carbon and oxygen in the gas phase involves a mechanism referred to as a two-phase chemical reaction: a gas phase and a solid phase. Gas molecules of oxygen are adsorbed on the surface of the solid carbon, resulting in a transition state that involves the reconfiguration of electrons of the two reactants. The formation of such a transition state depends on the probability of the two reactant species interacting and assuming a required orientation in space, which will permit the necessary reconfiguration of electrons to form new chemical bonds. The transition state is the precursor to the final product species being formed. This is an example of the *mechanistic aspect* of the mechanistic-thermodynamic paradigm, with the heat of combustion providing the *thermodynamic segment*; this constitutes the two elements of molecular reactivity.

The magnitude of the heat of reaction for the combustion of coal is dependent on bond energies that are broken and reformed in the course of the reaction's mechanism. The heat content of reactants and product are formally expressed as heats of formation, or *enthalpy*, **H**. For our purposes, enthalpy, **H,** may be approximated by **E** in Equation A-1; that is, both may be considered as *heat content,* as the difference between the two is attributable to process conditions that are irrelevant to this discussion. For a given process, the net heat of reaction, **ΔH**, is expressed as the difference in the heats of formation between the products formed and the reactants.

However, the change in enthalpy, **ΔH**, representing the net change in *heat content* resulting from the process, is only one of two distinct thermodynamic parameters that contribute to and determine the spontaneity of the specific process. The other influencing factor is related to probability, more specifically, the thermodynamic parameter entropy, represented by the symbol **S**. The entropy change for a given process is represented by **ΔS**.

Probability is a dynamic force of *molecular reactivity* and may be illustrated by the requirement that reacting species must achieve a definitive, favorable orientation in space in order to finalize a specific mechanistic process. The reaction of gaseous oxygen on a solid carbon surface during the combustion of coal has been provided as an example. This probability is defined by the chance occurrence of a specified three-dimensional configuration of carbon and oxygen atoms occupying a particular volume of space necessary to permit the formation of a transitional configuration of reactants enroute to forming products. Such an event permits bond energy transitions to occur and the formation of a temporary intermediate species, or a transition state. This phase is followed by the formation of new bonds representative of the product species. It will be shown that this probability factor is related to the thermodynamic parameter entropy. The change in entropy (*probability*) for a process plus the change in enthalpy (*energy content*) influence the spontaneous *direction* of physical and chemical processes, that is, whether the process is thermodynamically favorable to actually take place. Thus, whether a reaction or process actually occurs spontaneously depends on both the change in the heat content and the probability of favorable mechanistic events occurring.

Note that when a metal bar with one end hot and the other end cold is transformed to a bar of uniform temperature, the change in heat content **ΔH** for the system is zero. Consequently, the spontaneity of the process depends solely on the probability or entropy parameter. In contrast, the combustion of coal results from a mechanism that liberates heat and thus involves a change in heat content **ΔH,** in addition to a probability or entropy effect as previously described. The solid carbon is transformed to gaseous carbon dioxide, which constitutes a more disorderly state and an increase in entropy, **ΔS**, for the system. This combination of a heat content change and a probability or entropy influence for the process is defined by and referred to as the thermodynamic parameter *free energy* **G**. Thus, the spontaneity of a thermodynamic process is measured by its free energy change, **ΔG,** and is mathematically represented at a constant temperature **T** as:

$$\Delta G = \Delta H - T\Delta S \qquad \text{Equation A-2}$$

For a process to be spontaneous and irreversible, its free energy change is required to be a negative quantity. The system undergoes change as a result of, and at the expense of, free energy. The thermodynamic parameter free energy represents the fundamental driving force in nature and determines whether physical and chemical processes conducted by nature and society will take place. This is Mother Nature's intrinsic dynamic force, representing the probability parameter entropy and the energy content parameter enthalpy.

The representation of entropy as a thermodynamic probability parameter is a very powerful tool in characterizing the processes of society and nature. The entropy of a specific state of a system may be related to the probability of the existence of that state. The usual chemical and physical processes of life involve large numbers of molecules or *objects,* thus permitting the application of statistical methods, and as a result of entropy's relationship to probability, entropy is often referred to as the *statistical law*. The process of the molecular diffusion of vapors in the atmosphere is an appropriate illustration.

If one opens a bottle of perfume in a large room, the odor's presence becomes evident throughout the room even though the number of particles necessary to permit human detection of the odor is statistically large. Particles of the perfume quickly diffuse throughout the room's volume. As the particles randomly migrate to every corner of the room, it is reasonable to assume that any one specified particle motion is equally probable to any other specified motion. Given the large number of different motions that are possible, the total number of motions that produce uniform occupancy of the total volume available is significantly greater than the probability for total occupation in any given fractional volume of the totality of available space. Thus, the room becomes uniformly filled with particles of the perfume. The probability of this uniform filling of the room is proportional to the total number of motions that would produce this outcome, and this probability is overwhelming, greater than for any other possible set of outcomes. The driving force for this diffusion of perfume particles to occupy the total available space is the *probability of uniform distribution* (i.e., the spontaneous and irreversible tendency toward increased randomness of the system of perfume molecules to achieve the equilibrium condition). Thus, the reference is found to entropy as the *statistical law*.

It is noted that this diffusion process proceeds spontaneously from a more orderly arrangement of perfume molecules, in a relatively smaller volume within the bottle, to a more disorderly state occupying a larger volume of the whole room. The driving force for this process is solely a probability phenomenon, and the system is spontaneously and irreversibly driven from a more orderly state to a more disorderly state. This illustration provides meaning to the statement that *entropy is a measure of the randomness or the chaos of a system* and the thermodynamic principle that spontaneous processes of society and nature produce a more disorderly system.

Recall that the general equation for representing the spontaneity of nature's driving force or process is free energy; that is, $\Delta G = \Delta H - T\Delta S$. For this molecular diffusion process, there is no change in temperature of the system; thus, the heat content of the system of perfume molecules is constant and $\Delta H = 0$. Consequently, the free energy change or the driving force for the spontaneity of the molecular diffusion process is $\Delta G = -T\Delta S$. This demonstrates that the driving force for such an isothermal process may be represented by the entropy change alone; that is, the increase in the number of possible arrangements of *objects* in space and the occupancy of a larger volume that becomes available for the *objects*. For this example, ΔS is a positive quantity corresponding to an entropy increase for a more disorderly system, which in turn determines that ΔG will become a negative quantity, as is indicative of spontaneous processes.

In general, the entropy of a system can be defined in terms of the number of possible arrangements of *things* (e.g., molecules that comprise the system). Any resulting change in entropy for the system from one defined state to another is therefore related to the probability of a transition actually occurring from one arrangement to another. The significance of the entropy concept, particularly its mathematical expressions in terms of probabilities that represent physical states, cannot be overemphasized. Many of the world's most respected thinkers, scientists and nonscientists, have recognized the power of the entropy concept and its governance of the universe. Einstein referred to it as the premier law of the sciences,[1] Sir Arthur Eddington called it the supreme metaphysical law of the universe,[2] and the writer and dramatist Vaclav Havel stated, "Just as the constant increase of entropy is the basic law of the universe, so it is the basic law of life to be ever more highly structured and to struggle against entropy."[3] In addition, compelling aspects of the entropy concept appear in the popular press in a lighter vein, such as, "Stupidity is like entropy: It always increases and you have to do work to get rid of it" or "Entropy isn't what it used to be."

Given entropy's relationship to probability and statistics as the driving force for such physical processes as the uniform diffusion of perfume particles throughout an enclosure, it is useful to examine probability as a factor in a society's functional processes of existence. This requires the establishment of a linkage between probability as a thermodynamic parameter and the consumption of national wealth in socioeconomic and political processes. Pursuing this objective first necessitates addressing the following issues: How and to what extent do entropy and probability play a role in the random everyday proceedings of a society as its economic, social, and political events unfold? What is the relationship of these fundamental processes of societal existence to the investment and consumption of national wealth? To resolve such issues, it is useful to begin with a generic definition of entropy in terms of probability that is applicable to the functions of a culture's effort to survive and prosper as a society.

To begin this task, it is necessary to recognize that regardless of whether a system takes the physical form of small particles, atoms, molecules, or electron energy configurations, *the definition of entropy does not require a structural model.* It is sufficient to treat the system as simply a large number of small generic "things." However, *it is necessary to have a structural model in order to ascribe physical meaning to the probability of a given state* and thus provide meaning to the resulting quantification of entropy for that specified state of the system. As an illustration, to define entropy in terms of probability and generic energy levels does not require a model of actual distributions of electron energy levels of atoms. However, to actually perform a calculation of the probability of a specified electronic configuration for a particular atom would require a model for the actual electron configuration for that element.

Consequently, the *definition of entropy in terms of probability* may be illustrated by devising a box consisting of four cells, with each cell being capable of holding one ball:

|○○ ○○|

If two identical balls, ☺, are placed in the box and shaken or randomized, six possible arrangements of the two balls are possible:

The probability of obtaining one ball in both halves of the box is four chances in six, whereas the probability of obtaining both balls in one half of the box is two chances of six possible arrangements.

The relationship of such probabilities to the entropy of a system in a given state is based on the number of potential arrangements for the system that coincides with a defined state. This relationship has been shown to be **S = k ln Ω** where **ln** is the natural logarithm, **k** is the Boltzmann constant, and **Ω** is the number of arrangements of the system that will produce a specifically defined state of the system. For the defined state of both balls in the same half of the cell, **Ω = 2,** while for the state defined as being anywhere in the cell, $\Omega = 6$. If the number of cells increases to a very large number, the probability of finding one ball in each half of a cell approaches one chance in two as compared to four chances in six for the situation of only four cells. For two balls, the *number of arrangements* Ω is $\Omega = N(N-1) \div 2$ where N is the number of cells in the box.

Consider an isothermal expansion of a gas in the piston-cylinder arrangement previously discussed and visualize the expansion process as a larger volume becoming available for the occupancy of a fixed number of molecules. This is analogous to the two balls (now representing two molecules) in the above box illustration having more cells (now representing volume) available for occupancy. For example, four cells, or 6 possible arrangements for two balls, are defined as an initial state for a transition (i.e., gas expansion) to a second state, defined as having sixty-four cells, or 2,016 possible arrangements for the two balls. That is, the two molecules are involved in a hypothetical process where the total volume available for occupancy increases in volume from four cells to sixty-four cells.

Thus, the entropy change for the system of two balls representing two molecules is:

$$\Delta S = k \ln \Omega_2 / \Omega_1 = k \ln 2016/6 = k \ln 336$$

This is a positive quantity, as expected, for this simulated process of a gas expanding to occupy a larger volume that corresponds to a more disorderly condition.

Perfume molecules diffusing throughout a room to assume the maximum available volume project the vision of the most probable states, as do the gas molecules uniformly occupying the total volume of space available in the piston cylinder arrangement, e.g., two balls possessing 2,016 possible arrangements. Each corresponds to an equilibrium condition as the maximum entropy state for the conditions specified for the system. A system in this energy state or condition lacks any driving force to spontaneously depart from the equilibrium position. Molecular diffusion of perfume throughout a room is a *natural phenomenon* that exhibits directionality that represents the most probable sequence of events.

The treatment of more substantive, realistic science-related topics such as the determination of the probability of a valence electron of a given atom absorbing energy and migrating to a given higher energy level employs the same approach as outlined above for the determination of the probability of various arrangements of the balls in the cells. Given appropriate data, the physicist can calculate the probability of an electron being found in a particular excited state. Obviously, entropy, or the driving force of probability, has more significant ramifications beyond these applications.

When applied to human behavior and socioeconomic systems, probability is a major, but not the exclusive, driving force. Human actions are influenced by intellect, values, and ambition and are capable of altering this statistical driving force by reducing some elements of chance. In the ball and cell illustration, if a pereson blocks one or more cells, restricting or preventing the ball from occupying a particular cell, the probability of achieving a given arrangement of balls would undergo a man-made perturbation and would alter the probability of some events occurring. Likewise, human actions may alter the normal affairs of nature and society that are influenced by probability, including the functions of socioeconomic processes. The ramifications of human judgment and action may consciously or unconsciously alter the probabilities of events and outcomes in a positive or negative fashion regardless of the wisdom of humankind. Nevertheless, entropy principles place inherent restrictions on socioeconomic initiatives that may retard attempts to affect cultural change. Such restrictions control the evolution of a culture, just as entropy limits the maximum possible efficiency of the operation of a steam engine. Regardless of the quality and wisdom of individual leaders and a nation's policies, entropy dictates that complexity and disorder will be one of the by-products of living despite what might appear to be creative ideas and programs capable of progressive solutions to problems and issues. However, the almost infinite number of implementation steps involved in progressive solutions will encounter probability as potentially affecting the course of society's activities.

Thus, the fundamental laws of thermodynamics may be extended beyond generally appreciated applications in science and engineering to the realm of a society's political, social, and economic activities that are a fundamental component of cultural development. In the early 1900s, Frederick Soddy recognized this universal applicability of the laws of science to economics and to social stability and equity. More specifically, he refers to the laws of thermodynamics as the "laws of common sense" as he applies these principles to public finance, economics, politics, and social issues. This will constitute the subject matter of subsequent chapters that consider the fortunes of ancient Rome and Greece, Imperial Spain, and the British and American empires, and the twenty-first-century Asian resurgence. This history reflects Soddy's appreciation and interpretation of thermodynamics that with the passage of time, social, political, and economic systems, like wealth, "rot with old age" and are "consumed in the process of living."[4] Thus, the discipline and management of Mother Nature, as manifest by the laws of science, are not restricted to applications within the science and engineering disciplines. Mother Nature's scope of influence extends to and governs a culture's social, economic, and governmental processes that, while initially created for survival, ultimately are utilized for pursuing and securing riches, glory, territory, and power that ultimately lead to cultural degradation.

> **The second law of thermodynamics tells us that the cosmos as a whole is breaking down structurally and running down dynamically; matter is becoming less organized and energy more uniformly diffused. But in a tiny sector of the cosmos, namely in living material systems, the direction of the cosmic process is reversed: Matter becomes more highly organized and energy more concentrated. Life is a building up process ... must draw upon free energy in non-living systems, capture it, and put it to work in the maintenance of the vital process. All life is a struggle for free energy ... a movement toward greater organization, greater differentiation of structure, increased specialization of function, higher levels of integration, and greater degrees of energy concentration.[5]**

References and Notes

1. Rifkin, Jeremy. *Entropy*. New York: Viking Press. 1980. 6.
2. Ibid.
3. Havel, Vaclav: Czech playwright and president of Czechoslovakia (1989-92), b. 1936.
4. Soddy, Frederick. *Wealth, Virtual Wealth and Debt: The Solution of the Economic Paradox*. London: George Allen & Unwin. 1983. 70.
5. White, Leslie A. *The Science of Culture*. Toronto: Doubleday Canada. 1969. 367.

Index

A

adenosine triphosphate (ATP) xv
Afghanistan and Iraq wars 241, 361, 370
Alfonso V 254
Alternative Minimum Tax (AMT) 345
American gross national debt 361
American public higher education 392–393
American Society of Civil Engineers 375
America's celebrity culture 377
America's public school system 391
America's teaching profession 386
America's trade deficit 358
Ancient Egypt 202, 231, 431, 465
Aragon 248, 252–255, 257–264
Atlantic States Marine Fisheries Commission 281, 282
available degrees of motion 478

B

barbarian interregnum 465
Bell, Daniel A 439
Berman, Morris 324
Bernard L. Madoff Investment Securities 94
Brander, B. G. xxvi, 22, 227, 268, 318, 439, 468
Braudel, Fernand 147, 177, 243
  cultural traditions 138, 139
  intermediate scale of "conjunctures" 139
  quasi-immobile time 138, 139
  social structures 136, 138, 139, 152
  time frames for cultural processes 139
British Empire 7, 21, 22, 125, 159, 320, 349, 463
bureaucratic stasis 88, 89
bureaucratic syndrome 88

C

Camarota, Steven 398
Cardinal Virtues 433
Carnot, Sadi 102
Castile 70, 248–250, 254–265
Catalan 248, 253, 255
Catalonia 248, 252, 258
catastrophic flooding of New Orleans 372
Catholic Kings 249, 252, 255–261
Center for Immigration Studies 398, 399
CEOs of the 350 major U.S. companies 378
Chesapeake Bay vii, xiii, 270–273, 275–280, 282–289
Chesapeake Bay Foundation 275, 277
China's escalating productivity 421
China's rural population 423
China's urbanized industrial economy 423
Chinese gross domestic product 422

Christian Crowns 248, 255
civilization of the Nile Valley 222
Clausius, Rudolf 102
Cleveland, Cutler J. 46, 49
collateralized debt obligations (CDOs) 322
Colonial American values 332
commercial capitalism 310
Commission for the Conservation of Atlantic Tunas 283
concept of steady state 140
Confucian doctrine 423
Confucianism 10, 430–436, 439
corporate entropy production 365
Cortes 253, 254, 257, 258
crab population 275
credit default swaps 322
Cretan civilization 222
cultural
  adolescent growth 200
  assimilation in the American culture 348
  breakdown, stagnation, and deterioration xv, 445
  crystallization mechanism 169
  cultural dynamics 2, 130, 164
  displacement of primitive cultures 229
  dynamic equilibrium 144, 146, 147, 164, 178, 185, 186, 201, 205, 236, 237, 309, 450, 451, 453
  genesis and initial development 154
  homeostasis 193, 194
  incoherence 112, 327
  integration, crystallization, and stratification 447
  interregnum 465
  productivity 27, 180, 181, 197, 198, 200, 201, 203, 205, 209, 220, 222, 233, 248
  rejuvenation 101, 185, 222, 223, 226, 248, 337, 425, 459
  self-destruction xiii, 65, 138
  stages of decline and disintegration 187
culture
  concept of xxii, xxv, 4–20, 53–58, 65, 104, 130–140, 143–151, 163–167, 192–197, 204–209
  integration and stratification 114, 119, 354
  morphological structure 194
  structural differentiation 136, 142

D

Darwin, Charles 103
Dash, Eric 397
Dawson, Christopher. 10, 415
Defense Department's annual "Base Structure Report" 369
degree of crystallization 450, 456
degree of integration 139, 140, 146, 445, 453
degree of societal economic effectiveness 27
degree of stratification 182
degrees of freedom of a system 55, 469
demise of America's public school system 391
Department of Homeland Security 371
dietary globalization 356
differentiated patterns of social organization and function 135
Durant, Will xiv, 96, 317, 398, 402, 467
Durkheim, Emile. 21, 161
dynamic equilibrium condition 142, 146, 179, 186, 236

E

Earned Income Tax Credit 351
ecological management 282
ecology of innovation 336, 364
Economic
  globalization 310, 338, 356, 409
  stratification 114, 119, 144, 182, 340, 354, 447
  utility 35, 47, 59, 92, 155
  value xxiii, 27–29, 37, 47, 78, 109
economic wealth-energy resource 89
Egyptiac Society 208
Egyptian Empire 222
Eisenstadt, S. N 19, 61, 98
  premises of sociological analysis 19, 98
  transformative properties of social systems 19
Elliot, J. H 69
Emergency Management Agency (FEMA) 374
Employee Retirement Income Security Act 362
Endocrine disrupter chemicals (EDCs) 278
energy-based complexity model 461
energy-efficiency technology 417
energy return on investment, EROI 46
Enthalpy 472, 480, 482, 483
Entropy xii, xiii, 3, 4, 51–76, 83, 86, 88–95, 101, 102, 104, 105, 106, 107, 109, 110, 111, 112, 117, 142, 150, 169, 176, 177, 186, 200, 205, 206, 213–214, 220–225, 245–246, 270–272, 278, 280, 285, 299, 316, 325, 365, 408, 420, 445, 450, 478–488
  randomness or the chaos of a system 54, 484
  the directional property of 52
  the statistical law 54, 484, 488
  time's arrow 51, 83, 478
Entropy Law and the economic process 48, 73, 96
Environmental societal entropy production 270
environmental stress 276, 278
Ernst, Howard 285

F

Far Eastern Society in China 209
federal budget deficit 359
federal debt 359
Federal Reserve 330
first law of thermodynamics 75, 473, 477
"flowering of intellectuo-aesthetic culture 179
Forbes World richest people 378
free energy 142, 483, 484, 489
Friedman, Thomas L 439
Fried, Morton H. 18, 130, 132, 137, 196
  enculturation 137, 138
  situational learning 137
  social learning 132, 137
  social learning mechanisms 132
  symbolic learning 137, 138, 196
fuel of socioeconomic processes 39

G

Galbraith, John Kenneth 28, 75, 80–94, 105, 108, 368, 434
  debilitating private and public expenditures for public services 369
  economic value of "a steady flow

of purchasing power" 28
excessive public services 266, 368, 434
"human satisfaction" and "social achievement" 80
investment balance 88, 105, 394
social balance 90, 105, 204, 394
stabilization of the flow of aggregate demand 28, 29
gap in individual wealth 379
Geddes, Patrick 25
General Motors 120–126, 334, 335, 363–369, 379, 412
genetically based human nature 132–134
Georgescu-Roegen, Nicholas 27
Gergen, David 398, 402
Gibbon, Edward 73
global
  economic empowerment 355
  economic expansion 254, 313, 408
  energy consumption 81, 418, 421
  food shortage 356
  marketplace 284, 333, 338, 364, 410
global diffusion 335, 341
  of industrialization and capitalism, 406, 408
  of knowledge 335, 341, 383, 409
  of socioeconomic principles 336
  of technology to less-developed countries 411
globalization 337, 341, 356
golden era of profitability 340
good enough technology 326
Graeco-Roman Western civilizations 69, 180, 192, 229, 290, 313, 440, 441, 449, 461
Granada 248, 255, 259, 260
Green Revolution 420

Gulf of Mexico 271

H

Habsburg reign of Charles V 252, 259
Hahm, Chaibong 439
Harappa Empire 222
Harrison, Lawrence H 14
harvesting of egg-bearing crabs 275
Helmholtz, Hermann 102
Helmholtzian cosmos 111, 112
Herbert, Bob 397, 400, 401
hereditary information system 195, 196
Hindu, Chinese, and Arabian cultures 217
Hindu civilization 222
Hinduism 10, 431, 432
Homer ix
Huntington, Samuel P. 6, 11, 14, 21, 22, 130, 308, 342, 351, 405
  American multiculturalists 406
  integration and assimilation 351
  modern American cultural degradation 342
  the fading of the West 308
Hurricanes Rita and Katrina 371

I

idealistic rationalism 304
imperfect societal efficiency and effectiveness 89
Imperial Spain 18, 68–71, 73, 159, 171, 188, 206, 243–251, 262, 266–270, 320, 349, 380, 463, 489
India's well-educated and trained labor force 420
Indic civilization 222

Industrialism 112, 115, 447
Industrial Revolution 57, 98, 99, 102, 112–118, 126, 158, 311–313, 333, 335, 357, 381, 383, 415, 432
inevitability of failed civilizations xi–xiv, 3, 8, 20, 31, 56, 102
institutionalized feudalism 310
integration of institutional structures 88
Internal Revenue Service 344
International Energy Agency 356, 417
intersex imposition 278
intrinsic value or wealth 34
Islamic civilization 405, 424, 426
  resurgence xiii, 12, 404, 414, 416, 424, 425
  terrorism 405
  transition from colonial domination 425

J

James, Lawrence 7, 10
Japanese cultural trends 419
Japan's modern cultural evolution 417
Japan's steel industry 417
Jobs Bank 366

K

Kennedy, Paul 240, 268, 269
  differentials in growth rates and technological change 242
  economic change and military conflict 240
kepone 272, 277
King Charles I 262
King Ferdinand and Queen Isabella 252

kinship-oriented socialization 134
Kristof, Nicholas 396, 402
Kroeber, A.L. ix, xxiv, 129, 140, 147–152, 166, 172–174, 179–183, 198, 210, 230, 308, 442
  "absorption," "displacement," and "adaptation" 230
  cultural "dying" 231
  cultural florescence 172
  "florescent growth" period 179
  growth configurations 152, 172, 173
  methodology and concept of "pulses" and "lulls." 179
  "patterns" as in multidimensional structures 140
  "pauses, lulls, regressions, or barren intervals" 183
Krugman, Paul 338

L

labor power 103–112
labor power and technology 104
Latin American immigration 347, 351, 352
Legal and illegal immigration 352
Lessnoff, Michael 115
liquid modernity 325

M

MacCurdy, George Grant 13, 149
  The degree of civilization of any epoch 13
MacLeod, H. D 42, 49
mahogany tide 276
Malthusian socioeconomics 45
Marshall, Alfred xxvi
Martinez-Alier, Juan 25, 26
Marx, Karl 93, 104, 107–110, 413

laboring activity and labor power 109
labor power as a commodity 108
minimizing "useless labor." 107
self-created alienation and class conflict 93
the principle of entropy at work in capitalism 110
maximum cultural productivity and breakdown 200
maximum useful work 475, 476
mechanistic-thermodynamic paradigm xii–xxiv, xxiii, xxiv, 3, 4, 9, 13, 18, 24, 25, 50, 75, 84, 85, 93, 118, 141, 213, 231, 238, 243, 246, 286, 308, 440–442, 452, 457, 464, 482
Mediterranean civilization 180, 222, 224
Melko, Morton 18, 130, 139–141, 146–152, 167, 192, 442–453, 461, 465
   a disintegration mechanism (reversion) 141
   a formation mechanism (crystallization) 141
   cause-and-effect political history 150, 151, 442
   concept of "reconstitution," 167, 443
   concepts of both irreversibility and randomness 451
   crystallized social order 446
   "determinism" or human control of destiny 443
   economic factors, referred to as "fluxes" 465
   Feudal phase 450, 451, 452, 453, 454, 463, 464
   Imperial phase 451, 453, 454
   integration" model 442
   limitation of man 456
   model of civilization development 443, 445, 450, 456, 457
   reconstitution 167–168, 442–445
   State phase 447, 448
   transitional mechanism 457
   transition model 464
Menhaden 281
Mesopotamian civilization 222
Mill, John Stuart 49
Minoan period 217, 222, 404
Modern urbanization 141
Mogul Empire 222
Money, a form of national debt 41
morphological structure 194
Mother Nature's principles 25, 126, 194, 285
Moynihan, Daniel Patrick 16
MSX 272
Muslim modernity 425
Mycenaean Empire 222
Mycobacteriosis 279

N

narrative coherence 326, 327, 328
National Academy of Engineering 336
National Assessment of Educational Progress 386
National Football League (NFL) 378
National Research Council 350
national wealth xv–xxiv, 2, 12, 14, 20, 27, 29, 31, 32, 35–45, 53, 58, 60, 67, 76–92, 100, 106, 109, 121, 125, 126, 149, 159, 180, 188, 200, 203, 207, 208, 239–242, 246, 247, 250, 285, 305, 306, 322–327, 331, 334,

338, 346, 354, 358, 368–
370, 382, 394, 422, 423,
429, 430, 434, 435, 458,
466, 473, 476, 477, 485, 486
Navarre 248, 255
Neumann, Carl 59, 104, 105
No Child Left Behind Act 388
Nye, Joseph S. Jr. 22
  hard power of military and economic strength 12
  soft power of cultural appeal 12

O

open thermodynamic system 164,
    183, 186, 204, 236, 237, 450
optimal utilization of available
    energy 475
organizational dynamics 365
ossification xv, 141, 168, 181, 183,
    184, 188, 202, 208, 229,
    424, 443–445, 450–455
Ostwald, Friedrich Wilhelm 13
oxygen-depleted marine "dead
    zones" 271
Oyster population 273, 274, 277

P

PCBs 277
perceived consumer needs 108
periodicity of the Graeco-Roman
    Western civilizations 192
perishable wealth 38
permanent wealth 38, 43, 90
Persian Empire 222
Phillip II 241, 252, 263, 264
philosophical and spiritual tenants
    of Confucianism 430
photosynthesis xxiv, 2, 51, 68
polycyclic aromatic hydrocarbons
    (PAHs) 278

pork barrel projects 360
poultry waste regulations 278
probability of uniform distribution
    53, 484
profile of modern U.S. immigrants
    351
progressive Japanese national energy
    policy 418
properties of value symbolize a
    wealth-energy commodity 34
public fishery 274
public services 56, 87–90, 101,
    208, 251, 266, 348–351,
    368–369, 392–394, 423, 434
purchasing power 28–29, 41–43

Q

Quigley, Carroll 8, 13–14, 130,
    148, 238, 308
  Age of Conflict 310–312
  Age of Crisis 308
  Age of Expansion 311, 312
  institutionalized feudalism 310
  state mercantilism 310, 311
  Western cultures as consisting of
      three "expansion" periods
      308

R

Rabinbach, Anson 59, 106–111,
    120
  social energy as related to the labor
      force 59
  the devastating effects of entropy
      59, 111
Ramadan 428
rate of cultural productivity 180,
    181, 198, 201, 203, 205, 233
rates of cultural growth xii, xx, 14,
    20, 39, 58, 76, 81, 84, 144,

168, 178–185, 200, 218, 238–244, 261, 324–331, 416, 446, 452–458
rebuilding of the American social order 337
religious militant leaders 428
restoration of the Bay 280
reversion 141, 446–452
Rifkin, Jeremy 48, 490
River and Harbor Act of 1965 277
Robinson, Joan 97
Rockfish 279
Roman conquest of Greece 297
Ruskin, John 23, 27, 61, 73, 441
  concept of community or national wealth 40
  hypotheses to thermodynamics principles and to economic theory 24

S

Schumpeter, Joseph 4, 8, 63–67, 89, 93, 122, 138, 368
  economic foundation of creative destruction 65
  tendency of successful societies to self-destruct 63
  the culture and ethics of economics 65
  the realities of science-based economics 124
second law of thermodynamics xii, 45, 76, 90, 110, 251, 443, 477, 478, 489
social and economic stability 34, 59, 99, 117, 143, 150, 197, 208, 218, 222, 323, 362, 367, 415, 426, 458, 466
social energy 59, 103–105
societal differentiation 175
societal disorder 53–62, 66–71, 100, 109, 112, 170, 177, 200, 214, 221, 243, 300, 473
societal entropy xii, 3, 55, 58–69, 76, 86–95, 101, 106, 110, 176, 186, 200, 205, 221, 224, 245, 270, 280, 285, 299, 316, 325, 365, 408, 420, 445
societal entropy production 3, 4, 55, 58, 61, 65, 66, 69, 86–95, 101, 102, 106, 110, 112, 221–225, 245–270, 285, 325, 365, 408, 420
Society of Civil Engineers 375
society's economic effectiveness 27, 59
sociocultural transition 1, 168, 181, 187, 221, 224, 320, 330, 405
Soddy, Frederick 20, 23–29, 37–42, 66–67, 90–93, 106, 250, 322, 441, 467
  community wealth 42
  science-based concepts of wealth 20
  virtual wealth 28, 43, 48, 49, 97, 127, 490
  wealth as an agent of production 38
  wealth commodity or wealth capital 37
  Wealth II commodities (i.e., agents of production) 38, 39
  Wealth II, permanent wealth 38, 43, 90
  Wealth I, representing both value and stored energy 38
  Wealth I, the fuel of socioeconomic processes 39
Sorokin, Pitirim A 10, 19, 21, 22, 53, 73, 84, 93, 96, 140, 148, 163, 189, 190, 201, 209–

218, 230–232, 305–306,
    431, 438, 448–449
altruistic ethics 431
categorizations of truth, thought,
    and ethics 304
continuum of cultural mentalities
    213, 410, 457
dominant system of truth 304,
    314, 437
Ethics of Happiness 6, 305, 306,
    307, 323, 324, 330
Ethics of Love 6, 307
Ethics of Principles 6, 211, 301,
    305–307, 323, 330, 374
generic cycle of the "triple rhythm"
    217
idealistic rationalism 304
idealistic supersystem 212
ideational supersystems 212
"logico-meaningful method" of
    ordering chaos 306
model of societal maturation 6
principle of immanent change
    215, 220
principle of the limited possibili-
    ties 217, 219
Supersystems of Culture": Sensate,
    Idealistic and Ideational 306
systems of the truths of faith, rea-
    son, and the senses 304
theologico-supersensory mentality
    449
triple rhythm of sociocultural
    transitions 314, 405
species management 282
Spengler, Oswald xxiv, 8, 129,
    148–152, 192, 194, 239,
    430, 442, 456
cultural soul 137
culture as being "morphological"
    192

cyclical nature of history 151
"organism" model 192
spiritual and cultural hybridization
    407
spontaneous molecular diffusion
    480
stable thermodynamic state 142,
    143
strategy of permanent innovation
    326, 327, 364
stratification of organizational and
    functional processes 182
structural integration and stratifica-
    tion 354

T

Tainter, Joseph A xx, 2, 90–93,
    110, 191, 204, 232, 441,
    455–469
consequences of excessive cultural
    complexity 120
energy flow and sociopolitical
    organizations 440, 464
fatal overinvestment in societal
    complexity 191
hierarchical specialization xx, 90,
    191, 204, 232, 459
marginal return on investment
    in cultural complexity 458,
    461, 466
sociopolitical complexity 121,
    126, 248
Tax Policy Center 345
technological evolution of weaponry
    335
The Agrarian Revolt (145 to 78 b.c.)
    298
the ecology of innovation 336
The Elizabeth River 278
the ethos of globalization 337
The Law of Civilization and Decay

224, 228
The National Academy of Sciences 351
The Organization for Economic Cooperation and Development 341
Theory of Credit 42, 49
theory of faulty behavior and failure to adapt 4
thermal equilibrium condition 478
thermodynamic efficiency 27, 52, 149, 155, 205, 240, 367, 476, 477
The United Nations Food and Agriculture Organization 283
Toynbee, Arnold J. ix, xv, xxi–xxv, 4, 129, 148, 151, 170–177, 181–188, 192–202, 206–209, 230–239, 308, 424, 431, 447
  abortive civilizations 236
  arrested civilizations 236
  "barbarian interregnum" 465
  "challenge-and-response" thesis 176
  creative minority 182
  creative personalities and creative mimesis 186, 196
  cultural "mimeses" 202
  cultural "rejuvenation" 358
  cultural self-destruction xiii, 138
  "discords" and "schisms" 178
  dominant minority 182, 233, 234
  economic reorientation of the priorities 337
  élan vital 185
  final transitional stages 323
  forces of action 175
  four stages of civilization development 418, 426
  four stages of genesis to deterioration 192
  interregnum 181, 202, 234, 235, 465
  ossification xv, 141, 168, 188, 202, 208, 229, 424, 443–445, 450–455
  petrifaction 168, 181, 208
  rout-and-rally concept 236
  self-determination as a dynamic mechanism 176
  social schisms 178
  societal "breakdown" phase 177
  societal "disillusionment" 206
  soul-fractures 234
  survival instincts and perceived needs 133, 333
  the universal state as "a temporary shelter 234
  times of trouble 239
  transfer mechanism 41
  universal church 234, 235
trend of new oil discoveries relative to oil consumption 355
tributyltin 272, 277, 278
triple rhythm of prosperity to failure 441
tuna 284
twenty-first-century India 419

U

Union of the Crowns 248
universal health care 385
unsustainable level of societal complexity 456
U.S. Army Corps of Engineers 277
U. S. Congressional Budget Office (CBO) 241
U.S. federal spending 359
U.S. oil consumption 355
U.S. Pension Benefit Guaranty Corporation (PBGC) 362

U.S. research and development
	spending 381

V

Valencia 248, 252
value of faith 34
value of substance 34
Virginia General Assembly 281
Virginia Marine Resources Commission 275
Virginian-Pilot 281
Virginia Public Access Project 281
virtual wealth 28, 41–43, 322

W

Wall Street's financial giants 320
wasting disease 279
Watt, James 33, 157
wealth-energy capital 12, 18–20, 27, 36, 38–64, 89, 106–107, 117, 125, 136, 142, 155–158, 171–178, 184–187, 195–198, 204–210, 221, 238–255, 264–266, 300, 315, 367, 414–417, 445, 455–462
wealth- energy consumption-societal entropy production cycle 101
wealth transfer mechanism 41
Weber, Max x, 10, 26, 115, 119, 120, 415, 432
  concept of rationalization 120
  Protestant Puritan philosophy 430
  Protestant Puritan work ethic 116, 415, 432
  secularization 119
  spirit of capitalism 115, 128
West's 2003 invasion of Iraq 427
White, Leslie A. xxvi, 3, 5, 12, 127, 129–132, 139, 149, 490
  functional specialization 136
  influence of cultural context 132
  integrative view of culture 136
  structural differentiation 136, 142
  vision of culture 149
Wicomico River 279
Will, George 400
Winslow, Francis 274, 287
winter dredging 275
world's grain production 357
worldwide coal consumption 357
World Wildlife Fund (WWF) 283

Y

Yamada, Masahiro 419

Z

Zuckerman, Mortimer B 397, 398, 399, 400, 403, 438

Printed in the United States
148683LV00001B/146/P